JavaScript from Zero to Superhero: Unlock your web development superpowers

First Edition

Copyright © 2023 Cuantum Technologies

First edition: May 2024

Published by Cuantum Technologies LLC.

Plano, TX.

ISBN 979-8-89379-858-6

"Artificial intelligence is the new electricity."

- Andrew Ng, Co-founder of Coursera and Adjunct Professor at Stanford University

Who we are

Welcome to this book created by Cuantum Technologies. We are a team of passionate developers who are committed to creating software that delivers creative experiences and solves real-world problems. Our focus is on building high-quality web applications that provide a seamless user experience and meet the needs of our clients.

At our company, we believe that programming is not just about writing code. It's about solving problems and creating solutions that make a difference in people's lives. We are constantly exploring new technologies and techniques to stay at the forefront of the industry, and we are excited to share our knowledge and experience with you through this book.

Our approach to software development is centered around collaboration and creativity. We work closely with our clients to understand their needs and create solutions that are tailored to their specific requirements. We believe that software should be intuitive, easy to use, and visually appealing, and we strive to create applications that meet these criteria.

This book aims to provide a practical and hands-on approach to starting with **Mastering the Creative Power of AI**. Whether you are a beginner without programming experience or an experienced programmer looking to expand your skills, this book is designed to help you develop your skills and build a **solid foundation in Generative Deep Learning with Python**.

Our Philosophy:

At the heart of Cuantum, we believe that the best way to create software is through collaboration and creativity. We value the input of our clients, and we work closely with them to create solutions that meet their needs. We also believe that software should be intuitive, easy to use, and visually appealing, and we strive to create applications that meet these criteria.

We also believe that programming is a skill that can be learned and developed over time. We encourage our developers to explore new technologies and techniques, and we provide them with the tools and resources they need to stay at the forefront of the industry. We also believe that programming should be fun and rewarding, and we strive to create a work environment that fosters creativity and innovation.

Our Expertise:

At our software company, we specialize in building web applications that deliver creative experiences and solve real-world problems. Our developers have expertise in a wide range of programming languages and frameworks, including Python, AI, ChatGPT, Django, React, Three.js, and Vue.js, among others. We are constantly exploring new technologies and techniques to stay at the forefront of the industry, and we pride ourselves on our ability to create solutions that meet our clients' needs.

We also have extensive experience in data analysis and visualization, machine learning, and artificial intelligence. We believe that these technologies have the potential to transform the way we live and work, and we are excited to be at the forefront of this revolution.

In conclusion, our company is dedicated to creating web software that fosters creative experiences and solves real-world problems. We prioritize collaboration and creativity, and we strive to develop solutions that are intuitive, user-friendly, and visually appealing. We are passionate about programming and eager to share our knowledge and experience with you through this book. Whether you are a novice or an experienced programmer, we hope that you find this book to be a valuable resource in your journey towards becoming proficient in **Generative Deep Learning with Python**.

YOUR JOURNEY STARTS HERE...

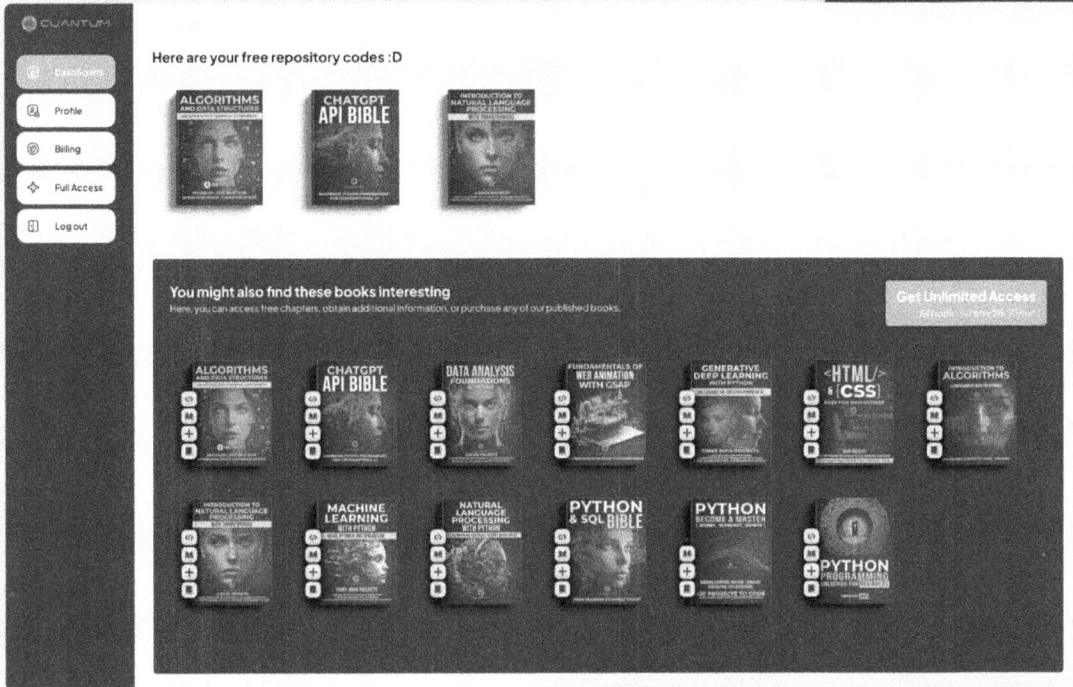

Here are your free repository codes :D

You might also find these books interesting

Here, you can access free chapters, obtain additional information, or purchase any of our published books.

Get access to all the benefits of being one of our valuable readers through our new **eLearning Platform:**

1. Free code repository of this book

2. Access to a **free example chapter** of any of our books.

3. Access to the **free repository code** of any of our books.

4. Premium customer support by writing to **books@cuantum.tech**

And much more...

HERE IS YOUR
FREE ACCESS

www.cuantum.tech/books/javascript-from-zero-to-superhero/code/

TABLE OF CONTENTS

Introduction

Welcome to "**JavaScript from Zero to Superhero: Unlock your web development superpowers**," a comprehensive guide designed to take you on a transformative journey from a novice to a proficient JavaScript developer. This book is not just about learning JavaScript; it's about mastering it to the extent that you can creatively and effectively build web applications that might just change the world.

JavaScript is the language of the web. As of today, it remains the only programming language native to web browsers, essential for adding interactivity and functionality to web pages. It has grown from a simple tool for making web pages interactive to a powerhouse capable of driving complex applications across numerous platforms. This ubiquity makes JavaScript an indispensable skill for any aspiring web developer.

However, diving into JavaScript can be daunting. The language has evolved rapidly, incorporating features that accommodate complex architectures and advanced functionalities. This book demystifies these complexities by breaking down concepts into manageable, understandable pieces. Whether you're a complete beginner or someone with basic programming knowledge, our goal is to equip you with the skills and confidence needed to forge your path in the dynamic world of web development.

"**JavaScript from Zero to Superhero**" is structured to provide a logical progression from fundamental concepts to advanced topics. We start by setting up your development environment—a space where you can write, test, and debug code. Understanding the tools and setups before diving into coding is crucial, as it ensures that you are well-prepared and focused purely on learning JavaScript.

From there, we delve into the core of JavaScript. You'll learn about variables, data types, and operators—basic building blocks that form the foundation of any JavaScript application. Control structures and functions will follow, teaching you to write succinct and effective code. Each chapter builds on the previous one, gradually increasing in complexity.

To reinforce learning, each chapter concludes with practical exercises that challenge you to apply what you've learned in real-world scenarios. At the end of each major section, a quiz will help consolidate your knowledge, ensuring that key concepts are mastered before moving on to more challenging topics.

As we venture into intermediate and advanced topics, you will learn about the Document Object Model (DOM), which allows JavaScript to manipulate the web pages users interact with. You'll explore advanced functions like callbacks, promises, and async/await for handling asynchronous operations—a critical skill in modern web development.

In the latter parts of the book, we introduce you to server-side JavaScript with Node.js, and guide you through developing single-page applications using modern frameworks such as React, Vue, and Angular. By this stage, you will be comfortable using JavaScript not just to enhance web pages but to build sophisticated front-end and back-end systems.

The capstone of your learning journey will be the three comprehensive projects included throughout the book. These projects are designed to synthesize all the concepts covered, providing you with hands-on experience in building and deploying real-world applications.

This book also serves as a reflection of the broader technology landscape, encouraging you to think critically about code and its impact on users and society. Ethical considerations, performance implications, and accessibility are woven through the discussions, preparing you not just to be a coder, but a conscientious developer.

By the end of this book, you will have the knowledge, skills, and confidence to use JavaScript effectively in a variety of contexts. You'll be able to architect robust applications, solve complex programming challenges, and contribute to projects that excite you, armed with a deep understanding of one of the world's most popular programming languages. Welcome to the beginning of your journey to unlocking your web development superpowers with JavaScript.

How to use this book

"**JavaScript from Zero to Superhero: Unlock your web development superpowers**" is designed to be both a comprehensive guide and a practical toolkit for aspiring and seasoned developers alike. To get the most out of this book, it's important to approach it not just as a reader, but as an active participant in your own learning journey. Here's how you can effectively utilize this book to maximize your learning experience.

1. Set Up Your Development Environment

Before you dive into the first chapter, take the time to set up your development environment as outlined in the Preface. This setup includes choosing a code editor, configuring a browser for testing, and understanding basic tools like the JavaScript console. A proper setup is crucial as it will be your workspace for the numerous code examples and projects you will undertake.

2. Progress Linearly

While it might be tempting to jump around based on what topics immediately interest you, JavaScript is a language built on foundational concepts that build upon each other. For the best understanding, proceed chapter by chapter. The structure of this book is deliberately planned to gradually escalate in complexity.

3. Engage with Practical Exercises

At the end of each chapter, there are practical exercises designed to test and reinforce what you have learned. Do not skip these exercises. They are crucial for transitioning from theoretical knowledge to practical skills. If an exercise seems challenging, revisit the concepts in the chapter, or even attempt to re-write the examples provided before tackling the exercise again.

4. Take Quizzes Seriously

Quizzes are provided at the end of each major section and are essential for assessing your understanding of the material covered. Treat these quizzes as both a learning tool and a diagnostic instrument to identify areas where you may need more focus or review.

5. Work on the Projects

This book includes three major projects spaced strategically throughout the book. Each project is designed to consolidate your knowledge and simulate real-world JavaScript development tasks. By working on these projects, you gain confidence in handling larger codebases and complex problem-solving scenarios.

6. Utilize the Appendices and Online Resources

The appendices and the resources listed at the end of the book are valuable for deepening your knowledge and staying updated with the latest in JavaScript development. These resources include additional readings, tools, community forums, and ongoing education opportunities.

7. Reflect on Ethical Considerations

As you learn, think about the broader implications of your code. Consider accessibility, user privacy, and the ethical use of software. We cover these topics throughout the book to ensure that as you grow into a skilled developer, you also become a conscientious one.

By following these guidelines, "**JavaScript from Zero to Superhero: Unlock your web development superpowers**" will not only teach you JavaScript but also how to think and solve problems like a seasoned developer. Whether your goal is to build dynamic websites, contribute to open source, or develop scalable enterprise applications, this book is your first step towards mastering the art and science of web development with JavaScript.

Tools and setup required

Embarking on your journey to learn JavaScript requires a solid foundation, not just in terms of skills and concepts, but also in the tools and environment you use for development. This section outlines the essential tools and setup required to effectively follow along with "**JavaScript from Zero to Superhero: Unlock your web development superpowers**." Ensuring you have a proper setup from the start will facilitate a smoother learning experience and allow you to focus more on learning and less on troubleshooting.

1. Code Editor

A good code editor is indispensable for programming. For JavaScript and web development, there are several popular options:

- **Visual Studio Code (VS Code)**: Highly recommended for its robust features, such as syntax highlighting, IntelliSense (code completion), and a vast ecosystem of extensions. It's lightweight, free, and available for Windows, macOS, and Linux.

- **Sublime Text**: Known for its speed and efficiency, Sublime Text is another excellent choice. It offers many of the same features as VS Code but can be quicker on older hardware.

- **Atom**: Developed by GitHub, Atom is a hackable text editor for the 21st Century. It's highly customizable, though sometimes slower than the other options.

Choose one that you find intuitive and comfortable, as you will spend many hours working with it.

For more information about how to install VS Code, please visit our detailed blog posts on this topic: https://www.cuantum.tech/post/stepbystep-guide-to-installing-visual-studio-code-vs-code-on-windows-mac-and-linux

2. Web Browser

A modern web browser is crucial for testing and running JavaScript. While most browsers support JavaScript, the following are recommended for their developer-friendly tools:

- **Google Chrome**: Widely used due to its extensive Developer Tools, which make debugging JavaScript straightforward.
- **Mozilla Firefox**: Offers robust Developer Tools with some unique features such as a CSS Grid inspector and a more privacy-focused browsing experience.
- **Microsoft Edge**: Built on the same engine as Chrome, Edge provides similar developer tools with additional integrations unique to the Windows operating system.

3. Node.js and npm

Node.js is a JavaScript runtime built on Chrome's V8 JavaScript engine. It allows you to run JavaScript on the server side or for developing server-side applications. npm (Node Package Manager) is included with Node.js and is essential for managing third-party packages you might use in your projects.

Download and install Node.js from the official website (https://nodejs.org/). This installation will also install npm, which is vital for handling project dependencies.

For more details on how to install Node.js, please refer to the following informative blog post: https://www.cuantum.tech/post/how-to-install-nodejs-on-windows-mac-and-linux-a-stepbystep-guide

4. Version Control System

Git is the most widely used system for version control. It helps you track changes, revert to previous states, and collaborate with others on projects. Download and install Git from the official website (https://git-scm.com/). Once installed, you can integrate it with your code editor to streamline your workflow.

For more information about version control, please visit the following informative and useful blog post: https://www.cuantum.tech/post/understanding-version-control-systems-what-they-are-and-why-you-need-one

5. GitHub Account

While not a tool in the traditional sense, having a GitHub account will be immensely useful. It allows you to manage your repositories, collaborate, and store your projects in the cloud. Create a free account at GitHub (https://github.com/) and start familiarizing yourself with its features.

6. Local Server Environment

While many simple JavaScript tasks can be run directly in a browser, developing more complex applications often requires a local server. For beginners, extensions in Visual Studio Code, like Live Server, can automatically reload your browser as you modify files, simulating a live environment.

7. Developer Extensions and Plugins

Consider installing additional browser extensions such as React Developer Tools or Vue.js devtools if you plan on working with those frameworks. These tools provide deeper insights into applications built with these technologies.

To know what is and how to install React Developer Tools and Vue.js please visit the following blog posts:

1. **React Developer Tools:** https://www.cuantum.tech/post/how-to-install-and-set-up-react-developer-tools-a-comprehensive-guide

2. **Vue.js:** https://www.cuantum.tech/post/how-to-install-and-set-up-vuejs-for-your-next-project

With these tools and settings, you will have a robust development environment that supports learning and building applications in JavaScript. Each tool contributes to a seamless development experience, allowing you to focus on mastering JavaScript and building impressive web projects.

Part I: Getting Started with JavaScript

Chapter 1: Introduction to JavaScript

Welcome to the introductory chapter of "**JavaScript from Zero to Superhero: Unlock your web development superpowers**." This chapter serves as your gateway into the captivating world of JavaScript, a programming language that has not only become a cornerstone of web development but also revolutionized the way we interact with the digital realm.

If you're an aspiring web developer, a seasoned programmer looking to diversify your skill set, or even a tech enthusiast with a passion for coding, gaining a solid understanding of JavaScript is an invaluable asset. The knowledge of JavaScript will open up a multitude of possibilities for you, paving the way for creativity, innovation, and problem-solving.

As we embark on this exciting journey, we'll start with a deep dive into the rich history of JavaScript. By understanding its origins, development trajectory, and evolution, we can appreciate why it holds such a pivotal and irreplaceable position in today's tech landscape. This historical overview will provide the context needed to understand the language's profound impact and its continuing relevance in an ever-evolving industry.

1.1 History of JavaScript

JavaScript, often abbreviated as JS, was created in 1995 by Brendan Eich while he was an engineer at Netscape. It was initially developed under the name Mocha, then later renamed to LiveScript, and finally to JavaScript. This renaming coincided with Netscape adding support for Java applets in its browser, leading to a common misconception that JavaScript is somehow an offshoot of Java. In reality, the similarities between the two languages are few, as they were developed independently of each other for different purposes.

The primary motivation behind creating JavaScript was to make web pages interactive. Before JavaScript, web pages were static, meaning that any interaction required a user to reload a page or submit data to the server. JavaScript enabled developers to add interactive elements to web pages that could respond to user actions without needing to reload the page. This capability fundamentally changed how developers and designers thought about building websites, paving the way for the dynamic web experiences we have today.

In 1997, JavaScript was taken to ECMA International, a standards organization, to carve out a standard specification which would ensure that different browsers could implement the language consistently. This led to the formation of ECMAScript, the official standard for JavaScript. ECMAScript has undergone many revisions to add new features and improvements, with major versions known as ES5 (ECMAScript 5), ES6 (also known as ECMAScript 2015), and so on.

The introduction of AJAX (Asynchronous JavaScript and XML) in the early 2000s was another pivotal moment for JavaScript. AJAX allowed web pages to request data from the server asynchronously without interfering with the display and behavior of the existing page. This not only improved the user experience by making web pages faster and more responsive but also gave rise to single-page applications (SPAs) — web applications that load a single HTML page and dynamically update that page as the user interacts with the app.

JavaScript's capabilities have been vastly extended over the years from a simple scripting tool to a powerful language capable of front-end and back-end development, thanks to Node.js. Introduced in 2009, Node.js is an open-source, cross-platform JavaScript runtime environment that allows developers to execute JavaScript code outside a browser. This innovation has been crucial in popularizing JavaScript even further, making it a versatile, all-encompassing programming language.

Example Code: Basic JavaScript Interaction

Let's look at a simple example to illustrate JavaScript's ability to add interactivity to a web page. Consider a web page with a button that, when clicked, displays the current date and time:

```
<!DOCTYPE html>
<html>
<head>
    <title>Simple JavaScript Example</title>
</head>
<body>
    <button onclick="displayDate()">What's the time?</button>
    <p id="time"></p>

    <script>
        function displayDate() {
            document.getElementById("time").innerHTML = new Date().toLocaleString();
        }
    </script>
</body>
```

```
</html>
```

In this example, clicking the button triggers the displayDate() function, which modifies the content of the paragraph element to display the current date and time. This is a fundamental example of how JavaScript can interact with the HTML structure of a page to dynamically modify content.

The initial interaction with code can seem overwhelming, but there's no need to worry. We will simplify everything step by step in this book. For now, let's start by breaking down the code:

1. Building the Page Structure (HTML):

- The code starts with <!DOCTYPE html>, which tells the web browser this is an HTML document.
- <html> and </html> tags define the main structure of the web page.
- Inside the <html>, we have a <head> section containing the page title displayed on the browser tab. Here, the title is "Simple JavaScript Example".
- The most important part for our purpose is the <body> section. This is where the content that appears on the webpage is written.

2. Button and Display Area (HTML):

- Inside the <body>, we first create a button using the <button> tag. The text displayed on the button is "What's the time?"
- An important attribute for the button is onclick. This tells the browser what action to perform when the button is clicked. In this case, the value is set to displayDate(), which refers to a function we'll define later.
- Next, we have a paragraph (<p>) element with an id of "time". This paragraph will be used to display the current date and time.

3. Making it Work with JavaScript:

- The <script> tag tells the browser that the code within it is JavaScript.
- Inside the <script>, we define a function called displayDate(). This function will be executed whenever the button is clicked.
- The function uses document.getElementById("time") to find the paragraph element with the id "time" on the webpage.
- The magic happens with .innerHTML. This property allows us to change the content displayed within the paragraph element.

- In our case, we set the content to new Date().toLocaleString(). Let's break this down further:
  - new Date(): This creates a new instance of the built-in JavaScript Date object, which represents the current date and time.
  - .toLocaleString(): This is a method of the Date object that converts the date and time into a human-readable format, depending on your browser's language settings.

Summary:

This code creates a simple web page with a button. Clicking the button triggers a JavaScript function that retrieves the current date and time, formats it in a user-friendly way, and displays it on the webpage within the paragraph element.

1.1.1 Influential Figures

While Brendan Eich is widely recognized as the original creator of JavaScript, having developed it during his time at Netscape in the 1990s, there have been several other influential figures who have played pivotal roles in its ongoing evolution and increasing sophistication:

Douglas Crockford: Crockford is well-known in the development community for his substantial work on JSON (JavaScript Object Notation). This lightweight data-interchange format, which is easy for humans to read and write and easy for machines to parse and generate, is now universally used for data interchange on the web. Beyond his work with JSON, Crockford has also written extensively about JavaScript. His influential book, "JavaScript: The Good Parts," has been widely read and has helped many developers to understand and utilize the more robust parts of the language effectively, leading to a greater appreciation of JavaScript's capabilities and potential.

Ryan Dahl: As the creator of Node.js, Dahl has had a significant impact on expanding the capabilities of JavaScript, extending its reach beyond the browser. By enabling server-side programming, Dahl's work with Node.js has truly transformed JavaScript into a full-stack development language. This has allowed it to handle everything from front-end interactions to back-end database operations, greatly increasing the versatility and utility of JavaScript in the world of web development.

1.1.2 Community and Culture

JavaScript's extensive and vibrant community is undoubtedly one of its most valuable attributes, fostering a rich culture of innovation, creativity, and support that is essential for the language's continued growth and development:

Open Source Projects: The JavaScript community has developed and continues to maintain a vast array of libraries and frameworks. These include the likes of jQuery, AngularJS, React, and Vue.js, to name just a few. These projects, driven by the community's pioneering spirit, are continually pushing the boundaries of what can be achieved with JavaScript. They have played a substantial role in shaping the language's capabilities and have led to significant advancements in web development.

Forums and Learning Platforms: Knowledge sharing is a key element of the JavaScript community, facilitated by various online platforms. Websites like Stack Overflow, Mozilla Developer Network (MDN), and freeCodeCamp provide invaluable resources where both new learners and seasoned developers can find ample support. These sites offer comprehensive tutorials, in-depth articles, and a platform for community insights and discussions, thereby promoting a conducive learning environment.

Conferences: The JavaScript community also organizes annual conferences such as JSConf and Node.js Interactive. These events serve as vital hubs for knowledge exchange, industry networking, and showcasing new technologies and techniques within the JavaScript ecosystem. They bring together developers from around the world, fostering a sense of unity and collaboration, which in turn contributes to the collective growth and development of the JavaScript language.

1.1.3 JavaScript on the Web Today

JavaScript, a powerful and versatile scripting language, is virtually omnipresent in today's world of web development. With almost all web browsers incorporating dedicated JavaScript engines to process and interpret it efficiently, JavaScript has become an integral part of the digital landscape:

Statistics and Adoption: According to the latest surveys and research findings, JavaScript is used by over 95% of all websites worldwide. This ubiquitous presence not only highlights its critical role in modern web development but also underscores its importance as a key technology in the digital world. JavaScript's widespread adoption is a testament to its versatility and robustness, making it a cornerstone of web technology.

Frameworks and Tools: The advent of modern JavaScript frameworks such as React, Angular, and Vue has revolutionized web development, making it easier than ever to build complex and high-performing web applications. These innovative tools abstract many of the complexities and challenges associated with raw JavaScript, providing developers with powerful and sophisticated capabilities to create responsive and dynamic user experiences.

They allow developers to focus on creating engaging and intuitive interfaces, all while ensuring optimal performance and responsiveness. This new generation of JavaScript tools has ushered in a new era of web development, empowering developers to create more efficient, effective, and user-friendly web applications.

1.1.4 Controversies and Challenges

JavaScript, despite its widespread use and immense popularity in the realm of web development, has had its fair share of criticisms. These criticisms primarily revolve around security issues, performance concerns, and the complexity of best practices.

One of the most significant criticisms is related to **Security Issues**. JavaScript's ability to interact directly with web browsers can be a double-edged sword. While it provides a high level of interaction and user experience, it can also be exploited for malicious purposes if not appropriately managed. A common vulnerability that JavaScript applications need to be aware of and guard against is Cross-site scripting (XSS). This is where malicious scripts are injected into trusted websites, which can then be used to steal sensitive information.

Another area of concern that has been raised over the years is **Performance Concerns**. In its earlier iterations, JavaScript was considerably slower, which consequentially limited what could be done efficiently. However, the advent of modern JavaScript engines like V8, used in Google Chrome, and SpiderMonkey, used in Firefox, has dramatically enhanced execution speeds, making JavaScript much more efficient.

The growing complexity of JavaScript and the associated best practices are also areas of criticism. As JavaScript has evolved and grown in functionality and use cases, so too has the **Complexity of Best Practices** associated with it. This growing complexity can be daunting for beginners trying to get a foothold in JavaScript programming and is a frequent topic of debate within the developer community.

1.1.5 Future Prospects

As we look towards the future, the role of JavaScript in web and application development only seems to be expanding and becoming even more prominent for several reasons:

ECMAScript Evolution: JavaScript is a dynamic language that continues to evolve with the addition of new features and improvements. These are added to the JavaScript standard, known as ECMAScript. Future versions of ECMAScript are expected to enhance capabilities around modularity, performance, and syntactic sugar. These updates and revisions are designed to make the language more powerful, versatile, and user-friendly, catering to the requirements of modern web development.

Beyond the Web: The use of JavaScript is extending beyond traditional web development. It's proving its versatility and adaptability by reaching out into new areas like mobile app development. Notably, through frameworks like React Native, JavaScript is making its mark in this space. Moreover, JavaScript is also extending into the realm of IoT (Internet of Things), demonstrating its capacity to adapt and stay relevant in a rapidly changing technological landscape.

WebAssembly: The introduction of WebAssembly has been a game-changer for web applications. WebAssembly allows for running code written in languages other than JavaScript at near-native speed.

This significant advancement can complement and enhance JavaScript's capabilities in web applications, providing a powerful combination of speed and functionality. This means that JavaScript can work hand in hand with other programming languages, thereby boosting the performance of web applications and providing a richer user experience.

1.2 What Can JavaScript Do?

Once upon a time, JavaScript was simply conceived as a straightforward scripting language. It was designed with a singular purpose in mind: to enhance the functionality of web browsers by infusing them with dynamic capabilities. However, as time passed and technology evolved, JavaScript has grown significantly, transforming into a potent tool with immense power. Today, it does not merely serve to add dynamism to web browsers. Instead, it powers complex, intricate applications and spans across a multitude of environments, far beyond the confines of its original purpose.

To truly appreciate the adaptability of JavaScript and the reason behind its emergence as a cornerstone in the realm of modern web development, one must unravel and understand its expansive capabilities. As a language, JavaScript has evolved and adapted, much like a living organism, to meet the demands of a rapidly evolving digital landscape, proving its worth time and again.

Let's delve deeper and explore the fascinating world of JavaScript's capabilities. This section will take us beyond the realm of conventional web development, introducing us to a vast range of domains where JavaScript plays a crucial role and leaves a significant, enduring impact.

Note: Some examples in this section may seem overwhelming or confusing. They are used merely to illustrate the capabilities of JavaScript. Don't worry if you don't fully understand them yet. Throughout the book, you will gain the knowledge to master JavaScript and its uses.

1.2.1 Interactivity in Web Pages

JavaScript, at its core, plays a crucial and game-changing role in enhancing user experiences on the digital platform. It cleverly achieves this by providing web pages with the ability to respond in real-time to user interactions, effectively eliminating the tedious process of reloading the entire page, thus ensuring a more seamless and efficient user experience.

This element of instant responsiveness goes beyond being just a mere feature or add-on, it is indeed an inherent and fundamental capability in the landscape of contemporary web design. It serves as a testament to the power of modern technologies and how they can greatly improve the way we interact with digital platforms. This dynamic nature of JavaScript, which allows for the creation and implementation of interactive content, has the potential to transform static and potentially monotonous web pages into lively, engaging, and interactive platforms.

Therefore, the importance of JavaScript is not just significant, but rather, it is monumental in today's digital age. In a world where user interaction and experience are paramount, where the digital experience can make or break a business or service, the role of JavaScript is pivotal. Its ability to create a more interactive and engaging platform is a key driver in the success of modern web design, and as such, its importance cannot be overstated.

Example: Interactive Forms Consider a sign-up form where JavaScript is used to validate input before it is sent to the server. This immediate feedback can guide users and prevent errors in data submission.

```html
<!DOCTYPE html>
<html>
<head>
    <title>Signup Form Example</title>
</head>
<body>
    <form id="signupForm">
        Username: <input type="text" id="username" required>
        <button type="button" onclick="validateForm()">Submit</button>
        <p id="message"></p>
    </form>
```

```
    <script>
        function validateForm() {
            var username = document.getElementById("username").value;
            if(username.length < 6) {
                document.getElementById("message").innerHTML = "Username must be at
least 6 characters long.";
            } else {
                document.getElementById("message").innerHTML = "Username is valid!";
            }
        }
    </script>
</body>
</html>
```

In this example, the JavaScript function validateForm() checks the length of the username and provides immediate feedback on the same page, enhancing the user's experience by providing immediate, useful feedback without reloading the page.

Code breakdown:

1. Building the Signup Form (HTML):

- Similar to the previous example, we start with the basic HTML structure.
- This time, the title is "Signup Form Example".
- Inside the <body>, we create a form element using the <form> tag. The form element is used to collect user input. We assign it an id of "signupForm" for easier referencing later.

2. Username Input and Submit Button (HTML):

- Within the form, we have a label "Username:" followed by an <input> element. This is where the user will enter their username.
    - The type attribute is set to "text", indicating it's a text field for entering characters.
    - The id attribute is set to "username" to identify this specific input field.
    - The required attribute ensures the user enters a username before submitting the form.
- Next, we have a button element with the text "Submit". However, this time, the type attribute is set to "button" instead of "submit". This means clicking the button won't submit the form by default, allowing us to control the submission process with JavaScript.

- The button has an onclick attribute set to validateForm(). This calls a JavaScript function we'll define later to validate the username before submission.
- Finally, we have a paragraph element with the id "message". This paragraph will be used to display any messages related to username validation.

3. Validating Username with JavaScript:

- The <script> tag remains the same, indicating the code within is JavaScript.
- We define a function called validateForm(). This function will be executed whenever the "Submit" button is clicked.
- Inside the function:
  - We use var username = document.getElementById("username").value; to retrieve the value entered by the user in the username field.
    - document.getElementById("username") finds the element with the id "username" on the webpage (the username input field).
    - .value extracts the actual text the user typed in that field.
  - We use an if statement to check if the username length is less than 6 characters using .length.
    - If the username is too short, we display a message using document.getElementById("message").innerHTML. We set the message content to inform the user about the minimum username length requirement.
  - Otherwise (else block), we display a success message indicating the username is valid.

Summary:

This code creates a signup form with a username field and a submit button. When the button is clicked, a JavaScript function validates the username length. If the username is less than 6 characters, an error message is displayed. Otherwise, a success message is shown.

1.2.2 Rich Internet Applications (RIAs)

JavaScript is the fundamental backbone, the essential building block, of single-page applications (SPAs) that are a common feature in the landscape of modern web applications. We see examples of these sophisticated applications in popular platforms such as Google Maps or Facebook. In these highly interactive, feature-rich applications, JavaScript holds an incredibly pivotal role where it is tasked with handling a myriad of responsibilities. These range from data requests, managing front-end routing, and controlling page transitions, to a host of other tasks that are vital for the application's performance.

JavaScript's functionalities go well beyond the simple execution of tasks. It is instrumental in providing a seamless, almost desktop-like experience directly within the confines of a web browser. This significant enhancement to the user's experience is achieved by making the interface more smooth and interactive. It successfully replicates the fluidity and responsiveness one would naturally expect from a full-fledged desktop application, thereby bridging the gap between web and desktop application experiences.

Without the power and flexibility of JavaScript, these single-page applications would not be able to deliver the kind of seamless, immersive user experience they are known for. It is JavaScript that breathes life into these applications, making them more than just static web pages, transforming them into dynamic, interactive digital experiences that engage and delight users.

Example: Dynamic Content Loading

JavaScript can dynamically load content into a page without a full reload. This is used extensively in SPAs where user actions trigger content changes directly.

```javascript
document.getElementById('loadButton').addEventListener('click', function() {
  fetch('data/page2.html')
    .then(response => response.text())
    .then(html => document.getElementById('content').innerHTML = html)
    .catch(error => console.error('Error loading the page: ', error));
});
```

In this snippet, when a button is clicked, JavaScript fetches new HTML content and injects it into a content div, updating the page dynamically.

Code breakdown:

1. Triggering the Action (JavaScript):

- This code snippet uses JavaScript to add functionality to a button.

2. Finding the Button and Adding a Listener (addEventListener):

- The first line, document.getElementById('loadButton'), finds the button element on the webpage using its id, which we can assume is set to "loadButton" in the HTML code (not shown here).
- .addEventListener('click', function() {...}) is a powerful function that allows us to attach an event listener to the button.

- o In this case, the event we're listening for is "click". So, whenever the user clicks this button, the code within the curly braces ({...}) will be executed.

3. Fetching External Content (fetch):

- Inside the function triggered by the click event, we use the fetch function. This is a modern and powerful way to retrieve data from the server.
- In our case, fetch('data/page2.html') tries to fetch the content of a file named "page2.html" located in a folder called "data" (relative to the current HTML file).

4. Processing the Fetched Data (then):

- The fetch function returns a promise. A promise is a way of handling asynchronous operations (operations that take time to complete) in JavaScript.
- Here, we use the .then method on the promise returned by fetch. This allows us to define what to do with the data once it's successfully fetched.
    - o Inside the .then method, we receive a "response" object. This object contains information about the fetched data.
- We use another .then method on the "response" object. This time, we call the response.text() method. This extracts the actual text content from the response, assuming it's HTML in this case.

5. Updating the Page Content (innerHTML):

- We receive the fetched HTML content (text) from the previous .then method.
- We use document.getElementById('content') to find the element with the id "content" on the webpage (presumably a container element where we want to display the loaded content).
- We set the innerHTML property of the "content" element to the fetched HTML content (html). This essentially replaces the existing content within the "content" element with the content from "page2.html".

6. Handling Errors (catch):

- The fetch operation might fail for various reasons, like server errors or network issues.
- To handle potential errors, we use the .catch method on the initial fetch call.
- The .catch method receives an "error" object if the fetch operation fails.
- Inside the .catch block, we use console.error('Error loading the page: ', error) to log the error message to the browser's console. This helps developers identify and troubleshoot any issues during development.

Summary:

This code demonstrates how to load content from an external HTML file using the fetch API and update the current webpage dynamically based on user interaction (clicking a button). It also introduces the concept of promises for handling asynchronous operations and error handling using .catch.

1.2.3 Server-Side Development

Node.js signifies a remarkable evolution in JavaScript's capabilities, liberating it from the constraints that once relegated it strictly to the confines of the browser. With the advent of Node.js, developers can now harness the power of JavaScript to engineer a myriad of new applications, including the construction of server-side applications.

One such application is the handling of HTTP requests, an integral feature of any contemporary web application. This functionality allows developers to create a more interactive and responsive user experience. It enables real-time updates and asynchronous communication, thereby transforming static web pages into dynamic platforms for user engagement.

Moreover, Node.js also facilitates interaction with databases. It provides a seamless interface for querying and retrieving data, further showcasing its versatility and strength as a robust tool for back-end development. This capability makes Node.js an ideal choice for building applications that require real-time data handling, such as chat applications, collaborative tools, and data streaming platforms.

The expanded capabilities of Node.js open up a universe of opportunities for developers, enhancing the power and flexibility of JavaScript as a programming language. This evolution redefines JavaScript's role in web development, elevating it from a simple scripting language to a versatile tool for building complex, scalable applications.

Example: Simple HTTP Server This example uses Node.js to create a basic HTTP server that responds to all requests with a friendly message:

```javascript
const http = require('http');

const server = http.createServer((req, res) => {
    res.writeHead(200, {'Content-Type': 'text/plain'});
    res.end('Hello, welcome to our server!');
});

server.listen(3000, () => {
    console.log('Server is running on <http://localhost:3000>');
```

```
});
```

When you run this script with Node.js, it starts a web server that sends "Hello, welcome to our server!" to any client requests.

Code breakdown:

1. Entering the Node.js World:

- This code snippet is written in JavaScript, but it's specifically designed to run in a Node.js environment. Node.js allows JavaScript to be used for server-side development, meaning it can create web servers and handle requests from users.

2. Including the HTTP Module (require):

- The first line, const http = require('http');, is essential for working with HTTP in Node.js.
    - const is used to declare a variable.
    - require('http'); imports the built-in HTTP module provided by Node.js. This module gives us the tools to create an HTTP server.

3. Creating the Server (http.createServer):

- The next line, const server = http.createServer((req, res) => {...});, creates the actual HTTP server.
    - http.createServer is a function from the imported HTTP module.
    - It takes a callback function (the part between parentheses) that defines how the server will respond to incoming requests.
    - Inside the callback function, we receive two arguments:
        - req (request): This object represents the incoming HTTP request from a client (like a web browser).
        - res (response): This object allows us to send a response back to the client.

4. Sending a Simple Response:

- Inside the callback function:
    - res.writeHead(200, {'Content-Type': 'text/plain'}); sets the response headers.
        - The first argument, 200, is the status code indicating a successful response.

- The second argument is an object defining the response headers. Here, we set the Content-Type to text/plain, indicating the response content is plain text.
  - res.end('Hello, welcome to our server!'); sends the actual response content as a string to the client.

5. Starting the Server (server.listen):

- The last two lines, server.listen(3000, () => {...});, start the server and log a message to the console.
  - server.listen is a method on the server object. It takes two arguments:
    - The first argument, 3000, is the port number on which the server will listen for incoming requests.
    - The second argument is a callback function that executes once the server starts listening successfully.
  - Inside the callback function, we use console.log to print a message indicating the server is running and accessible at http://localhost:3000. This includes the "http://" part because it's a web server and "localhost" refers to your own machine.

Summary:

This code creates a basic HTTP server in Node.js. It demonstrates how to handle incoming requests, set response headers, send content back to the client, and start the server on a specific port. This is a fundamental building block for creating more complex web applications with Node.js and JavaScript.

1.2.4 Internet of Things (IoT)

JavaScript is not just a versatile and widely-used programming language but has significantly expanded its reach into the rapidly growing and innovative field of the Internet of Things (IoT). In this cutting-edge sector, the use of JavaScript extends beyond traditional web and mobile applications, allowing developers to control a variety of hardware devices, gather data from a multitude of diverse sources, and perform a host of other critical functions that are pivotal in the technologically advanced digital age we live in today.

Supporting JavaScript's foray into IoT are numerous frameworks, one of the most prominent being Johnny-Five. This particular framework significantly enhances the capabilities of JavaScript, transforming it from a mere scripting language to an invaluable, powerful tool that is extensively used in the prototyping and building of robust, efficient, and scalable IoT applications.

These applications are not limited to a specific domain but span across a wide variety of sectors. They include everything from home automation systems that improve the quality of life by automating routine tasks, to industrial IoT applications that streamline and optimize complex industrial processes. These diverse applications perfectly showcase the immense flexibility and power that JavaScript wields in the ever-evolving landscape of IoT.

1.2.5 Animation and Games

JavaScript, a remarkably versatile programming language, is far from being confined to the usual realm of static web pages and simplistic data management. In reality, it is much more expansive, being extensively employed in the complex design and meticulous implementation of animations and game development. Its usage adds a significant layer of dynamism and interactive components to various digital platforms, thereby enhancing user engagement and experience.

Developers can leverage the powerful capabilities of JavaScript in combination with robust libraries such as Three.js and Phaser. These libraries provide a rich set of tools and functionalities that empower developers to not only build but also intricately design complex 3D animations and interactive games, adding a new dimension to digital platforms.

The use of these libraries transcends the traditional boundaries of programming. They provide the tools necessary to breathe life into otherwise static digital scenes. With their aid, developers have the ability to transform these static scenes into immersive virtual realities, interactive games, and visually stunning web interfaces that captivate the audience. These tools open up new avenues for creativity and innovation in the digital world, making it possible to create engaging and visually appealing experiences for users.

1.2.6 Educational and Collaborative Tools

JavaScript, a powerful and versatile programming language, serves as the backbone of many modern educational platforms and real-time collaboration tools. It is this multifaceted language that breathes life into a wide variety of advanced features that, in our current digital age, have become almost second nature to us.

Take, for example, the function of document sharing. This technological advancement has completely transformed both our professional and educational landscapes by providing a platform for the seamless exchange of information and fostering a collaborative environment. And the unsung hero behind this revolution? None other than JavaScript.

In addition, consider the tool of video conferencing. In our current era, marked by remote work and distance learning, video conferencing has proven to be an invaluable resource. It allows us to maintain a semblance of normalcy, facilitating face-to-face interactions despite geographical barriers. And the technological wizardry that makes this possible is, once again, JavaScript.

Lastly, let's look at real-time updates. This feature, often overlooked, guarantees that we always have access to the most up-to-date and accurate information. Whether it's the latest news headlines, stock market fluctuations, or simply the score in a live sports event, real-time updates keep us informed and in the loop. And it is JavaScript, with its robust capabilities, that provides this feature.

In summary, JavaScript, with its power and versatility, lies at the heart of the digital tools and platforms we often take for granted. Its influence spans across various aspects of our digital lives, enabling functionalities that have become integral to our everyday routines.

1.2.7 Progressive Web Applications (PWAs)

JavaScript plays a crucial role in the development of Progressive Web Applications (PWAs), which are an important aspect of modern web development. PWAs use the latest web capabilities to provide an experience that feels very much like using a native app, yet they run in a web browser.

This is a powerful combination that offers numerous advantages to users, including the ability to work offline, perform well even on slow networks, and be installed on the user's home screen just like a native app. A key part of this is the use of JavaScript, which is responsible for managing service workers.

Service workers are essentially scripts that your browser runs in the background, separate from a web page, opening the door to features which don't need a web page or user interaction. Among other things, they enable features such as push notifications and background sync, both of which significantly enhance user experience.

These service workers are a key feature of PWAs, and it is JavaScript that controls their operation.

Example: Registering a Service Worker

```javascript
if ('serviceWorker' in navigator) {
    navigator.serviceWorker.register('/service-worker.js')
    .then(function(registration) {
        console.log('Service Worker registered with scope:', registration.scope);
```

```
    }).catch(function(error) {
        console.log('Service Worker registration failed:', error);
    });
}
```

This example shows how to register a service worker using JavaScript, which is fundamental for enabling offline experiences and background tasks in PWAs.

Code breakdown:

1. Checking for Service Worker Support (if statement):

- This code checks if the browser supports service workers.
    - if ('serviceWorker' in navigator) is the conditional statement that starts it all.
        - 'serviceWorker' in navigator checks whether the navigator object (which provides information about the browser) has a property named serviceWorker. This property indicates if service workers are supported in that particular browser.

2. Registering the Service Worker (navigator.serviceWorker.register):

- If service workers are supported (if condition is true), the code proceeds to register a service worker using navigator.serviceWorker.register('/service-worker.js').
    - navigator.serviceWorker.register is a method provided by the navigator object to register a service worker script.
    - /service-worker.js is the path to the JavaScript file containing the service worker logic. This file likely resides in the same directory (or a subdirectory) as the HTML file where this code is placed.

3. Handling Registration Success and Failure (then and catch):

- The .register method returns a promise. A promise is a way of handling asynchronous operations (operations that take time to complete) in JavaScript.
    - .then(function(registration) {...}) defines what to do if the service worker registration is successful.
        - The function receives a registration object as an argument. This object contains information about the registered service worker.
    - .catch(function(error) {...}) defines what to do if the service worker registration fails.

- The function receives an error object as an argument, which contains details about the encountered error.

4. Logging Registration Status:

- Inside both the .then and .catch blocks, we use console.log to log messages to the browser's console.
  - In the success case, we log a message indicating successful registration along with the service worker's scope using registration.scope. The scope determines the URLs the service worker can control.
  - In the failure case, we log a message indicating registration failure and the specific error details from the error object.

Summary:

This code snippet registers a service worker script if the browser supports them. It leverages promises to handle the asynchronous nature of the registration process and logs success or failure messages to the console for debugging and informative purposes. Service workers are powerful tools for enhancing web applications with offline capabilities, push notifications, and background functionality. This code provides a foundational step for utilizing them in your web projects.

1.2.8 Machine Learning and Artificial Intelligence

With the advent of advanced libraries such as TensorFlow.js, developers who specialize in JavaScript now have the ability to seamlessly incorporate machine learning capabilities directly into their web applications.

This opens up a whole new realm of possibilities and enables the integration of sophisticated and cutting-edge features such as image recognition - which allows the application to identify and process objects in images, natural language processing - a technology that enables the application to understand and interact in human language, and predictive analytics - a feature that uses data, statistical algorithms and machine learning techniques to identify the likelihood of future outcomes.

All these can be achieved without the developers needing to have a specialized background in machine learning or artificial intelligence. This is a significant step forward in making machine learning more accessible and widely used in web application development.

Example: Basic TensorFlow.js Model

```
async function run() {
    const model = tf.sequential();
    model.add(tf.layers.dense({units: 1, inputShape: [1]}));
    model.compile({loss: 'meanSquaredError', optimizer: 'sgd'});

    const xs = tf.tensor2d([1, 2, 3, 4], [4, 1]);
    const ys = tf.tensor2d([1, 3, 5, 7], [4, 1]);

    await model.fit(xs, ys, {epochs: 500});
    document.getElementById('output').innerText = model.predict(tf.tensor2d([5], [1,
1])).toString();
}

run();
```

This script sets up a simple neural network model that learns to predict output based on input data, showcasing how JavaScript can be used for basic AI tasks directly in the browser.

Code breakdown:

1. Entering the Machine Learning World:

- This code dives into the world of machine learning using TensorFlow.js, a popular library that allows you to train machine learning models directly in the browser with JavaScript.

2. Defining an Async Function (async function run):

- The code starts with async function run() {...}, which defines an asynchronous function named run. Asynchronous functions allow us to handle code that takes time to complete without blocking the main thread. This is important for machine learning tasks that often involve training models on data.

3. Building the Machine Learning Model (tf.sequential):

- Inside the run function:
    - const model = tf.sequential(); creates a sequential model object using TensorFlow.js (tf). This object will hold the layers and configuration of our machine learning model.
    - model.add(tf.layers.dense({units: 1, inputShape: [1]})); adds a dense layer to the model.

- Dense layers are a fundamental building block for neural networks. They perform linear transformations on the data.
- Here, units: 1 specifies the layer has one output unit.
- inputShape: [1] defines the expected input shape for this model. In this case, it expects a single number as input.

4. Configuring the Model (model.compile):

- model.compile({loss: 'meanSquaredError', optimizer: 'sgd'}); configures the training process for the model.
  - loss: 'meanSquaredError' defines the loss function used to measure how well the model's predictions match the actual values. Mean squared error is a common choice for regression problems.
  - optimizer: 'sgd' specifies the optimizer algorithm used to adjust the model's weights during training. SGD (stochastic gradient descent) is a popular choice.

5. Preparing the Training Data (tf.tensor2d):

- const xs = tf.tensor2d([1, 2, 3, 4], [4, 1]); creates a 2D tensor named xs using TensorFlow.js. This tensor represents our training data for the model's inputs.
  - The data is an array of numbers: [1, 2, 3, 4].
  - [4, 1] defines the shape of the tensor. It has 4 rows (representing 4 training examples) and 1 column (representing the single input value for each example).
- const ys = tf.tensor2d([1, 3, 5, 7], [4, 1]); creates another 2D tensor named ys for the training data's target outputs (labels).
  - The data is an array of numbers: [1, 3, 5, 7].
  - The shape again matches [4, 1], corresponding to the 4 target values for each input in the xs tensor.

6. Training the Model (model.fit):

- await model.fit(xs, ys, {epochs: 500}); trains the model asynchronously.
  - model.fit is the method used to train the model. It takes three arguments:
    - xs: The input training data tensor (xs).
    - ys: The target output (label) tensor (ys).
    - {epochs: 500}: An object defining training options. Here, epochs: 500 specifies the number of training iterations (epochs) to perform. During each epoch, the model will go through all the training examples and adjust its internal weights to minimize the loss function.

7. Making a Prediction (model.predict):

- document.getElementById('output').innerText = model.predict(tf.tensor2d([5], [1, 1])).toString(); uses the trained model to make a prediction.
    - model.predict(tf.tensor2d([5], [1, 1])) predicts the output for a new input value of 5. It creates a new tensor with the shape [1, 1] representing a single input value.
    - .toString() converts the predicted value (a tensor) to a string for display.
    - Finally, we set the innerText property of the element with the id "output" (presumably a paragraph element) to display the predicted value on the webpage.

1.2.9 Accessibility Enhancements

JavaScript plays an absolutely pivotal role in the enhancement of web accessibility, a crucial aspect of modern web design. Its capabilities extend far beyond mere functionality, as it can dynamically update and alter web content in real time to meticulously comply with accessibility standards.

This not only ensures that the web content is compliant with international guidelines, but it also vastly improves the overall user experience. This dynamic nature of JavaScript is particularly beneficial for users with disabilities, providing them with far better navigation and interactivity options.

By doing so, it allows them to engage with the Web in a much more inclusive and user-friendly manner, thereby making the digital world a more accessible place.

Example: Enhancing Accessibility

```javascript
document.getElementById('themeButton').addEventListener('click', function() {
    const body = document.body;
    body.style.backgroundColor = body.style.backgroundColor === 'black' ? 'white' :
'black';
    body.style.color = body.style.color === 'white' ? 'black' : 'white';
});
```

This example shows how JavaScript can be used to toggle high contrast themes, which are helpful for users with visual impairments.

Code breakdown:

1. Triggering the Theme Change (Event Listener):

- This code snippet uses JavaScript to add interactivity to a button.
- The first line, document.getElementById('themeButton').addEventListener('click', function() {...});, sets up an event listener for the button with the id "themeButton".
    - .addEventListener('click', function() {...}) is a powerful function that allows us to attach an event listener to the button.
    - In this case, the event we're listening for is "click". So, whenever the user clicks this button, the code within the curly braces ({...}) will be executed.

2. Toggling Background and Text Color:

- Inside the function triggered by the click event, we define the logic for changing the theme (background and text color).
- const body = document.body; retrieves a reference to the <body> element of the webpage, where we want to apply the theme changes.
- The next two lines: body.style.backgroundColor = body.style.backgroundColor === 'black' ? 'white' : 'black'; body.style.color = body.style.color === 'white' ? 'black' : 'white'; use a clever technique to toggle between two color schemes (black background with white text and white background with black text) based on the current background color.

JavaScript

```
body.style.backgroundColor = body.style.backgroundColor === 'black' ? 'white' :
'black'; body.style.color = body.style.color === 'white' ? 'black' : 'white';
```

- .style.backgroundColor and .style.color access the CSS style properties for background-color and color of the body element, respectively.
- The assignment uses a ternary operator (? :). This is a shorthand way of writing an if-else statement. Here's how it works:
    - body.style.backgroundColor === 'black' checks if the current background color is black.
    - If it's black (=== 'black'), then the background color is set to 'white' (switch to white theme).
    - Otherwise (using the : after the first condition), the background color is set to 'black' (switch to black theme).
- The same logic applies to the color property, toggling between 'white' and 'black' based on the current text color.

Summary:

This code demonstrates how to listen for user interaction (button click) and dynamically change the webpage's theme (background and text color) using JavaScript's DOM manipulation techniques and clever use of the ternary operator for conditional assignments. This is a great example of adding user interactivity and basic styling control to a webpage.

1.3 Practical Exercises

At the end of this chapter, we provide practical exercises that allow you to apply what you've learned about JavaScript's history and capabilities. These exercises are designed to reinforce your understanding and help you gain hands-on experience with JavaScript.

If you find some of the exercises overwhelming, don't worry. You can return to them after completing the following chapters.

Exercise 1: Historical Quiz

- Q1: Who created JavaScript and in what year?
- Q2: What was JavaScript originally called?
- Q3: Describe one major milestone in the evolution of JavaScript.

Exercise 2: Form Validation

Create a simple HTML form for user registration that includes fields for username and email. Use JavaScript to validate the form so that:

- The username must be at least 6 characters long.
- The email must include an "@" symbol. Display error messages next to each field if the validation fails.

Solution:

```
<!DOCTYPE html>
<html>
<head>
    <title>Registration Form Validation</title>
    <script>
        function validateForm() {
            var username = document.getElementById("username").value;
            var email = document.getElementById("email").value;
            var errorMessage = "";
```

```
            if(username.length < 6) {
                errorMessage += "Username must be at least 6 characters long.\\\\n";
                document.getElementById("usernameError").innerText = errorMessage;
            } else {
                document.getElementById("usernameError").innerText = "";
            }

            if(email.indexOf('@') === -1) {
                errorMessage += "Email must include an '@' symbol.\\\\n";
                document.getElementById("emailError").innerText = errorMessage;
            } else {
                document.getElementById("emailError").innerText = "";
            }

            if(errorMessage.length > 0) {
                return false;
            }
        }
    </script>
</head>
<body>
    <form id="registrationForm" onsubmit="return validateForm()">
        Username: <input type="text" id="username" required>
        <span id="usernameError" style="color: red;"></span><br>
        Email: <input type="text" id="email" required>
        <span id="emailError" style="color: red;"></span><br>
        <button type="submit">Register</button>
    </form>
</body>
</html>
```

Exercise 3: Content Loading

Using JavaScript, write a function to load content from a text file into a div element on your webpage when a button is clicked. Assume the text file is called "content.txt".

Solution:

```
<!DOCTYPE html>
<html>
<head>
    <title>Dynamic Content Loading</title>
    <script>
        function loadContent() {
            fetch('content.txt')
            .then(response => response.text())
            .then(data => {
                document.getElementById('contentDiv').innerHTML = data;
            })
```

```
            .catch(error => {
                console.log('Error loading the content:', error);
                document.getElementById('contentDiv').innerHTML = 'Failed to load
content.';
            });
        }
    </script>
</head>
<body>
    <button onclick="loadContent()">Load Content</button>
    <div id="contentDiv"></div>
</body>
</html>
```

Exercise 4: Theme Switcher

Write a JavaScript function to toggle the color theme of a webpage between light (white background with black text) and dark (black background with white text) modes.

Solution:

```
<!DOCTYPE html>
<html>
<head>
    <title>Theme Switcher</title>
    <script>
        function toggleTheme() {
            var body = document.body;
            body.style.backgroundColor = body.style.backgroundColor === 'black' ?
'white' : 'black';
            body.style.color = body.style.color === 'white' ? 'black' : 'white';
        }
    </script>
</head>
<body>
    <button onclick="toggleTheme()">Toggle Theme</button>
</body>
</html>
```

These exercises provide practical scenarios to help you apply and deepen your understanding of JavaScript. By attempting these exercises, you'll enhance your ability to solve real-world problems using JavaScript.

Chapter 1 Summary

In this opening chapter of "JavaScript from Scratch: Unlock your Web Development Superpowers," we embarked on a journey to understand JavaScript, a language that stands as a cornerstone of modern web development. This chapter laid the foundational knowledge necessary to appreciate JavaScript's capabilities and its pivotal role in both the historical and current context of the web.

We began by delving into the history of JavaScript, which was developed by Brendan Eich in 1995. Originally conceived under the name Mocha, then renamed to LiveScript, and finally to JavaScript, this language was created to add interactivity to web pages—a novel concept at the time. JavaScript's evolution was significantly marked by its standardization as ECMAScript, which ensured consistent interpretation across different web browsers. This standardization has been crucial for the development community, fostering a reliable environment where JavaScript could thrive and expand its capabilities.

Further, we explored the extensive functionalities of JavaScript. Initially designed to make static HTML pages interactive, today JavaScript powers complex applications across multiple platforms. Its capabilities extend from simple page enhancements to managing back-end services via Node.js, and even to applications in artificial intelligence and the Internet of Things. The examples provided, from interactive forms to dynamic content loading, illustrated JavaScript's ability to enhance user experience and streamline web functionalities.

In practical terms, JavaScript enables developers to create rich internet applications (RIAs) such as single-page applications that offer seamless user experiences akin to desktop applications. We also touched upon JavaScript's role in server-side development, showcasing its versatility beyond client-side scripting.

The exercises at the end of the chapter were designed to reinforce the knowledge gained, with hands-on tasks ranging from form validation to theme switching. These exercises not only helped cement your understanding of JavaScript's basic functionalities but also encouraged you to think creatively about how JavaScript can be used to solve real-world problems.

As we conclude this chapter, you should have a solid understanding of what JavaScript is, where it came from, and the multitude of tasks it can perform. The historical insights, coupled with practical applications, set the stage for deeper exploration in the subsequent chapters. With this foundation, you are now better prepared to delve into the more complex aspects of JavaScript, from DOM manipulation to modern frameworks that are shaping the future of web development.

This chapter is just the beginning of your journey with JavaScript. As we move forward, each chapter will build upon this foundation, introducing more sophisticated concepts and techniques. Your path from learning the basics to mastering advanced JavaScript functionalities will be filled with exciting challenges and opportunities for growth.

Chapter 2: Fundamentals of JavaScript

Welcome to Chapter 2. This chapter is designed to provide you with an in-depth exploration of JavaScript's fundamental concepts, laying the essential groundwork for more complex topics and diverse applications that you'll encounter later in your coding journey.

Understanding these basic concepts is not just an academic exercise, but a crucial step in your development as a programmer. They form the building blocks of any JavaScript program, and a thorough grasp of them will enable you to write more efficient, effective code.

From the seemingly simple, like variables and data types, to the more nuanced, like operators and control structures, each concept will be explored thoroughly. Our goal is to provide you with a solid and unshakeable foundation in JavaScript programming.

We begin this chapter by introducing the very basics—variables and data types. These are essential components that you will use in every JavaScript program you write. This knowledge is not just foundational but also vital for understanding how JavaScript interprets and processes data. By the end of this chapter, you should have a clear understanding of these concepts, ready to apply them to your own coding projects.

2.1 Variables and Data Types

In JavaScript, a variable serves as a symbolic name or identifier for a value. The role of variables is central to programming as they are used to store data, which forms the backbone of any program. This stored data can be of various types, be it numbers, strings, or more complex data structures, and can be modified, manipulated, and utilized at different points throughout the execution of the program.

One of the key characteristics of JavaScript is that it is a dynamically typed language. What this means is that you do not have to explicitly declare the type of the variable when you initialize it, unlike statically typed languages where such a declaration is mandatory. This grants a significant amount of flexibility, allowing for rapid scripting, and makes JavaScript an accessible language for beginners due to its less stringent syntax.

However, this flexibility also demands a strong understanding and careful handling of the various types of data that JavaScript can deal with. Without a clear understanding of data types, there is a risk of unexpected behavior or errors in the program. Therefore, while the dynamic nature of JavaScript can speed up the process of scripting, it also places an emphasis on the importance of a thorough knowledge of data types.

2.1.1 Understanding Variable Declarations

JavaScript, provides three distinct keywords for declaring variables: var, let, or const. Each of these has its own unique characteristics and scopes.

Variable var

The keyword var has been a longstanding element in the realm of JavaScript, used traditionally for the purpose of variable declaration. It's a feature that has been woven into the language since its very birth. The scope of a var variable, which refers to the context in which the variable exists, is its current execution context. This context could either be the function in which it is enclosed or, in cases where the variable is declared outside the realms of any function, it is given a global scope.

To put it in simpler terms, a var variable can only be seen or accessed within the function within which it was declared. However, when a var variable is declared outside the boundaries of any specific function, its visibility spreads throughout the entire program. This universal visibility, spanning the breadth of the whole program, thereby assigns the variable a global scope. This means it can be accessed and manipulated from any part of the code, making var variables extremely versatile in their use.

Example:

The var keyword is used to declare a variable in JavaScript. Variables declared with var have function scope or global scope (if declared outside a function).

```javascript
// Global scope
var globalVar = "I'm a global variable";

function example() {
  // Function scope
  var functionVar = "I'm a function variable";
  console.log(functionVar); // Output: "I'm a function variable"
```

```
}

example();
console.log(globalVar); // Output: "I'm a global variable"
```

Code breakdown:

1. Global Variable:

- The code starts with var globalVar = "I'm a global variable";. This line declares a variable named globalVar and assigns the string value "I'm a global variable" to it.
  - The var keyword is used for variable declaration (older way in JavaScript, modern way uses let or const).
  - Since there's no let or const before it, and it's not inside any function, globalVar is declared in the global scope. This means it's accessible from anywhere in your code.

2. Function Scope:

- The code then defines a function named example().
- Inside the function:
  - var functionVar = "I'm a function variable"; declares another variable named functionVar with the value "I'm a function variable".
  - Here, functionVar is declared with var within the function, so it has **function scope**. This means it's only accessible within the example function and not outside of it.
- The function also includes console.log(functionVar); which prints the value of functionVar to the console, and you'll see the expected output "I'm a function variable".

3. Accessing Variables:

- After the function definition, the code calls the function with example();. This executes the code inside the function.
- Outside the function, there's another line: console.log(globalVar);. This attempts to print the value of globalVar. Since globalVar was declared globally, it's accessible here, and you'll see the output "I'm a global variable".

Summary:

This code shows the difference between global variables and function-scoped variables. Global variables can be accessed from anywhere in your code, while function-scoped variables are only accessible within the function where they are declared.

Variable let

let – Introduced in ECMAScript 6 (ES6), also known as ECMAScript 2015, let provides a contemporary and advanced way to declare variables in JavaScript. This is a step up from the traditional var declaration.

The key difference between the two lies in their scoping rules. Unlike the function-scoped var, let is block-scoped. Block scoping means that a variable declared with let is only visible within the block where it is declared, as well as any sub-blocks contained within. This is a significant improvement over var, which is function-scoped and can lead to variables being visible outside their intended scope.

As a result, using let for variable declaration enhances code readability and maintainability, as it offers more predictable behavior and reduces the risk of accidentally declaring global variables. This makes let an ideal choice when dealing with variable data that may change over time, particularly in larger codebases where managing scope can be challenging.

Example:

The let keyword is used to declare a variable with block scope ({ }). Variables declared with let are limited in scope to the block they are defined in.

```javascript
function example() {
  if (true) {
    // Block scope
    let blockVar = "I'm a block variable";
    console.log(blockVar); // Output: "I'm a block variable"
  }
  // console.log(blockVar); // Error: blockVar is not defined
}

example();
```

Code breakdown:

1. Block Scope with let:

- The code defines a function named example().
- Inside the function, there's an if statement: if (true) {...}. The condition is always true, so the code within the curly braces ({...}) will always execute.
- Within the if block:
  - let blockVar = "I'm a block variable"; declares a variable named blockVar using the let keyword and assigns the string value "I'm a block variable".
  - Here's the key point: let creates a block scope, meaning blockVar is only accessible within the code block where it's declared (the if block in this case).

2. Accessing blockVar:

- Inside the if block, there's console.log(blockVar);. This line can access blockVar because it's declared within the same block. You'll see the output "I'm a block variable" as expected.

3. Trying to Access Outside the Block (Error):

- Notice the commented line, // console.log(blockVar); // Error: blockVar is not defined. If you uncomment this line and try to run the code, you'll get an error message like "blockVar is not defined".
- This is because blockVar is declared with let within the if block, and its scope is limited to that block. Once the code execution moves outside the block (after the closing curly brace of the if statement), blockVar is no longer accessible.

Summary:

This code demonstrates block scope using let. Variables declared with let are only accessible within the block they are defined in, promoting better code organization and reducing the risk of naming conflicts between variables with the same name in different parts of your code.

Variable const

Introduced in ES6, const is a specific type of variable declaration that is utilized for variables which are not intended to undergo any sort of change after their initial assignment. const shares the block-scoping characteristics of the let declaration, meaning that the scope of the const variable is limited to the block in which it is defined, and it cannot be accessed or used outside of that particular block of code.

However, the const declaration brings an additional layer of protection to the table. This added protection ensures that the value assigned to a const variable remains constant and unalterable throughout the entirety of the code. This is a crucial feature because it prevents the value of the const variable from being unintentionally altered or modified at any point in the code, which could potentially lead to bugs or other unintended consequences in the program.

In essence, the const declaration is an important tool in the JavaScript language that helps programmers maintain the integrity of their code by ensuring that certain variables remain constant and unchangeable, thereby preventing potential errors or bugs that could occur as a result of unwanted or inadvertent changes to these variables.

Example:

The const keyword is used to declare a constant variable. Constants must be assigned a value at the time of declaration, and their values cannot be reassigned.

```javascript
const PI = 3.14159; // Constant value
console.log(PI); // Output: 3.14159

// PI = 3.14; // Error: Assignment to constant variable

const person = {
  name: "John Doe"
};
console.log(person.name); // Output: "John Doe"

person.name = "Jane Smith"; // Allowed, but modifies the object property
console.log(person.name); // Output: "Jane Smith"
```

Code breakdown:

1. Constant Variables with const:

- The first line, const PI = 3.14159;, declares a constant variable named PI using the const keyword. It's assigned the value 3.14159, representing the mathematical constant pi.
  - const is used to create variables whose values cannot be changed after they are assigned. This ensures the value of pi remains consistent throughout your code.
- The next line, console.log(PI);, prints the value of PI to the console, and you'll see the output 3.14159.
- The commented line, // PI = 3.14; // Error: Assignment to constant variable, attempts to reassign a new value to PI. This will result in an error because constants cannot be changed after their initial assignment.

2. Objects and Modifying Properties:

- The code then defines a constant variable named person using const. However, in this case, const doesn't mean the entire object itself is immutable. It means the reference to the object (person) cannot be reassigned to a new object.
- person = { name: "John Doe" }; creates an object with a property named name and assigns the value "John Doe" to it.
- console.log(person.name); prints the value of the name property of the object referenced by person, and you'll see the output "John Doe".
- Here's the key distinction:
    - While person itself is constant (its reference cannot change), the object it references can still be modified.
- That's why the next line, person.name = "Jane Smith";, is allowed. It modifies the value of the name property within the object that person references.
- Finally, console.log(person.name); again prints the name property, but this time you'll see the updated value "Jane Smith".

Summary:

This code demonstrates constant variables with const and the difference between constant variable references and mutable object properties. While const prevents reassignment of the variable itself, it doesn't prevent modifications of the data within the object it references if the object is mutable (like an array or another object).

In order to write clean, efficient, and error-free JavaScript code, it's essential to fully understand the differences between the three methods of variable declaration - var, let, and const. Each of these methods has its own unique characteristics and quirks, and they're each suited to different situations.

The nuances of these methods might seem subtle, but they can have a significant impact on how your code behaves. By carefully choosing the right method of variable declaration for each situation, you can make your code more intuitive and easier to read, which in turn makes it easier to debug and maintain.

In the long run, this understanding can save you, and any others who might work with your code, a significant amount of time and effort.

2.1.2 Data Types

In the world of JavaScript, variables act as an absolutely essential and fundamental component for programming. They are the containers that hold different types of data, serving as the backbone of numerous operations within any given piece of code. The beauty of these variables lies in their ability to accommodate a wide range of data types, from the most simple and straightforward numbers and text strings to intricate and complex data structures like objects.

In addition to these, the nature of JavaScript's variables is such that they are not strictly confined to holding these specific types of data. On the contrary, their functionality extends to encompass a much broader array of data types.

This aspect ensures JavaScript's variables provide maximum flexibility to programmers, allowing them to dynamically alter the type of data a variable holds in accordance to the changing needs of their code. This gives programmers the freedom to manipulate their variables in a manner that best suits the particular requirements of their programming context.

Here are the basic data types in JavaScript:

Primitive Types:

String: In the world of JavaScript programming, a "string" is a critical data type that is used to represent and manipulate a sequence of characters, forming textual data. For instance, a simple string could look like this: 'hello'. This could represent a greeting, a user's name, or any other piece of text that the program might need to store and access at a later point. The versatility and utility of the string data type make it a staple in a vast majority of JavaScript code.

Number: This represents a specific data type within programming that signifies both integer and floating-point numbers. An integer is a whole number without a fractional component, like 10, whereas a floating-point number includes a decimal component, as seen in 20.5. The number data type is extremely versatile and crucial in programming as it can represent any numerical value, making it integral for calculations and data manipulation.

Boolean: In the realm of computer science, a Boolean is a specific type of logical data type that can only adopt one of two possible values, namely true or false. This particular data type derives its name from George Boole, a mathematician and logician. The Boolean data type plays a pivotal role in a branch of algebra known as Boolean algebra, which forms the backbone of digital circuit design and computer programming.

It is frequently employed for conditional testing in programming, where it proves invaluable in decision-making structures such as if-else statements, enabling the program to choose different courses of action based on various conditions. In essence, the Boolean data type is a simple yet powerful tool in the hands of programmers, allowing them to mirror the binary nature of computer systems effectively.

Undefined: This is a special and unique data type in programming that is specifically assigned to a variable that is declared but has not been given a value yet. It's a state of a variable that signifies its existence, but it's yet to have an associated value or meaning assigned to it.

An undefined value is an indication or a straightforward suggestion that while the variable has been recognized and exists in the memory, it's still devoid of a defined value or hasn't been initialized. In essence, it refers to the scenario when a variable is declared in the program but has not been assigned any value, hence it is undefined.

Null: Null is a unique and special data type that is utilized in programming to signify a deliberate and intentional absence of any specific object value. In other words, it is used to represent 'nothing' or 'no value'. It holds a significant place in various programming languages because of its ability to denote or verify the non-existence of something.

For instance, it can be used in situations where an object doesn't exist or the default value for unassigned variables. It is a fundamental concept that programmers use to manage the state and behavior of their programs.

Symbol: Introduced in ES6, the Symbol is a unique and immutable data type. It's distinctive in its uniqueness as no two symbols can have the same description. This characteristic makes it particularly useful in creating unique identifiers for objects, ensuring no accidental alterations or duplications. This provides developers with a powerful tool for maintaining data integrity and control over object properties, thus improving the overall robustness of the code.

Objects:

In the expansive world of JavaScript programming, there exists a number of key concepts that are absolutely crucial for developers to grasp. Among these, perhaps one of the most significant is the notion of objects.

At a basic level, objects can be thought of as organized collections of properties. To elaborate, each property is a unique pair, composed of a key (also known as a name) and a corresponding value. This straightforward yet efficient structure forms the essence of what we call an object.

The key or name, within this pair, is invariably a string. This ensures a consistent method of identification across the object. On the other hand, the value associated with this key can be of any data type. Whether it's strings, numbers, booleans, or even other objects, the possibilities are virtually limitless.

This remarkable feature of objects, which facilitates the structuring and accessing of data in a highly versatile manner, is what makes them an indispensable component in JavaScript programming. By effectively utilizing objects, developers are able to manage data in a structured and coherent way, thereby enhancing the overall quality and efficiency of their code.

Example: Data Types

```javascript
let message = "Hello, world!"; // String
let age = 25; // Number
let isAdult = true; // Boolean
let occupation; // Undefined
let computer = null; // Null

// Object
let person = {
    name: "Jane Doe",
    age: 28
};
```

2.1.3 Dynamic Typing

JavaScript, a popular programming language, is known for being a dynamically typed language. This particular attribute points to the fact that the type of a variable is not checked until the program is executing - a phase also referred to as runtime. While this characteristic brings about certain flexibility, it can also potentially lead to unexpected behaviors, which can be challenging for developers.

As a software developer or programmer, having a deep understanding of JavaScript's dynamic typing nature is imperative. This is because it can generate bugs that are incredibly difficult to identify and rectify, particularly if you're not cognizant of this feature. Dynamic typing, although offering versatility, can be a double-edged sword, causing elusive bugs that could lead to system crashes or incorrect results, adversely affecting the overall user experience.

Therefore, as you embark on creating your applications or working on JavaScript projects, it is crucial to be particularly vigilant about this characteristic. Ensuring that you handle variables correctly, understanding the potential pitfalls and the ways to circumnavigate them, will not only help in reducing the risk of bugs but also improve the efficiency and performance of your applications.

Example: Dynamic Typing

```
let data = 20; // Initially a number
data = "Now I'm a string"; // Now a string
console.log(data); // Outputs: Now I'm a string
```

Code breakdown:

1. Dynamic Typing in JavaScript:

- JavaScript is a dynamically typed language. This means that the data type (like number or string) of a variable is not explicitly declared, but rather determined by the value assigned to it at runtime.
- The code demonstrates this concept:
    - let data = 20; declares a variable named data using let and assigns the number 20 to it. Here, data has the number data type.
    - In the next line, data = "Now I'm a string";, the same variable data is reassigned a new value, which is a string. JavaScript automatically understands that data now refers to a string value.

2. Reassigning Variables with Different Data Types:

- Unlike some other programming languages where variables have a fixed data type, JavaScript allows you to reassign variables with different data types throughout your code. This provides flexibility but can sometimes lead to unexpected behavior if you're not careful.

3. The Output:

- The last line, console.log(data);, prints the current value of data to the console. Since it was last assigned a string value, you'll see the output "Now I'm a string".

In Summary:

This code snippet highlights dynamic typing in JavaScript. Variables can hold different data types throughout the code, and their type is determined by the assigned value at runtime. This flexibility is a core characteristic of JavaScript, but it's essential to be aware of it to write predictable and maintainable code.

2.1.4 Type Coercion

Type coercion stands out as a distinctive feature of JavaScript, where the language's interpreter takes it upon itself to automatically convert data types from one form to another when it deems it necessary.

This is often observed during comparisons where, for instance, a string and a number may be compared, and JavaScript will automatically convert the string to a number to make a meaningful comparison. While on one hand, this can be quite helpful and adds to the flexibility of the language, especially for beginners who may not be fully versed in handling different data types, it can also lead to unexpected and often puzzling results.

This is because the automatic conversion may not always align with the programmer's intent, leading to bugs that can be hard to detect and fix. Therefore, while type coercion can be a useful tool, it's also important to understand its implications and use it judiciously.

Example: Type Coercion

```
let result = '10' + 5; // The number 5 is coerced into a string
console.log(result); // Outputs: "105"
```

In order to avoid the unanticipated results that can occur due to type coercion in JavaScript, it is highly recommended to always use the strict equality operator, denoted as ===. This operator is considered superior to the standard equality operator, represented by ==, due to its stricter evaluation criteria.

The strict equality operator doesn't just compare the values of the two operands, but also takes into account their data type. This means that if the value and data type of the operands do not match exactly, the comparison will return false. This level of strictness helps to prevent bugs and errors that can arise from unexpected type conversions.

Example: Avoiding Type Coercion

```
let value1 = 0;
let value2 = '0';

console.log(value1 == value2);  // Outputs: true (type coercion occurs)
console.log(value1 === value2); // Outputs: false (no type coercion)
```

2.1.5 Const Declarations and Immutability

In the realm of JavaScript, the utilization of the term const carries a significant distinction that is frequently misunderstood. Many individuals commonly interpret const as a clear indication of complete immutability, an assertion that the value in question is unchanging and fixed. However, this interpretation isn't entirely accurate. In reality, the primary function of const is to prohibit the reassignment of the variable identifier to a new value. It's essential to note that it does not guarantee the immutability of the value itself to which the variable reference is pointing.

To illustrate this, let's consider an example. If you declare an object or an array as a const, it's crucial to comprehend that the const keyword will not extend its protective shield over the contents of that object or array from being modified or manipulated. What this implies is that while the variable identifier itself is safeguarded from reassignment, the object or array it refers to can still have its properties or elements altered, changed, or modified.

In essence, the const keyword in JavaScript ensures that the binding between the variable identifier and its value remains constant. However, the contents of the value, especially when dealing with complex data types like objects and arrays, can still be subject to alteration.

Example: Const and Immutability

```
const person = { name: "John" };
person.name = "Doe"; // This is allowed
console.log(person); // Outputs: { name: "Doe" }

// person = { name: "Jane" }; // This would cause an error
```

Code breakdown:

1. Creating a Constant Object (const person)

- The code starts with const person = { name: "John" }. Here, we're using the const keyword to declare a constant variable named person.
  - Remember, const means the value of the variable itself cannot be changed after it's assigned.
- But in this case, the value we're assigning is an object literal ({ name: "John" }). This object stores a property named name with the value "John".

2. Modifying Object Properties (Allowed!)

- Even though person is a constant, the code proceeds to person.name = "Doe". This line updates the value of the name property within the object that person references.
  - It's important to understand that const prevents you from reassigning the variable person itself to a new object. But it doesn't freeze the entire object referenced by person.
- Objects are mutable in JavaScript, meaning their properties can be changed after they are created. So, here we're allowed to modify the name property.

3. Trying to Reassign the Entire Object (Error!)

- The commented line, // person = { name: "Jane" }, demonstrates what's not allowed. This line attempts to reassign a completely new object to the person variable.
- Since person is declared with const, this reassignment would violate the constant rule. You'll get an error if you try to run this line because you cannot change the reference person points to after the initial assignment.

4. Looking at the Output (console.log(person))

- The final line, console.log(person);, logs the value of the person variable to the console. Even though we modified the name property, it's still the same object referenced by person. So, you'll see the updated object: { name: "Doe" }.

Summary:

This code showcases how constant objects work in JavaScript. While you can't reassign the entire object referenced by a constant variable, you can still modify the properties within that object because objects themselves are mutable. This distinction between constant variable references and mutable object properties is essential to understand when working with const and objects in JavaScript.

2.1.6 Using Object.freeze()

One effective strategy to ensure the unchangeable nature of objects or arrays within your codebase is through the use of a specific JavaScript method known as Object.freeze(). This method serves an integral role in preserving the state of objects and arrays, as it effectively prevents any potential modifications that could be made.

The crux of Object.freeze()'s usefulness comes from its ability to maintain a constant state of the object or array throughout the execution of the program, regardless of the conditions it may encounter. By invoking this method, you put a stop to any changes that could alter the state of the object or array.

This feature of immutability provided by Object.freeze() can be significantly advantageous in software development, chiefly in the prevention of bugs. More specifically, unexpected mutations within objects and arrays are a common source of bugs in JavaScript. These can lead to a variety of issues, from minor glitches to major functional problems within the application.

By using Object.freeze(), you can prevent such mutations from occurring, thereby enhancing the stability of your program and reducing the likelihood of encountering any mutation-related bugs. Thus, the Object.freeze() method offers a robust, efficient solution to enforcing immutability and consequently preventing potential issues that could arise from unwanted mutations.

Example:

```javascript
const frozenObject = Object.freeze({ name: "John Doe", age: 30 });

// Trying to modify the object
frozenObject.name = "Jane Smith"; // This won't have any effect
console.log(frozenObject.name); // Outputs: "John Doe"

// Trying to add a new property
frozenObject.gender = "Male"; // This won't work
console.log(frozenObject.gender); // Outputs: undefined

// Trying to delete a property
delete frozenObject.age; // This won't work
console.log(frozenObject.age); // Outputs: 30
```

Code breakdown:

1. Creating a Frozen Object (const frozenObject):

The code starts with const frozenObject = Object.freeze({ name: "John Doe", age: 30 });. Here, we're using the Object.freeze() method to create a frozen object which cannot be modified, and we're storing the frozen object in a constant variable named frozenObject.

2. Trying to Modify the Object (No effect):

The code proceeds to frozenObject.name = "Jane Smith";. This line attempts to change the value of the name property within the frozen object.

Since Object.freeze() was used, this operation has no effect. The object remains as it was when it was frozen.

3. Attempting to Add a New Property (Won't work):

The next line, frozenObject.gender = "Male";, tries to add a new property gender to the frozen object.

Again, because the object is frozen, this operation does not succeed.

4. Attempting to Delete a Property (No effect):

The code then tries to delete a property with delete frozenObject.age;. This operation attempts to remove the age property from the frozen object.

5. Looking at the Outputs (console.log() statements):

The various console.log() statements in the code print the state of the object after each operation.

As you can see, none of the operations alter the state of the frozen object. The outputs confirm that the object remains as it was when it was first created and frozen.

Summary:

This code demonstrates how the Object.freeze() method works in JavaScript. Once an object is frozen, it cannot be modified, extended or reduced in any way. This immutability extends to all properties of the object, safeguarding the object's integrity.

2.1.7 Handling Null and Undefined

In JavaScript, null and undefined are both special data types that represent the absence of value. However, they are not completely interchangeable and are typically used in different contexts to convey different concepts:

undefined usually implies that a variable has been declared in the code, but it hasn't been assigned a value yet. It's a way of telling the programmer that this variable exists, but it doesn't have a value at this point in time. This could be because the variable is yet to be initialized or because it is a function parameter that was not provided when the function was called.

On the other hand, null is used explicitly to denote that a variable is intentionally set to have no value. When a programmer assigns null to a variable, they are clearly stating that the variable should have no value or object assigned to it, possibly indicating that the value or object it pointed to previously is no longer needed or relevant. It's a conscious declaration by the programmer that the variable should be empty.

Example: Handling Null and Undefined

```
let uninitialized;
console.log(uninitialized); // Outputs: undefined

let empty = null;
console.log(empty); // Outputs: null
```

This example demonstrates the difference between uninitialized and null variables. The variable 'uninitialized' is declared but not assigned a value, hence its value is 'undefined'. The variable 'empty' is assigned the value 'null', which is a special value representing no value or no object.

2.1.8 Using Template Literals for Strings

Introduced as part of the sixth edition of the ECMAScript standard, known as ES6, template literals have emerged as a powerful tool for handling string manipulation tasks. They provide a significantly simplified method for creating complex strings in JavaScript.

Unlike traditional string concatenation methods, template literals allow for the creation of multi-line strings without resorting to concatenation operators or escape sequences, thus making the code cleaner and more readable.

Additionally, they feature the ability to embed expressions within the string. These embedded expressions are then processed, evaluated, and ultimately converted into a string. This functionality can greatly streamline the process of integrating variables and computations within a string.

Example: Template Literals

```
let name = "Jane";
let greeting = `Hello, ${name}! How are you today?`;
console.log(greeting); // Outputs: "Hello, Jane! How are you today?"
```

In this example we declared a variable named "name" and assigns it the string value "Jane". We then declare another variable named "greeting" and assigns it a string value that uses a template literal to include the value of the "name" variable. The phrase "Hello, Jane! How are you today?" is created using this template literal. The last line of the code outputs this greeting to the console.

2.2 Operators

In the realm of JavaScript, operators serve as indispensable tools that empower you to carry out a plethora of operations on variables and values. They make it possible for you to perform everything from the most basic arithmetic to the immensely complex logical comparisons.

The mastery of how to utilize these operators correctly and efficiently is absolutely crucial for effective programming and is often a distinguishing factor in the success of a project.

This section is dedicated to uncovering and exploring the diverse range of operators present within JavaScript, providing you with a comprehensive understanding of their specific functionalities and capabilities.

In addition, we will be diving into examples that are drawn from real-world scenarios, thereby enabling you to grasp how these operators can be appropriately applied and leveraged. This practical and applied approach ensures that you are not just understanding these concepts theoretically, but also

2.2.1 Arithmetic Operators

In the field of programming, arithmetic operators play a crucial role as they are used to conduct mathematical calculations. The basic arithmetic operators, which are the foundation of any mathematical computation, include addition (+), subtraction (-), multiplication (*), and division (/). Each of these operators carry out their respective mathematical operations on numerical values.

Addition (+) combines two numbers, subtraction (-) takes away one number from another, multiplication (*) multiplies two numbers together, and division (/) divides one number by another.

In addition to these fundamental arithmetic operators, JavaScript, a widely-used programming language, includes a few other operators to enhance its mathematical capabilities. One of them is the modulus operator (%). This operator is used to obtain the remainder of a division operation, which can be useful in various programming scenarios.

Furthermore, JavaScript includes the increment (++) and decrement (--) operators. These operators are used to increase or decrease a numerical value by one, respectively, and are frequently utilized in loop structures and various other programming constructs. The increment (++) operator adds one to its operand, while the decrement (--) operator subtracts one.

Example: Using Arithmetic Operators

```
let a = 10;
let b = 3;

console.log(a + b);  // Outputs: 13
console.log(a - b);  // Outputs: 7
console.log(a * b);  // Outputs: 30
console.log(a / b);  // Outputs: 3.3333333333333335
console.log(a % b);  // Outputs: 1

let counter = 0;
counter++;
console.log(counter);  // Outputs: 1
counter--;
console.log(counter);  // Outputs: 0
```

This is a simple JavaScript (JSX) example demonstrating basic arithmetic operations. Here, two variables 'a' and 'b' are declared with the values 10 and 3, respectively. Then, the code logs the result of addition, subtraction, multiplication, division, and modulus (remainder of division) operations performed on 'a' and 'b'.

Following that, a 'counter' variable is declared with the value 0. The 'counter' is then incremented by 1 using 'counter++', which results in a new value of 1. The 'counter' is then decremented by 1 using 'counter--', bringing it back to its initial value of 0.

2.2.2 Assignment Operators

In JavaScript programming, assignment operators play a crucial role as they are used to assign values to variables. The most commonly used assignment operator is the simple equal sign (=), which assigns the value on its right to the variable on its left.

However, JavaScript also includes a variety of compound assignment operators, which are capable of performing an operation and an assignment in a single step, thus simplifying the code and improving readability. Some of these compound assignment operators include +=, -= , *=, /= and %=.

These operators, respectively, add, subtract, multiply, divide, or calculate the modulus of the current value of the variable and the value on the right, then assign the result to the variable. They not only make the code cleaner and easier to understand, but also increase efficiency by reducing the amount of code required to perform these operations.

Example: Using Assignment Operators

```
let x = 10;
x += 5;   // Equivalent to x = x + 5
console.log(x);   // Outputs: 15

x *= 3;   // Equivalent to x = x * 3
console.log(x);   // Outputs: 45
```

In this example, the code initially defines a variable 'x' and assigns it a value of 10. The operator '+=', known as an addition assignment, adds the number 5 to the current value of 'x'. So, after executing this statement, the value of 'x' becomes 15. This is then logged out to the console.

The operator '*=', known as a multiplication assignment, multiplies the current value of 'x' by 3. After executing this, the value of 'x' becomes 45, which is then logged out to the console.

2.2.3 Comparison Operators

In JavaScript, comparison operators play a critical role in comparing two values and subsequently returning a Boolean value that is either true or false. They are integral to control flow and decision-making structures in the code. The different types of comparison operators that JavaScript encompasses are as follows:

- The first one is 'Equal to' which is represented by ==. It evaluates whether two values are equal in value irrespective of their type. Alongside, there is 'Strictly equal to' denoted by ===. It's stricter in the sense that it checks both the value and the type of the two entities being compared.
- The 'Not equal to' operator is represented by !=. It returns true if the two values being compared are not equal in value, regardless of their type. Strictly not equal to, on the other hand, which is represented by !==, checks both the value and the type, returning true only if either one or both of these are not equal.
- The 'Greater than' operator (>) and the 'Less than' operator (<) are self-explanatory. They compare two values and return true if the value on the left side of the operator is greater than or less than the one on the right, respectively.

- Lastly, we have the 'Greater than or equal to' (>=) and 'Less than or equal to' (<=) operators. These operators return true not just when the value on the left is greater than or less than the one on the right, but also when both values are equal.

Example: Using Comparison Operators

```
let age = 30;
console.log(age == 30);   // Outputs: true
console.log(age === '30');  // Outputs: false (strict comparison checks type)
console.log(age != 25);   // Outputs: true
console.log(age > 20);   // Outputs: true
```

In this example, we demonstrate various types of comparison operators.

- "age == 30" checks if the variable 'age' is equal to 30 and returns true.
- "age === '30'" performs a strict comparison (checking both value and type), so it returns false because 'age' is a number, not a string.
- "age != 25" returns true because the 'age' is not equal to 25.
- "age > 20" checks if 'age' is greater than 20 and returns true.

2.2.4 Logical Operators

In the world of programming, logical operators command a vital position. They are instrumental in establishing the logic between variables or values, thereby playing a key role in defining the behavior and output of a code.

Logical operators, in essence, serve as the building blocks that help in formulating more complex and dynamic conditions, making them indispensable tools in every programmer's arsenal. JavaScript offers support for a number of logical operators that help in creating intricate logical constructs within the code.

These include the logical AND (&&), a powerful tool that returns true only if both operands it is evaluating are true. This operator is often used when multiple conditions need to be satisfied simultaneously.

Next is the logical OR (||), another commonly used operator, which returns true if at least one of the operands it is evaluating is true. This operator is typically used in scenarios where satisfying just one of many conditions is enough for the code to proceed.

Last but not least, we have the logical NOT (!), a unique operator that flips the truthiness of the operand it is applied to - if the operand was true, it turns it false, and vice versa. This operator is particularly useful for quickly negating conditions or for checking the opposite of a certain condition.

Mastering the use of these logical operators can open up new avenues of efficiency and reliability within a programmer's code. Proper utilization of these operators can lead to code that is not only easier to understand and maintain, but also more robust and less prone to bugs, thereby enhancing the overall quality and performance of the software.

Example: Using Logical Operators

```
let isAdult = true;
let hasPermission = false;

console.log(isAdult && hasPermission);  // Outputs: false
console.log(isAdult || hasPermission);  // Outputs: true
console.log(!isAdult);  // Outputs: false
```

In this example.

1. The '&&' operator returns true only if both operands are true. Here, 'isAdult' is true and 'hasPermission' is false, so the result is false.
2. The '||' operator returns true if at least one of the operands is true. Here, 'isAdult' is true, so the result is true regardless of 'hasPermission' value.
3. The '!' operator negates the value of the operand. Here, 'isAdult' is true, so '!isAdult' is false.

2.2.5 Conditional (Ternary) Operator

The conditional (ternary) operator, which is unique in JavaScript due to its requirement for three operands, stands as an exception to the common binary operators that usually take two operands.

This operator is frequently utilized as a more concise alternative to the standard if-else statement. It serves this role effectively due to its ability to evaluate conditions and return values based on the condition's outcome in a more succinct manner than traditional control flow structures.

Example: Using the Conditional Operator

```
let age = 18;
let beverage = (age >= 18) ? "Beer" : "Juice";
console.log(beverage);  // Outputs: "Beer"
```

In this example a variable 'age' is declared with a value of 18. Then a ternary operator is used to declare another variable 'beverage'. If the age is 18 or over, 'beverage' is assigned the value "Beer". If the age is less than 18, 'beverage' is assigned the value "Juice". Finally, the value of 'beverage' is logged to the console. In this case, since age is 18, "Beer" is logged to the console.

2.2.6 Bitwise Operators

Bitwise operators, as the term suggests, are operators that perform operations directly on the binary or bit level representation of numbers. These numbers are typically represented in a format that a computer can understand, such as binary or base 2.

Bitwise operators can be exceptionally useful in certain low-level programming tasks. Specifically, they shine in areas such as graphics programming or device control, where there is often a need to manipulate or control data right down to the individual bits.

These tasks often require a high degree of precision and control, which is exactly what bitwise operators provide. With them, programmers can easily manipulate data in ways that would be complex or impractical with more high-level operations.

Example: Using Bitwise Operators

```
let a = 5;  // binary 0101
let b = 3;  // binary 0011

console.log(a & b);  // AND operator, outputs: 1 (binary 0001)
console.log(a | b);  // OR operator, outputs: 7 (binary 0111)
console.log(a ^ b);  // XOR operator, outputs: 6 (binary 0110)
console.log(~a);     // NOT operator, outputs: -6 (binary 1010, two's complement)
```

In this example, we demonstrate the use of bitwise operators.

a & b uses the AND operator, which compares each bit of the first operand (a) to the corresponding bit of the second operand (b). If both bits are 1, the corresponding result bit is set to 1. Otherwise, the result is 0.

a | b uses the OR operator. It compares each bit of a to the corresponding bit of b. If either bit is 1, the corresponding result bit is set to 1. Otherwise, the result is 0.

a ^ b uses the XOR operator. It compares each bit of a to the corresponding bit of b. If the bits are not the same, the corresponding result bit is set to 1. Otherwise, the result is 0.

~a uses the NOT operator. It inverts the bits of the operand.

2.2.7 String Operators

In JavaScript, the + operator serves a dual purpose. Not only does it perform the standard mathematical function of addition when used with numerical values, but it is also capable of concatenating strings when used with string data types.

Concatenation, in the context of programming, refers to the process of joining two or more strings together to form a single, continuous string. This feature of the + operator is particularly useful in various programming scenarios, such as when you need to combine user input data or dynamically generate text.

Example: String Concatenation

```
let firstName = "John";
let lastName = "Doe";
let fullName = firstName + " " + lastName;

console.log(fullName);   // Outputs: "John Doe"
```

In this example, we start by declaring two variables, "firstName" and "lastName", and assign them the string values "John" and "Doe" respectively. We then declare another variable, "fullName", and assign it the combined value of "firstName", a space, and "lastName". Lastly, we print the value of "fullName" to the console, resulting in the output "John Doe".

2.2.8 Comma Operator

The comma operator, a somewhat underused feature in many programming languages, serves an interesting purpose. It provides a way for multiple expressions to be evaluated within a single statement, which can be incredibly useful in certain situations.

When this operator is deployed, the expressions are evaluated in sequence, from left to right, and the result of the last expression is then returned. This means that the value of the statement as a whole will always be the value of the last expression.

Despite its infrequent use, the comma operator can, when applied judiciously, make code more concise, cleaner, and more efficient. It's certainly worth understanding for those situations where it can provide a more elegant solution to a coding problem.

Example: Using the Comma Operator

```
let a = 1, b = 2, c = 3;
(a++, b = a + c, c = b * a);
console.log(a, b, c);   // Outputs: 2, 5, 10
```

In this example, variables a, b, and c are initially declared and assigned the values of 1, 2, and 3, respectively. Inside the parentheses, three operations occur simultaneously. First, 'a' is incremented by 1, making its value 2. Then, the sum of 'a' and 'c', which equals 5, is assigned to 'b'. Finally, the product of 'b' and 'a', which equals 10, is assigned to 'c'. The 'console.log' statement outputs the new values of 'a', 'b', and 'c', which are now 2, 5, and 10, respectively.

2.2.9 Nullish Coalescing Operator (??)

The nullish coalescing operator (??), which was introduced in the ES2020 version of JavaScript, plays an essential role as a logical operator in programming. This operator functions by returning its right-hand side operand, but only in cases where its left-hand side operand is either null or undefined. In all other scenarios, it will return the left-hand side operand.

The advantage of this operator is primarily seen in its ability to assign default values. This is especially useful in scenarios where a variable can be null or undefined. Instead of writing a conditional statement to check if the variable has a value, you can use the nullish coalescing operator to assign a default value, streamlining your code and making it more readable.

Example: Nullish Coalescing Operator

```
let userComment = null;
let defaultComment = "No comment provided.";

let displayComment = userComment ?? defaultComment;
console.log(displayComment);   // Outputs: "No comment provided."
```

In this example we declare two variables, 'userComment' and 'defaultComment'. 'userComment' is initially set to null, and 'defaultComment' is a string that says "No comment provided."

The '??' operator is the nullish coalescing operator. It returns the right-hand side operand (which is 'defaultComment' here) if the left-hand side operand (which is 'userComment') is null or undefined.

The variable 'displayComment' is set to the value of 'userComment' if it's not null or undefined. If 'userComment' is null or undefined, then 'displayComment' is set to the value of 'defaultComment'.

Finally, the value of 'displayComment' is logged to the console. In this case, since 'userComment' is null, "No comment provided." is logged to the console.

2.2.10 Optional Chaining Operator (?.)

In ES2020, another exciting feature was introduced: the optional chaining operator (?.). This powerful tool simplifies the process of accessing properties deeply nested within an object structure. Without this operator, you would normally have to manually check each reference in the chain to ensure it is not nullish (i.e., null or undefined).

This can be a tedious and error-prone process, especially for complex structures. However, with the optional chaining operator, you can now safely navigate through these structures, and the operator will automatically return undefined whenever it encounters a nullish reference.

This helps to prevent runtime errors and makes your code cleaner and more readable.

Example: Optional Chaining

```
let user = {
    name: "John",
    address: {
        street: "123 Main St",
        city: "Anytown"
    }
};

let street = user.address?.street;
console.log(street);  // Outputs: "123 Main St"

let zipcode = user.address?.zipcode;
console.log(zipcode);  // Outputs: undefined (safely handled)
```

In this example we use optional chaining (?.). The optional chaining operator allows you to read the value of a property located deep within a chain of connected objects without having to check that each reference in the chain is valid.

The 'user' object contains a nested 'address' object. The variables 'street' and 'zipcode' are assigned the values of the corresponding properties in the 'address' object. If the property does not exist, instead of causing an error, the expression short-circuits, returning undefined.

2.3 Control Structures (if, else, switch, loops)

Control structures play a pivotal role in JavaScript programming. They serve as the backbone of your scripts, allowing you to control how and when specific segments of code are executed based on a variety of conditions. This control over the flow of your program is what makes your scripts dynamic and responsive, enabling them to adapt to different inputs and situations.

In JavaScript, there are several types of control structures that you can use depending on the specific requirements of your code. These structures allow you to add complexity and functionality to your scripts, making them more efficient and effective.

In this section, we'll delve deeper into these control structures. We will focus on three main types: conditional statements, switch statements, and loops. Each of these structures serves a unique purpose and can be used in different scenarios.

Conditional statements, such as the if-else statement, allow you to execute different pieces of code based on whether a certain condition is true or false. This provides a great deal of flexibility and can make your scripts much more dynamic.

Switch statements, on the other hand, let you choose between several blocks of code to execute based on the value of a variable or expression. This can be particularly useful when you have multiple conditions to check.

Finally, loops offer a way to repeatedly execute a block of code until a certain condition is met. This can be incredibly useful for tasks that require repetition, such as iterating over an array.

Throughout this section, we'll not only explain how to use these control structures, but also offer detailed examples of each. These examples will serve to illustrate how these structures work in practice, thereby enhancing your understanding and helping you become a more proficient JavaScript programmer.

2.3.1 Conditional Statements (if, else)

Conditional statements serve as the cornerstone of logical programming, allowing us to check specified conditions and perform different actions depending on the results of these checks. The simplest and most basic form of these conditional statements is the if statement.

The if statement tests a given condition, and if the result of this test is true, it then executes a specific block of code associated with this condition. This allows for greater control and flexibility in the code. To further enhance this flexibility, we can also add else blocks to our conditional statements.

These else blocks are designed to handle scenarios where the initial condition tested in the if statement is not met or is false. In this way, we can ensure that our program has a robust and comprehensive response mechanism to various situations, further improving its functionality and effectiveness.

Example: Using if and else

```
let score = 85;

if (score >= 90) {
    console.log("Excellent");
} else if (score >= 75) {
    console.log("Very Good");
} else if (score >= 60) {
    console.log("Good");
} else {
    console.log("Needs Improvement");
}
```

In this example, a program evaluates a score and prints a corresponding message. It uses a simple yet effective method to handle multiple conditions. The program starts by initializing a variable named "score" with a value of 85. It then uses an if-else structure to print different messages based on the "score" value. If the score is 90 or above, it prints "Excellent". For scores between 75 and 89, it prints "Very Good". If the score is between 60 and 74, it prints "Good". For scores below 60, it outputs "Needs Improvement".

2.3.2 Switch Statements

In programming, when you encounter a situation where there are multiple conditions that all depend on the same variable, using a switch statement can become a more efficient and cleaner method than resorting to multiple if statements. The switch statement is a multi-way branch statement.

It provides an easier method of sequentially checking each condition of our variable. It starts by comparing the value of a variable against the values of multiple variants or cases. If a match is found, the corresponding block of code is executed. This enhances readability and efficiency of your code, making it a preferred choice in such scenarios.

Example: Using switch

```javascript
let day = new Date().getDay(); // Returns 0-6 (Sunday to Saturday)

switch(day) {
    case 0:
        console.log("Sunday");
        break;
    case 1:
        console.log("Monday");
        break;
    case 2:
        console.log("Tuesday");
        break;
    case 3:
        console.log("Wednesday");
        break;
    case 4:
        console.log("Thursday");
        break;
    case 5:
        console.log("Friday");
        break;
    case 6:
        console.log("Saturday");
        break;
    default:
        console.log("Invalid day");
}
```

This JavaScript code generates a variable called 'day' that identifies the current day of the week as a number (0-6, representing Sunday to Saturday). It then employs a switch statement to output the corresponding day name. If the day number falls outside the 0-6 range, it prints "Invalid day" to the console.

2.3.3 Loops

In programming, loops are incredibly useful tools that allow a block of code to be repeated multiple times. This repetition can be utilized to iterate through arrays, perform calculations multiple times, or even to create animations.

JavaScript, a versatile and widely-used programming language, supports several types of loops. These include the for loop, which is often used when you know the exact number of times you want the loop to run.

The while loop, on the other hand, continues to run as long as a specified condition is true. And finally, the do...while loop is similar to the while loop but it ensures that the loop will run at least once, as it checks the condition after executing the loop's code block.

For Loop

This is an ideal loop structure to utilize when the total number of iterations is known beforehand, prior to the commencement of the loop. The 'For Loop' provides a concise way to write a loop that needs to execute a specific number of times, making it particularly useful in scenarios where you need to iterate through elements of an array, or perform repetitive operations a certain number of times.

Example:

```
for (let i = 1; i <= 5; i++) {
    console.log("Iteration number " + i);
}
```

This loop prints the iteration number five times. It's a basic for loop that begins with an index (i) of 1 and runs until i is less than or equal to 5. During each iteration, it displays the phrase "Iteration number " followed by the current iteration number on the console.

While Loop

This is a programming concept that comes into play when there is uncertainty about the precise number of loop iterations required before the loop initiates. It is a mechanism that continuously runs a specific block of code as long as a given condition holds true. This condition forms the backbone of the loop and as long as the condition stays true, the loop will keep running, executing the block of code within it over and over.

Once the condition evaluates to false, the loop is brought to a halt. This makes this type of loop an optimal choice for situations where the number of iterations is not fixed but depends on dynamic factors or inputs which may change during the course of program execution. Hence, it provides a lot of flexibility and control, making it a valuable tool in the programmer's arsenal.

Example:

```
let i = 1, n = 5;
while (i <= n) {
    console.log("Iteration number " + i);
    i++;
}
```

This method achieves the same outcome as a for loop but is commonly used when the termination condition depends on something other than a basic counter. This program sets two variables, i and n, to the values of 1 and 5 respectively. The while loop then runs as long as i is less than or equal to n. Inside the loop, it logs the current iteration number and increments i by one for each iteration. Consequently, it prints "Iteration number 1" to "Iteration number 5" on the console.

Do...While Loop

The do...while loop is a control flow statement that functions in a manner akin to the while loop, but it has a significant distinction. The main characteristic of the do...while loop is that it first executes the block of code enclosed within it, and only after this execution is the condition for the loop checked. This ensures that the block of code is always run a minimum of one time, irrespective of whether the condition is true or false.

This is in contrast to the while loop where the condition is evaluated before the execution of the code block, and if the condition is false from the outset, the code block may not run at all. Therefore, the do...while loop provides an advantage in specific scenarios where it's necessary for the code block to execute at least once before the loop condition is evaluated.

This might be applicable in cases where an operation or a method needs to be performed before a condition can be tested or a certain value can be obtained for testing. Thus, understanding the do...while loop can be an essential tool in the programmer's toolkit to handle such scenarios efficiently.

Example:

```
let result;
do {
    result = prompt("Enter a number greater than 10", "");
} while (result <= 10);
```

This loop will repeatedly prompt the user until they enter a number greater than 10. It uses a do-while loop to continuously prompt the user to enter a number. This loop will keep repeating until the user enters a number greater than 10. The input is stored in the variable 'result'.

2.3.4 Nested Control Structures

Control structures are fundamental building blocks in programming that can be nested within each other to create more intricate and sophisticated decision-making processes along with finely detailed flow control.

This nesting ability provides the programmer with the flexibility to precisely dictate how a program should function and respond under different circumstances. An illustrative example of this can be seen when working with nested loops.

These are particularly useful, and in many cases necessary, when operating with multi-dimensional arrays or more complex data structures. The nested loop allows for the traversal of these more intricate structures, enabling the manipulation, analysis, or display of their data in a detailed and comprehensive manner.

Example: Nested for Loops

```
for (let i = 0; i < 3; i++) {
    for (let j = 0; j < 3; j++) {
        console.log(`Row ${i}, Column ${j}`);
    }
}
```

This example employs nested for loops to traverse a 3x3 grid, which might represent the rows and columns of a game board or a pixel grid in image processing. The outer loop executes thrice, iterating i values from 0 to 2. For each i iteration, the inner loop also runs thrice, iterating j values from 0 to 2. Each inner loop iteration generates a console log statement displaying the current row (i) and column (j). This produces a total of 9 console log statements, one for each pair of i and j values.

2.3.5 Using Conditional Statements with Logical Operators

In the realm of programming, it is crucial to underline the significance of using conditional statements in harmony with logical operators, such as '&&' (which represents 'and') or '||' (which represents 'or'). This can lead to a code structure that is not merely more streamlined and efficient but also more intelligible and maintainable.

The value of this approach becomes particularly apparent when one is tasked with the evaluation of multiple conditions within a single 'if' statement. By harnessing the power of this combination, it becomes possible to achieve a range of benefits.

Firstly, the readability of your code can be substantially improved. This makes it easier for others to understand your work, which is an often overlooked but critically important aspect of professional programming.

Secondly, the manageability of your code can be enhanced. A well-structured codebase can be more easily navigated, updated, and debugged, thereby reducing the likelihood of errors and making your work more reliable.

Lastly, the performance of your code can be significantly improved. By simplifying the structure of your code and eliminating potential redundancies, you can reduce its complexity. This can lead to faster execution times and less strain on system resources, which is particularly important in environments where efficiency is paramount.

The use of conditional statements and logical operators can be a powerful tool in the programmer's arsenal, providing a range of benefits that can improve the quality, readability, manageability, and performance of your code.

Example: Combining Conditions

```
let age = 25;
let resident = true;

if (age > 18 && resident) {
    console.log("Eligible to vote");
}
```

This example demonstrates the use of logical operators to streamline condition checks. It involves two variables: 'age', assigned a value of 25, and 'resident', assigned a value of true. The system then checks if the age is more than 18 and if the person is a resident. If both conditions are met, "Eligible to vote" is printed to the console.

2.3.6 Loop Control with break and continue

In programming, the break and continue statements are crucial as they allow you to control and modify the flow of loops:

The break statement serves as a potent tool in programming, it provides an immediate exit from the loop, completely disregarding any remaining iterations that may have been scheduled. This implies that as soon as the break statement is encountered in the flow of the program, the execution of the remaining portion of the loop is instantly stopped.

The program then exits the loop structure without any further delay, and it proceeds to execute the rest of the code that lies beyond the loop. This feature of the break statement allows programmers to have a significant degree of control over the flow of execution and can be particularly useful in numerous scenarios, such as when an error condition is detected within a loop or when a particular condition has been satisfied, thus making further iterations unnecessary.

The continue statement in programming languages holds a unique and significant role. Unlike the break statement that entirely breaks out of the loop, the continue statement only skips the remaining part of the current iteration and quickly moves on to the next iteration.

Therefore, when a program's execution flow encounters a continue statement, it doesn't terminate the entire loop. Instead, it bypasses the rest of the code in the current iteration and swiftly advances to the starting point of the next cycle in the loop.

This means that all the code after the continue statement in the current iteration will not be executed, but the loop itself will continue with its next iteration, making the continue statement a powerful tool to control the flow of loops in programming.

Example: Using break and continue

```
for (let i = 0; i < 10; i++) {
    if (i === 5) {
        break;  // Exits the loop when i is 5
    }
    if (i % 2 === 0) {
        continue;  // Skips the current iteration if i is even
    }
    console.log(i);  // This line will only run for odd values of i less than 5
```

```
}
```

In this example, break stops the loop early, and continue is used to skip even numbers, effectively filtering the output to odd numbers less than 5. This program uses a for loop to iterate from 0 to 9. Inside the loop, there are two conditional statements.

The first conditional statement breaks the loop when the value of i is equal to 5. This means that the loop will stop executing as soon as i reaches 5, and the code after the loop will start executing.

The second conditional statement uses the continue statement to skip the rest of the current loop iteration if i is an even number. This means that if i is an even number, the console.log(i) line will be skipped, and the loop will immediately move on to the next iteration.

Therefore, the console.log(i) line will only run for odd values of i that are less than 5 (i.e., 1 and 3 will be printed to the console).

2.3.7 Error Handling with Try-Catch in Loops

When running a loop, especially ones that deal with external data sources or engage in complex calculations, there are many instances where errors may occur. These errors could be due to a variety of reasons such as faulty data, bugs in the code, or unexpected inputs.

In such cases, it is crucial to have a mechanism in place that can handle these errors efficiently so that the entire loop does not fail due to a single error. One such efficient error handling mechanism is the try-catch structure.

By wrapping the loop or its body within this structure, the program can catch any errors that occur and deal with them accordingly, without causing the entire loop to fail. This also ensures that the rest of the loop can continue to function as expected even if one iteration encounters an error.

Example: Error Handling in Loops

```
for (let i = 0; i < data.length; i++) {
    try {
        processData(data[i]);
    } catch (error) {
        console.error(`Error processing data at index ${i}: ${error}`);
    }
}
```

This loop continues processing data even if an error occurs in processData, logging the error and moving on to the next iteration. This is a program where a for-loop is used to iterate over an array of data. For each item in the array, a function called 'processData' is called. If an error occurs during the processing of data, the error is caught and logged to the console with the index of the array where the error occurred.

2.4 Functions and Scope

In the world of JavaScript, functions represent one of the most fundamental building blocks of the language. They allow programmers to encapsulate pieces of code that can be reused and executed whenever necessary, bringing modularity and efficiency to your scripts. Having a firm understanding of how to define and utilize these functions effectively is a cornerstone skill for any JavaScript programmer, and a key element in writing clean, efficient code.

In addition to this, having a clear understanding of the concept of scope is equally as important. The scope essentially determines the visibility or accessibility of variables within your code. This concept is absolutely vital when it comes to managing data within your functions, as well as across your entire program. Scope management can dictate the structure of your code and directly impact its performance and efficiency.

This section is designed to take a deep dive into the mechanics of functions within JavaScript, exploring the details of how they are declared, how expressions are handled within them, and how scope management works. Through a clear understanding of these elements, you can write more efficient and effective JavaScript code.

2.4.1 Function Declarations

A function declaration is a fundamental concept in programming that sets the ground for creating a function with the defined parameters. It starts with the function keyword, which signals the start of the function definition.

This keyword is then followed by the name of the function, which is a unique identifier used to call the function in the program. After the function name comes a list of parameters, enclosed in parentheses. These parameters are the inputs to the function, and they allow the function to perform actions based on these provided values.

Lastly, a block of statements, enclosed in curly braces, follows the parameter list. These statements form the body of the function and define what the function does when it is called.

This entire structure forms the function declaration, which is a critical component in the structure of any program.

Example: Function Declaration

```
function greet(name) {
    console.log("Hello, " + name + "!");
}

greet("Alice");  // Outputs: Hello, Alice!
```

This example demonstrates a simple function that takes a parameter and prints a greeting message. The function 'greet' takes one input parameter named 'name'. When this function is called, it prints a greeting message in the console that includes the input name. The last line of the code calls the function 'greet' with the argument "Alice", which results in the output: "Hello, Alice!".

2.4.2 Function Expressions

A function expression is a potent and valuable concept in the world of programming. In essence, a function expression is a technique where a function is directly assigned to a variable. The function that is to be assigned to the variable could either be a named function with its own designated name, or it could be an anonymous function, which is a function without a specific, identified name attached to it.

This concept and technique of function expressions open up a considerable amount of flexibility and adaptability in the realm of programming. Once a function has been assigned to a variable, it can be passed around as a value within the code. This ability to transport and utilize the function throughout the code not only increases its usability but also enhances the fluidity with which the code operates.

What this means practically is that the function can be used in a myriad of different contexts and can be invoked at different points throughout the code, based entirely on the needs, requirements, and discretion of the programmer. This is a significant advantage as it allows the programmer to tailor the use of the function to best suit their specific objectives.

This ability to assign functions to variables and use them flexibly throughout the programming code is a testament to the complexity and dynamic nature of programming languages, such as JavaScript. It showcases the numerous ways in which these languages can be manipulated to create complex functionalities, adapt to different needs, and execute tasks in a more efficient and effective manner.

Example: Function Expression

```
const square = function(number) {
    return number * number;
};

console.log(square(4));  // Outputs: 16
```

Here, the function is stored in a variable square, and it calculates the square of a number. The code defines a function called 'square' that takes a number as an input and returns the square of that number. Then, it uses the console.log statement to print the result of the function square when the input is 4, which is 16.

2.4.3 Arrow Functions

Introduced in the sixth edition of ECMAScript (ES6), arrow functions brought a new and concise syntax to the JavaScript landscape. They were designed to provide a more compact and streamlined method of writing functions, particularly in comparison to traditional function expressions. With their less verbose and more readable syntax, they became an instant favorite among developers, especially when it comes to working with short, single-line expressions.

One of the standout features that sets arrow functions apart from their traditional counterparts is their unique ability to share the same lexical this as their surrounding code. In other words, they inherit the this binding from the enclosing context.

This is a significant departure from traditional functions which create their own this context. With arrow functions, this retains the same meaning inside the function as it has outside of it. This feature not only simplifies the code but also makes it easier to understand and debug. It eliminates common bugs and confusions that arise from the this keyword behaving differently in different contexts, thereby enhancing the overall programming experience.

Example: Arrow Function

```
const add = (a, b) => a + b;

console.log(add(5, 3));  // Outputs: 8
```

This example employs an arrow function for a simple addition operation. It establishes a constant function called 'add', which accepts two arguments 'a' and 'b', then returns their sum. The 'console.log' statement invokes this function using 5 and 3 as arguments, consequently outputting the number 8 to the console.

2.4.4 Scope in JavaScript

In JavaScript, the concept of scope is used to define the context where variables can be accessed. This is a fundamental concept that has a significant impact on how your code behaves. The scope in JavaScript can be divided into two main types:

Global Scope

When variables are defined outside the confines of any specific function, they are referred to as having a 'global scope'. This particular designation implies that these variables, termed as 'global variables', can be accessed from any part of the code, irrespective of the location or the context from which they are invoked or called upon.

This characteristic of global scope denotes a universal accessibility, enabling these variables to be available throughout your entire codebase. This availability persists throughout the lifecycle of the program, making global variables a powerful tool that should be used judiciously to avoid unanticipated side effects.

Local Scope

On the other hand, variables that are defined within the structure of a function have what is known as a local scope. In essence, this means that they are only accessible or 'visible' within the confines of that specific function in which they were originally declared. They cannot be called upon or accessed from outside that function.

This is a significant and deliberate restriction, as it prevents these locally scoped variables from interacting, interfering or colliding with other parts of your code that lie beyond the function's boundaries.

This design principle helps to maintain the integrity of your code, ensuring that functions operate independently and that variables don't unexpectedly alter in value due to interactions with other parts of the code.

Example: Global vs. Local Scope

```
let globalVar = "I am global";

function testScope() {
    let localVar = "I am local";
    console.log(globalVar);   // Accessible here
    console.log(localVar);    // Accessible here
}

testScope();
console.log(globalVar);       // Accessible here
// console.log(localVar);     // Unaccessible here, would throw an error
```

This example explains the distinction between global and local scope. A global variable named "globalVar" and a function named "testScope" are declared. Within the function, a local variable called "localVar" is declared. The global variable can be accessed both inside and outside the function, but the local variable can only be accessed within the function where it's declared. Trying to access the local variable outside of the function will cause an error.

2.4.5 Understanding let, const, and var

The introduction of let and const in ES6 brought about a significant change in JavaScript's handling of variable scope. Instead of being limited to function-level scope, as is the case with var, these new declarations introduced the concept of block-level scope.

This means that a variable declared with let or const is only accessible within the block of code in which it was declared. This differs from var, which is accessible anywhere within the function it was declared in, regardless of block boundaries.

The function-scope nature of var can be a source of confusion and unexpected results if not used with caution, particularly in loops or conditional blocks. Therefore, the use of let and const for block-level scope can lead to more predictable code and fewer bugs.

Example: Block Scope with let

```
if (true) {
    let blockScoped = "I am inside a block";
    console.log(blockScoped);   // Outputs: I am inside a block
}

// console.log(blockScoped);   // Unaccessible here, would throw an error
```

This shows the block-level scope of let, which limits the accessibility of blockScoped to the if block. It illustrates how to use the let keyword to declare a block-scoped variable, blockScoped, which is only accessible within its declaration block (between the curly braces). Trying to access it outside this block, as shown in the commented out line, results in an error due to it being out of scope.

2.4.6 Immediately Invoked Function Expressions (IIFE)

An Immediately Invoked Function Expression (IIFE) is a function that is declared and executed simultaneously. This is an important concept in JavaScript, and it is a pattern that programmers often use when they want to create a new scope. When an IIFE is used, the function is executed right after it is defined.

This unique characteristic is particularly beneficial for creating private variables and maintaining a clean global scope. By using an IIFE, we can prevent any unwanted access or modification to our variables, thus ensuring the integrity and reliability of our code.

In other words, it mitigates the risk of polluting the global scope, which is a common issue in JavaScript development. This makes IIFEs an essential tool in any JavaScript developer's toolbox.

Example: IIFE

```
(function() {
    let privateVar = "I am private";
    console.log(privateVar);  // Outputs: I am private
})();
// The variable privateVar is not accessible outside the IIFE
```

This example shows how IIFE helps in encapsulating variables, making them private to the function and inaccessible from the outside.

In this example, an IIFE is declared using the syntax (function() { ... })(). The outer parentheses (...) are used to group the function declaration, making it an expression. The trailing parentheses () cause the function expression to be immediately invoked or executed.

Inside the IIFE, there is a variable declaration let privateVar = "I am private";. This variable privateVar is local to the IIFE and cannot be accessed outside the function's scope. This is a technique for encapsulating variables and making them private, which is useful for preventing unwanted external access or modifications, and keeping a clean global scope.

After the variable declaration, there's a console log statement console.log(privateVar);, which outputs the string "I am private". This console log statement is within the IIFE's scope, so it has access to the variable privateVar.

Once the IIFE has executed, the variable privateVar goes out of scope and is not accessible anymore. As a result, if you try to access privateVar outside of the IIFE, you'll get an error.

2.4.7 Closures

A closure, in the world of programming, is a particular type of function that comes with its own unique set of abilities. What sets a closure apart from other functions is its inherent capacity to remember and access variables from the scope in which it was originally defined. This holds true regardless of where it is subsequently executed, making it an immensely powerful tool in the programmer's arsenal.

This distinctive feature of closures forms the basis for the creation of functions that are equipped with their own private variables and methods, effectively creating a self-contained unit of code. These variables and methods are not accessible externally, thereby enhancing the encapsulation and modularity of the code. This characteristic of closures is a boon to programmers, as it allows them to create sections of code that are both secure and self-reliant.

Within the sphere of functional programming, closures are not just an important concept, they are an essential tool. They offer a level of power and flexibility that can significantly boost the efficiency and simplicity of the code. Through the use of closures, programmers can streamline their code, reducing unnecessary complexity and making it easier to understand and maintain. All these factors combine to make closures an indispensable part of modern programming.

Example: Closure

```
function makeAdder(x) {
    return function(y) {
        return x + y;
    };
}

const addFive = makeAdder(5);
console.log(addFive(2));  // Outputs: 7
```

Here, addFive is a closure that remembers the value of x (5) and adds it to its argument whenever it's called. This JavaScript code defines a function named makeAdder which accepts a single argument x. This function returns another function that takes another argument y, and returns the sum of x and y.

In the code, makeAdder function is called with 5 as an argument, and the returned function is stored in the addFive variable. This makes addFive a function that adds 5 to whatever number it receives as an argument.

Finally, addFive is called with the argument 2, which results in 7 (because 5 + 2 equals 7), and this result is logged to the console.

2.4.8 Default Parameters

In the complex and intricate world of programming, the concept of default parameters assumes a critical role. Default parameters are a remarkable feature. They allow named parameters to be automatically initialized with default values in situations where no explicit value is provided, or in cases where undefined is passed.

Think of a scenario where you are juggling with numerous parameters in your function, and a handful of them often remain the same or are used repeatedly. In such instances, wouldn't it be convenient to have those parameters automatically assigned their usual values without having to explicitly specify them every single time? That's precisely what default parameters allow you to do.

This feature is particularly beneficial in situations where certain values are used frequently. For instance, you might have a function that pulls data from a database. Most of the time, you might be pulling from the same table, using the same user credentials. Instead of having to specify these parameters every time, you could set them as default parameters, simplifying your function calls significantly.

Moreover, the use of default parameters ensures that your code is not only simpler but also more robust. By automatically assigning values to parameters, you prevent potential errors that could arise from missing arguments. This strengthens your code, making it more resistant to bugs and errors, and ultimately leading to a more efficient and effective programming process.

Example: Default Parameters

```javascript
function greet(name = "Stranger") {
    console.log(`Hello, ${name}!`);
}

greet("Alice");  // Outputs: Hello, Alice!
```

```
greet();          // Outputs: Hello, Stranger!
```

This functionality provides flexibility and safety for functions, ensuring that they handle missing arguments gracefully. This is a program that defines a function named 'greet'. This function takes one parameter 'name' and if no argument is provided when calling the function, it defaults to 'Stranger'.

The function then logs a greeting to the console, including the provided name or the default value. When the function is called with "Alice" as the argument, it outputs "Hello, Alice!". When it's called without any argument, it outputs "Hello, Stranger!".

2.4.9 Rest Parameters and Spread Syntax

Rest parameters and spread syntax are both highly useful features in JavaScript that might seem similar at first glance, but they serve different, yet complementary, purposes:

Rest parameters, denoted by an ellipsis (...), provide a way to handle function parameters that allows a variable number of arguments to be passed to a function. What this means is that rest parameters allow you to represent an indefinite number of arguments as an array. This is especially handy when you don't know ahead of time how many arguments will be passed to a function.

On the flip side, **spread syntax**, also denoted by an ellipsis (...), performs the opposite function. It allows an iterable such as an array or string to be expanded in places where zero or more arguments (for function calls) or elements (for array literals) are expected.

This can be useful in a variety of scenarios, such as combining arrays, passing elements of an array as separate arguments to a function, or even copying an array.

Example: Rest Parameters

```
function sumAll(...numbers) {
    return numbers.reduce((acc, num) => acc + num, 0);
}

console.log(sumAll(1, 2, 3, 4));  // Outputs: 10
```

This is a JavaScript function named 'sumAll' which uses the rest parameter syntax ('...numbers') to represent an indefinite number of arguments as an array. Within the function, the 'reduce' method is used to add together all the numbers in the array and return the total sum. The 'console.log' line is a test of this function, passing in the numbers 1, 2, 3, and 4. The output of this test should be 10, as 1+2+3+4 equals 10.

Example: Spread Syntax

```
let parts = ["shoulders", "knees"];
let body = ["head", ...parts, "toes"];

console.log(body);  // Outputs: ["head", "shoulders", "knees", "toes"]
```

This is a program that uses the spread operator (...) to combine two arrays. The variable 'parts' contains the array ["shoulders", "knees"]. The variable 'body' creates a new array that combines the string 'head', the elements of the 'parts' array, and the string 'toes'. The console.log statement then prints the 'body' array to the console, outputting: ["head", "shoulders", "knees", "toes"].

2.5 Events and Event Handling

Events serve as the backbone of interactive web applications. They are pivotal in bringing life to static web pages by enabling users to interact with web elements in a variety of ways. With JavaScript's robust event-handling capabilities, users can interact with web pages through a wide range of actions such as clicking, keyboard inputs, mouse movements, and more.

It's not an overstatement to say that understanding how to correctly manage these events is absolutely critical to crafting responsive, intuitive, and user-friendly web interfaces. Without a good grasp of event handling, web applications risk becoming clunky and difficult to navigate.

In this comprehensive section, we will delve deeply into the world of events. We will explore what exactly events are in the context of web development, how they are handled in JavaScript--one of the most popular programming languages used in web development today, and provide detailed, step-by-step examples to effectively illustrate these concepts. Our aim is to provide a solid foundation upon which you can build your understanding and skills in JavaScript event handling.

2.5.1 What are Events?

In the digital landscape of the internet, the term 'events' refers to specific actions or occurrences that transpire within the confines of the web browser. These actions can then be detected and responded to by the web page. Events are an integral part of user interaction. They can be initiated by the user through various activities such as clicking on an element, scrolling through the page, or pressing a specific key.

Alternatively, these events can be triggered by the browser itself. This can occur through a multitude of scenarios such as when a web page is loaded, when a window is resized, or when a timer has elapsed. These are critical events that a well-designed web page should be prepared to handle.

JavaScript, a powerful programming language used widely in web development, is employed to respond to these events. It does so through the use of functions specifically designed to handle these instances, aptly named 'event handlers'. These event handlers are written into the JavaScript code of a web page and are set to execute when the event they are designed to handle occurs.

2.5.2 Adding Event Handlers

In the process of creating dynamic and interactive web pages, event handlers play a vital role and can be assigned to HTML elements through several methods:

HTML event attributes

These are unique attributes that are directly embedded within the HTML elements themselves. They are designed to respond immediately when a certain specified event occurs. Under these conditions, the event attribute will swiftly call upon the JavaScript code that has been assigned to it.

This method of embedding JavaScript is straight-forward and easy to understand, making it an accessible way for programmers to add interactive features to a website. However, one should exercise caution when using this method.

If HTML event attributes are used excessively or without careful organization, they can lead to a situation where the code becomes cluttered and unorganized. This can make the code difficult to read and debug, undermining the effectiveness and maintainability of the website.

Example: HTML Event Attribute

```
<button onclick="alert('Button clicked!')">Click Me!</button>
```

This example uses an HTML attribute to directly assign an event handler to a button. When the button is clicked, it triggers a JavaScript function that displays an alert box with the message 'Button clicked!'.

DOM Property Method

This method revolves around assigning the event handler directly to the DOM property of a specific HTML element. This assignment process takes place within the JavaScript code itself. This technique presents a distinct advantage compared to the HTML event attributes approach.

The main advantage is that it provides a much more organized and cleaner option for developers. This is because it separates the JavaScript code from the HTML markup, enhancing the readability and maintainability of the code. However, it's important to note a significant limitation associated with this method.

Only one event handler can be assigned to a specific HTML element for a particular event. This limitation can potentially restrict the functionality and flexibility of the application.

Example: DOM Property

```
<script>
    window.onload = function() {
        alert('Page loaded!');
    };
</script>
```

Here, an event handler is assigned to the window's load event using a DOM property. This script will display a pop-up alert with the message 'Page loaded!' once the web page has fully loaded.

Event listeners

These are powerful and dynamic methods that are invoked whenever a specific event occurs on the associated element. The primary advantage of using event listeners is their ability to handle multiple instances of the same event on a single element, which is distinct from other methods.

This means that you can assign multiple event listeners for the same event on a single element, without any interference between the different listeners. This unique feature makes the event listener method the most flexible and adaptable of the three methods, particularly when dealing with complex interactive functionality or when multiple events need to be tracked on a single element.

Example: Event Listener

```javascript
document.getElementById('myButton').addEventListener('click', function() {
    alert('Button clicked!');
});
```

This program attaches an event listener to the HTML element with the ID 'myButton'. When the button is clicked, a pop-up message saying 'Button clicked!' will appear.

This example uses addEventListener to attach an anonymous function to the button's click event, which is a more flexible method as it allows multiple handlers for the same event and more detailed configuration.

2.5.3 Event Propagation: Capturing and Bubbling

In the Document Object Model (DOM), events can propagate in two distinct ways, each serving a unique purpose in the overall structure and functionality of the application.

The first method is known as **Capturing**. In this phase, the event starts at the topmost element or the root of the tree, then trickles down through the nested elements, following the hierarchy until it reaches the intended target element. This top-down approach allows for specific interactions to be captured as the event moves through the lower levels of the DOM tree.

The second method is **Bubbling**. Contrary to capturing, bubbling initiates from the target element, then ascends up through the ancestors, moving from the lower level elements to the upper ones. This bottom-up approach ensures that the event doesn't remain isolated to the target element and can influence the parent elements.

Understanding both the capturing and bubbling phases of event propagation is crucial, especially for complex event handling scenarios. It allows developers to control how and when events are handled, providing flexibility in managing user interactions and overall application behavior.

Example: Capturing and Bubbling

```html
<div onclick="alert('Div clicked!');">
    <button id="myButton">Click Me!</button>
</div>
<script>
    // Stops the click event from bubbling
    document.getElementById('myButton').addEventListener('click', function(event) {
```

```
        alert('Button clicked!');
        event.stopPropagation();
    });
</script>
```

In this example, clicking the button triggers its handler and normally would bubble up to the div's handler. However, stopPropagation() prevents that, so only the button's alert is shown.

This program creates a button inside a 'div' element. When you click the button, it triggers an alert message saying 'Button clicked!'. Additionally, it stops the event propagation, meaning that the 'div' onclick event (which would trigger an alert saying 'Div clicked!') is not activated when the button is clicked. If you clicked anywhere else in the 'div' but not on the button, it would trigger the 'Div clicked!' alert.

2.5.4 Preventing Default Behavior

The Document Object Model (DOM), a crucial part of web technology, is composed of numerous elements that come equipped with their own inherent, or default, behaviors. These default behaviors, designed to streamline user interactions, are automatically activated or triggered when a user interacts with certain elements on a web page.

A quintessential example of this automatic triggering of default behaviors can be seen when a user clicks on a hyperlink embedded within a webpage. In this scenario, the default action for the web browser is to navigate to the URL or web address specified by the activated hyperlink.

However, there are circumstances where it might be deemed necessary to prevent, or override, this default action from taking place. A common reason for needing to halt the default behavior is when a developer opts to utilize the capabilities of JavaScript to control the process of navigation within a website, and load new content into the existing page without requiring a full page refresh.

This technique of dynamically loading new content without a full page reload is commonly employed in modern web development. It's an approach that not only enhances the overall user experience by delivering a more seamless and responsive interface, but also reduces the load on servers. Consequently, this can lead to improved website performance and potentially higher user satisfaction.

Example: Preventing Default Behavior

```
<a   href="<https://www.example.com>"   onclick="return   false;">Click   me   (going
nowhere!)</a>
```

Here, returning false from the onclick handler prevents the browser from following the link. The anchor tag <a> is used to create the hyperlink. The href attribute is set to a URL (https://www.example.com) which is normally where the user would be directed when they click the link. However, the onclick attribute is set to return false; which prevents the default action of the link and makes it so that the user is not redirected anywhere when they click this link.

2.5.5 Advanced Event Handling

Event handling in programming is not restricted to simple scenarios; it can also involve more complex situations. One of these situations includes handling events on elements that are created dynamically.

Dynamically created elements are those that are added to the webpage after the initial page load, and handling events on these elements can present unique challenges. Additionally, event handling can also involve optimizing performance for high-frequency events.

These are events that occur very frequently, such as resizing a window or scrolling through a webpage. Such events can potentially slow down the performance of a webpage if not handled correctly, so proper optimization is crucial.

Example: Event Delegation

```
document.getElementById('myList').addEventListener('click', function(event) {
    if (event.target.tagName === 'LI') {
        alert('List item clicked: ' + event.target.textContent);
    }
});
```

This is a program that adds an event listener to an HTML element with the id 'myList'. When any list item (LI) within this element is clicked, it triggers a function that opens an alert box displaying the text content of the clicked list item.

This is an example of event delegation, where a single event listener is added to a parent element instead of individual handlers for each child. It's particularly useful for handling events on elements that may not exist at the time the script runs.

2.5.6 Throttling and Debouncing

Managing high-frequency events such as resizing, scrolling, or continuous mouse movement can pose a significant challenge. This is because these actions can lead to performance issues due to the sheer number of event calls that they trigger. To mitigate this, two strategies are commonly employed: throttling and debouncing. These techniques are used to limit the rate at which a function gets executed, thereby preventing an overflow of calls that might bog down the system performance.

Throttling is a technique that ensures a function executes at most once every specified number of milliseconds. This method is particularly effective when dealing with high-frequency events because it allows us to set a maximum limit on the rate at which a function gets executed. By doing so, we can maintain a steady and predictable flow of function calls, and prevent any potential performance issues that may arise from too many function calls being executed in a short span of time.

On the other hand, **debouncing** is another technique that ensures a function executes only once after a specified amount of time has elapsed since its last invocation. This is particularly useful for events that keep firing as long as certain conditions are met, such as a user continuing to resize a window. By implementing a debounce, we can ensure that the function doesn't keep firing continuously, but instead only gets executed once after the user has stopped resizing the window for a certain period of time.

Example: Throttling

```
function throttle(func, limit) {
    let lastFunc;
    let lastRan;
    return function() {
        const context = this;
        const args = arguments;
        if (!lastRan) {
            func.apply(context, args);
            lastRan = Date.now();
        } else {
            clearTimeout(lastFunc);
            lastFunc = setTimeout(function() {
                if ((Date.now() - lastRan) >= limit) {
                    func.apply(context, args);
                    lastRan = Date.now();
                }
            }, limit - (Date.now() - lastRan));
        }
    };
}
```

```
window.addEventListener('resize', throttle(function() {
    console.log('Resize event handler call every 1000 milliseconds');
}, 1000));
```

Code breakdown:

1. Throttling Function Calls (throttle function):

- The code defines a function named throttle(func, limit). This function takes two arguments:
    - func: This is the function we want to throttle. It's the function whose calls we want to limit.
    - limit: This is a number (in milliseconds) that specifies the minimum time interval between allowed calls to the func function.
- Inside the throttle function, there are several variables and logic to control the call rate of the provided func.

2. Tracking Last Call Time (lastRan):

- The variable let lastRan; is declared to store a timestamp of the last time the func function was called through the throttled version.

3. The Throttling Logic (Inner Function):

- The throttle function returns another function. This inner function acts as the throttled version of the original func that gets passed as an argument.
    - Inside the inner function:
        - const context = this; captures the context (this) of the function call (important for some function types).
        - const args = arguments; captures the arguments passed to the throttled function.
        - The logic then checks if it's time to allow a call to the original func based on the limit:
            - If !lastRan (meaning there was no previous call or enough time has passed since the last call), the original func is called directly using func.apply(context, args). This ensures the function is called at least once immediately.
            - lastRan is then updated with the current timestamp using Date.now().

- Otherwise (if lastRan exists), a more complex throttling mechanism is used:
  - Any existing timeout (set to call func later) is cleared using clearTimeout(lastFunc).
  - A new timeout is created using setTimeout. This timeout will call another function after a delay.
    - The delay is calculated based on the limit and the time elapsed since the last call (Date.now() - lastRan). This ensures calls are spaced out by at least the limit time interval.
    - The inner function called after the timeout checks again if enough time has passed since the last call ((Date.now() - lastRan) >= limit). If so, it calls the original func and updates lastRan.

4. Applying Throttling to Resize Event (window.addEventListener):

- The last two lines demonstrate how to use the throttle function.
- window.addEventListener('resize', throttle(function() { ... }, 1000)); adds an event listener for the resize event on the window object.
  - The event listener function you want to call on resize is passed through the throttle function.
  - In this case, the throttled function logs a message "Resize event handler call every 1000 milliseconds" to the console.
  - The 1000 passed as the second argument to throttle specifies the limit (1 second or 1000 milliseconds) between allowed calls to the resize event handler function. This prevents the resize event handler from being called too frequently, improving performance.

Summary:

This code introduces function throttling, a technique to limit the rate at which a function can be called. The throttle function creates a wrapper function that ensures the original function is called at most once within a specific time interval (defined by the limit). This is useful for event handlers or any functions that you don't want to be called too often to avoid overwhelming the browser or causing performance issues.

Example: Debouncing

```
function debounce(func, delay) {
    let debounceTimer;
```

```
    return function() {
        const context = this;
        const args = arguments;
        clearTimeout(debounceTimer);
        debounceTimer = setTimeout(() => func.apply(context, args), delay);
    };
}

input.addEventListener('keyup', debounce(function() {
    console.log('Input event handler call after 300 milliseconds of inactivity');
}, 300));
```

Code breakdown:

1. Debouncing Function Calls (debounce function):

- The code defines a function named debounce(func, delay). This function takes two arguments:
    - func: This is the function we want to debounce. It's the function whose calls we want to control.
    - delay: This is a number (in milliseconds) that specifies the time to wait after the last call before actually executing the func function.
- Inside the debounce function, there's a variable and logic to handle the delayed execution of the provided func.

2. Debounce Timer (debounceTimer):

- The variable let debounceTimer; is declared to store a reference to a timeout timer. This timer is used to control the delay before calling the func function.

3. The Debouncing Logic (Inner Function):

- The debounce function returns another function. This inner function acts as the debounced version of the original func that gets passed as an argument.
    - Inside the inner function:
        - const context = this; captures the context (this) of the function call (important for some function types).
        - const args = arguments; captures the arguments passed to the debounced function.
        - The debouncing logic is implemented using a timeout:

- clearTimeout(debounceTimer); clears any existing timeout set by the debounce function. This ensures there's only one timeout waiting at any time.
- A new timeout is created using setTimeout. This timeout calls an anonymous function after the specified delay milliseconds.
  - The anonymous function calls the original func using func.apply(context, args). It also applies the function with the captured context and arguments.

4. Applying Debouncing to Keyup Event (input.addEventListener):

- The last two lines demonstrate how to use the debounce function.
- input.addEventListener('keyup', debounce(function() { ... }, 300)); adds an event listener for the keyup event on the input element (assuming there's an input element with this reference).
  - The event listener function you want to call on keyup is passed through the debounce function.
  - In this case, the debounced function logs a message "Input event handler call after 300 milliseconds of inactivity" to the console.
  - The 300 passed as the second argument to debounce specifies the delay (300 milliseconds) before calling the actual keyup event handler function. This prevents the event handler from being called for every keystroke. Instead, it waits for a pause of 300 milliseconds after the last keystroke before firing the function.

Summary:

This code showcases debouncing, a technique used to delay the execution of a function until a certain amount of time has passed since the last call. This is useful for situations like search bars or input fields where you don't want to perform an action (like sending a search request) after every single key press, but only after a short period of inactivity. Debouncing helps improve performance and user experience by avoiding unnecessary function calls.

2.5.7 Custom Events

JavaScript, a powerful and commonly used programming language in the world of web development, expands its range of functionality by providing the ability for developers to create their own custom events. This is achieved using the CustomEvent function, a feature that is built into the JavaScript language.

With CustomEvent, these tailor-made or 'custom' events can be dispatched or triggered on any element that exists within the Document Object Model (DOM). The DOM, in essence, is a representation of the structure of a webpage and JavaScript's CustomEvent function allows developers to interact with this structure in a highly customizable way.

This unique capability of JavaScript to create and dispatch custom events holds significant value, especially when dealing with complex interactions within web applications. The development of web applications often requires the management of numerous interactive elements and intricate user interfaces. In such cases, the ability to create custom events can greatly simplify the process of managing these interactions, thereby making the overall development process more efficient.

Furthermore, the use of custom events becomes even more critical when integrating with third-party libraries. In these scenarios, custom events serve as the necessary 'hooks' or connection points that enable successful interaction and integration with these external libraries. By providing these hooks, JavaScript's CustomEvent function can significantly enhance the overall functionality and user experience of the web application, making it more responsive, interactive, and user-friendly.

Example: Custom Events

```
// Creating a custom event
let event = new CustomEvent('myEvent', { detail: { message: 'This is a custom event'
} });

// Listening for the custom event
document.addEventListener('myEvent', function(e) {
    console.log(e.detail.message);  // Outputs: This is a custom event
});

// Dispatching the custom event
document.dispatchEvent(event);
```

This example uses the CustomEvent API. It first creates a new custom event named 'myEvent' with a detail object containing a message. It then sets up an event listener on the document for 'myEvent'. When 'myEvent' is dispatched on the document, the event listener is triggered and the message is logged to the console.

2.5.8 Best Practices in Event Handling

1. **Use Event Delegation**: This is a particularly useful technique when dealing with lists or content that is generated dynamically. Rather than attaching event listeners to each

individual element, it is more efficient to attach a single listener to a parent element. This approach minimizes the number of event listeners required for functionality, thus improving the performance and efficiency of your code.

2. **Clean Up Event Listeners**: It's important to always remove event listeners when they are no longer needed, especially when elements are removed from the Document Object Model (DOM). Keeping unnecessary event listeners can lead to memory leaks and potential bugs in your application. Ensuring you have a cleanup process in place will help maintain the health and performance of your application over time.

3. **Be Mindful of this in Event Handlers**: When working with event handlers, it's critical to be aware that the value of this refers to the element that received the event, unless it is bound differently. This can sometimes lead to unexpected behavior if not properly managed. To maintain control over the scope of this, consider using arrow functions or the bind() method, which allow you to specify the value of this explicitly.

2.6 Debugging JavaScript

Debugging is a critical and essential skill that every developer, regardless of their area of expertise, should master, and JavaScript programming is certainly not an exception to this rule. The ability to effectively and efficiently identify and rectify errors or bugs in your codebase is a pivotal factor in ensuring that your code not only performs its intended function correctly but also consistently maintains an overall high standard of quality.

This is crucial not only for the immediate task at hand but also for the long-term sustainability and reliability of your code. In this comprehensive section, we will embark on an in-depth exploration of a variety of different techniques and tools that are available for debugging in JavaScript.

The aim of this guide is to equip you with a robust and versatile toolkit that can be employed to tackle both common and complex issues that may arise during the course of development. This will provide you with the necessary skills and knowledge to ensure that your code is not only functional but also optimized and free of errors.

2.6.1 Understanding the Console

The console, an indispensable tool in a developer's toolkit, acts as the primary shield against potential bugs that could disrupt the smooth functioning of web applications. The console is an integral feature that is readily accessible in all modern web browsers, ranging from popular ones like Google Chrome and Mozilla Firefox to lesser-known ones. The JavaScript console, in particular, is an irreplaceable asset for any developer.

The console's functionality extends beyond just serving as a bug defense mechanism. It also provides a comprehensive platform for logging information, a vital component of debugging and monitoring the performance of web applications. The ability to log information provides developers with insights into the behavior of their applications, assisting them in identifying any irregularities that may be affecting performance.

Moreover, the JavaScript console bestows upon developers a unique capability: the ability to run JavaScript code in real-time. This real-time execution function can be an absolute game-changer in the field of web development. It grants developers the freedom to not just test their code, but also make modifications on the fly.

This flexibility can significantly boost productivity and efficiency, as it eliminates the need for time-consuming cycles of coding, testing, and debugging. Instead, developers can directly interact with their code in real-time, making adjustments as necessary to ensure optimal performance.

Example: Using console Methods

```
console.log('Hello, World!'); // Standard log
console.error('This is an error message'); // Outputs an error
console.warn('This is a warning message'); // Outputs a warning

const name = 'Alice';
console.assert(name === 'Bob', `Expected name to be Bob, but got ${name}`); //
Conditionally outputs an error message
```

This example uses the console object to print messages. 'console.log' is used for general output of logging information, 'console.error' is used for error messages, 'console.warn' is used for warnings. 'console.assert' is used for testing: if the condition inside the assert function is false, an error message will be printed. Here, it checks if the variable 'name' is 'Bob'. If not, it outputs an error message.

2.6.2 Using debugger Statement

In JavaScript, the debugger statement serves an instrumental and indispensable role to aid in debugging processes. It works by acting as a strategic breakpoint in the code, akin to a stop sign on a busy road. This essential feature prompts a well-timed pause in the execution of code within the browser's advanced debugging tool. This pause provides the developer with a golden opportunity to delve into the current state of the code and explore its operations on a deeper level.

This feature is a powerful tool in a developer's arsenal. It allows for a meticulous inspection of the values of variables at that specific point in time. This offers an insightful and illuminating view into the inner workings of how data is manipulated and transformed throughout the execution process. It's like a window into the soul of your code, shining a light on its innermost operations.

Moreover, the debugger statement allows for stepping through the code execution line by line. This is akin to following a roadmap, enabling the developer to trace the path of execution and identify potential issues or areas of improvement in the code. It's like having a tour guide through your code, pointing out areas of interest and potential red flags that could be causing problems or slowing down your execution.

The debugger statement in JavaScript is an essential tool for any developer looking to understand their code on a deeper level, troubleshoot potential problems, and streamline their code to be as efficient and effective as possible.

Example: Using debugger

```javascript
function multiply(x, y) {
    debugger; // Execution will pause here
    return x * y;
}

multiply(5, 10);
```

When running this code in a browser with the developer tools open, execution will pause at the debugger statement, allowing you to inspect the function's arguments and step through the multiplication process.

This is a simple JavaScript function named 'multiply'. It takes two parameters, 'x' and 'y', and returns the product of these two numbers. The 'debugger' command is used to pause the execution of the code at that point, which is helpful for debugging purposes. The function is then called with the arguments 5 and 10.

2.6.3 Browser Developer Tools

Modern internet browsers are now equipped with a suite of developer tools, designed to provide a robust set of features that are essential for debugging JavaScript. These tools are built directly into the browser, making it easier and more efficient for developers to debug their code. Here's a brief overview of the primary features these developer tools offer:

- **Breakpoints**: One of the key features is the ability to set breakpoints directly in the source code view. This allows you to pause the execution of your JavaScript code at a particular line, making it easier to inspect the state of your app at that point in time.
- **Watch Expressions**: Another useful feature is the ability to track expressions and variables. By using watch expressions, you can see how the values of specific variables change over time as your code runs, which can be invaluable in identifying unexpected behavior.
- **Call Stack**: The call stack view allows you to see the function call stack, giving you a clear understanding of how the execution of your code reached its current point. This is especially useful for tracking the flow of execution, particularly in complex applications with numerous function calls.
- **Network Requests**: Lastly, developer tools provide a way to monitor AJAX requests and responses. This feature is crucial for debugging server communication, as it allows you to see the data being sent and received, the duration of the request, and any errors that occurred during the process.

Example: Setting Breakpoints

1. Open the developer tools in your browser (usually F12 or right-click -> Inspect).
2. Go to the "Sources" tab.
3. Find the JavaScript file or inline script you want to debug.
4. Click on the line number where you want to pause execution. This sets a breakpoint.
5. Run your application and interact with it to trigger the breakpoint.

2.6.4 Handling Exceptions with Try-Catch

In the realm of programming, the try-catch statement stands out as a remarkably powerful tool. It provides an elegant and graceful means of dealing with errors and exceptions that might otherwise disrupt the smooth flow of execution.

The concept is simple but effective. When we pen our code, it is placed within the try block. This block is then executed sequentially, line after line, as if it were any other ordinary block of code. This process continues unimpeded, allowing the program to function as intended. However, should an error or exception arise in the process, the try-catch statement swings into action.

Contrary to what might happen without this provision, the occurrence of an error does not cause the execution to halt abruptly. Instead, the flow of control is immediately diverted to the catch block. This block serves as a safety net, catching the error before it can cause any serious disruption.

Once inside the catch block, you are given the opportunity to handle the error constructively. The options are manifold. You could correct it on the spot, if that is feasible. Alternatively, you might choose to log the error for future debugging, thereby providing valuable information that could assist in identifying patterns or recurring issues. You also have the option of communicating the error to the user in a format that they can understand, rather than confronting them with raw and often cryptic error messages.

By providing this safety net, the try-catch statement ensures the smooth execution of code, even in the face of unforeseen or exceptional situations. It is an invaluable tool for maintaining the robustness and reliability of your software.

Example: Using Try-Catch

```
try {
    let result = riskyFunction(); // Function that might throw an error
    console.log('Result:', result);
} catch (error) {
    console.error('Caught an error:', error);
}
```

This example uses the try-catch statement. The code inside the try block (riskyFunction) is executed. If an error occurs while executing this code, the execution is stopped, and control is passed to the catch block. The catch block then logs the error message.

2.6.5 Tips for Effective Debugging

- **Reproduce the Bug**: The very first step in the debugging process should be to ensure that you can consistently reproduce the issue before attempting to fix it. This will guarantee that you understand the problem well and that it isn't related to any external factors.
- **Isolate the Issue**: Once you've reproduced the bug, the next step is to isolate the issue. This means you should reduce your code to the smallest possible set that still produces the bug. By doing this, you can more easily identify the root cause of the problem.
- **Use Version Control**: One of the most critical tools in a developer's toolkit is version control. It allows you to keep track of changes you make to your code over time and revert to previous versions if necessary. This can be particularly helpful when trying to understand when and how the bug was introduced into the code base.
- **Write Unit Tests**: Finally, writing unit tests can be a very effective method of debugging. Tests can help narrow down where the problem lies and confirm it's fixed once you make changes. This not only helps solve the current issue but also prevents the same bug from reoccurring in the future.

By fully mastering these advanced debugging techniques, you will significantly improve your ability to maintain, troubleshoot, and enhance your JavaScript code. This not only helps in making your code more efficient but also saves considerable time and reduces potential frustration during the development process.

Effective debugging is an absolutely critical skill for any programmer to acquire. It enables better comprehension of the code, allows for quicker identification and rectification of issues, and ultimately results in the creation of more robust, reliable, and resilient code. This is essential for improving the overall quality and performance of your applications, thereby ensuring a better user experience.

Practical Exercises

Now that you've learned the fundamentals of JavaScript in Chapter 2, here are some practical exercises designed to test and reinforce your understanding of the concepts discussed. These exercises include challenges related to variables and data types, operators, control structures, functions, event handling, and debugging.

Exercise 1: Variable Manipulations

Create variables to store your name, age, and if you are a student (boolean). Print a greeting message using these variables.

Solution:

```
let name = "John Doe";
let age = 20;
let isStudent = true;

console.log(`Hello, my name is ${name}. I am ${age} years old and it is ${isStudent ?
'' : 'not '}true that I am a student.`);
```

Exercise 2: Using Operators

Calculate the area of a circle with a radius of 7 using the appropriate JavaScript mathematical operators. Output the result to the console.

Solution:

```
let radius = 7;
let area = Math.PI * radius * radius;
```

```
console.log("The area of the circle is:", area);
```

Exercise 3: Control Structure - Looping

Write a JavaScript for loop that counts from 1 to 10 but only prints odd numbers to the console.

Solution:

```
for (let i = 1; i <= 10; i++) {
    if (i % 2 !== 0) {
        console.log(i);
    }
}
```

Exercise 4: Functions - Prime Number Checker

Create a function to check whether a given number is a prime number or not. The function should return true if the number is prime, otherwise false.

Solution:

```
function isPrime(number) {
    if (number <= 1) return false;
    if (number <= 3) return true;

    if (number % 2 === 0 || number % 3 === 0) return false;

    for (let i = 5; i * i <= number; i += 6) {
        if (number % i === 0 || number % (i + 2) === 0) {
            return false;
        }
    }
    return true;
}

console.log(isPrime(29));  // Outputs: true
console.log(isPrime(10));  // Outputs: false
```

Exercise 5: Event Handling

Create a simple HTML button that changes its own text content from "Click me!" to "Clicked!" when clicked.

Solution:

```
<button id="clickButton">Click me!</button>
<script>
    document.getElementById('clickButton').addEventListener('click', function() {
        this.textContent = "Clicked!";
    });
</script>
Exercise 6: Debugging Challenge
Find and fix the error in the following code snippet:
function calculateProduct(a, b) {
    console.log(a * b);
}

calculateProuct(10, 2);
```

Solution:

```
function calculateProduct(a, b) {
    console.log(a * b);
}

calculateProduct(10, 2);  // Fixed the typo in the function call
```

These exercises provide practical applications of the concepts discussed in Chapter 2, helping you to deepen your understanding and proficiency with JavaScript. Completing these will further solidify your foundation in JavaScript, preparing you for more advanced topics and projects.

Chapter Summary

In Chapter 2 we explored the fundamental aspects of JavaScript, laying a solid foundation for building interactive and dynamic web applications. This chapter provided a thorough grounding in the essential concepts every JavaScript programmer must understand, including variables, data types, operators, control structures, functions, event handling, and debugging. Let's summarize the key points covered in each section to reinforce what you've learned and highlight how these elements interact to form the backbone of JavaScript programming.

Variables and Data Types

We began by understanding how to declare and initialize variables using var, let, and const, each serving different scopes and uses in JavaScript programming. We examined JavaScript's loosely typed nature, exploring various data types like strings, numbers, booleans, null, undefined, arrays, and objects. This knowledge is crucial for handling data effectively in your applications.

Operators

Next, we delved into operators, discussing how to manipulate data using arithmetic, assignment, comparison, logical, and conditional operators. These tools allow you to perform calculations, make decisions, and execute logic based on conditions, which are vital for creating dynamic behaviors in your scripts.

Control Structures

Control structures such as if, else, switch, and loops (for, while, do-while) were explored to demonstrate how you can control the flow of execution in your programs. Understanding these structures is essential for writing efficient and effective JavaScript code that responds to different conditions and repeats tasks multiple times.

Functions and Scope

We covered functions, one of the most powerful features of JavaScript, enabling you to encapsulate code into reusable blocks. This section emphasized the importance of scope—global and local—helping you manage where variables can be accessed and modified within your scripts.

Event Handling

Event handling was introduced to equip you with the ability to make your web pages interactive. We discussed how to respond to user actions like clicks, keyboard input, and other forms of interactions through event listeners and handlers, which are fundamental for engaging user experiences.

Debugging

Finally, the debugging section equipped you with strategies to identify and fix problems in your JavaScript code. Using the console, debugger statements, and browser developer tools, you learned how to systematically troubleshoot and refine your code, ensuring its reliability and functionality.

Throughout this chapter, practical examples and exercises were provided to help you apply what you've learned in a hands-on manner. These activities are designed not only to reinforce your understanding but also to encourage you to experiment and explore JavaScript's capabilities.

As we conclude this chapter, you should feel confident in your understanding of JavaScript's fundamental concepts. These basics will serve as stepping stones to more advanced topics in subsequent chapters, where you will build on this knowledge to create more complex and powerful web applications. Remember, mastery comes with practice, so continue experimenting with code and refining your skills.

Chapter 3: Working with Data

Welcome to an engaging journey that commences with Chapter 3. In this intricate and comprehensive chapter, we are about to delve deeply into the various riveting data structures and diverse types that lay the foundation for managing and manipulating data in a highly effective manner within the versatile programming language, JavaScript.

Understanding how to work efficiently with different types of data is a crucial skill set, fundamental for the development of applications that are not only efficient but also scalable, adaptable to growing needs and demands. In this chapter, we'll explore in detail the essential elements of JavaScript's data structures such as arrays, objects, and the popular data format, JSON, among others.

Our goal is to equip you with the necessary tools and knowledge you need to handle complex data operations with ease and proficiency, enhancing your programming capabilities. We'll break down each structure and type, providing practical examples and in-depth explanations to ensure a robust understanding.

Let's embark on this enlightening journey with one of the most versatile and widely used data structures in JavaScript—arrays, a powerful tool that any proficient JavaScript developer must master.

3.1 Arrays

In the realm of JavaScript, arrays play an indispensable role by providing the crucial functionality of storing multiple values within a single variable. This characteristic is not just useful, but incredibly beneficial for the task of managing and organizing diverse data elements. It ensures that these elements are kept in an orderly, systematic manner within a single container, thereby promoting the efficient handling of data.

The versatility of arrays is another aspect that sets them apart. These dynamic data structures have the capacity to hold elements of a wide array of data types. Whether it's numerical values, textual strings, complex objects, or even other arrays, JavaScript arrays are capable of storing them all. This unique ability to accommodate different types of data, without any restrictions, amplifies their functionality, making them an exceedingly powerful tool in the hands of developers.

Arrays are structured in nature, which makes them an ideal choice for storing and managing ordered data collections. This structured approach simplifies the task of data organization and manipulation. With arrays, managing data becomes less of a chore and more of a streamlined process.

This greatly simplifies many aspects of programming in JavaScript, allowing developers to write clean, efficient code. By helping to keep data organized and easily accessible, arrays play a pivotal role in making JavaScript a robust and versatile programming language.

3.1.1 Creating and Initializing Arrays

Arrays, a crucial and fundamental data structure available in a wide range of programming languages, are versatile structures that can be created through two primary methods.

The initial method of creating arrays is through the use of array literals. This method is simple yet effective, and it involves listing out the values that you want to include in your array within square brackets. This is a straightforward way to manually specify each element that you want in your array, and it's perfect for situations where you have a clear idea of the data you'll be working with.

The second method involves using the Array constructor, a special function that creates an array based on the arguments passed to it. This method is slightly more complex but offers greater flexibility, as it allows you to dynamically create arrays based on variable input. This is particularly useful for situations where the size or contents of your array may change based on user input or other factors.

Both of these methods for creating arrays are equally valid and useful, although the most appropriate method may vary depending on the specific context and requirements of your code. By understanding and utilizing both, you can ensure that you're using the most efficient method for your specific needs.

Example: Creating Arrays

```javascript
// Using an array literal
let fruits = ['Apple', 'Banana', 'Cherry'];
```

```
// Using the Array constructor
let numbers = new Array(1, 2, 3, 4, 5);

console.log(fruits);  // Outputs: ['Apple', 'Banana', 'Cherry']
console.log(numbers); // Outputs: [1, 2, 3, 4, 5]
```

This is JavaScript code that demonstrates two ways of creating an array. The first way is using an array literal represented by square brackets. The second way is using the Array constructor, which is a built-in JavaScript function for creating arrays. After creating the arrays, the 'console.log' function is used to print the contents of the arrays to the console.

3.1.2 Accessing Array Elements

Within the broad and complex world of programming, there exists a fundamental and indispensable concept, known as the usage of arrays. Arrays, in the simplest terms, are a structured collection of elements. Each of these elements is uniquely identified by a specific index, a numerical identifier that denotes its exact position within the array.

The process of indexing, which is a cornerstone in traditional programming languages, usually commences at the digit 0. This widely accepted convention implies that the very first element in any given array is denoted by the index 0, the subsequent one by 1, and so forth in a systematic sequence.

This methodical form of indexing provides not only a highly systematic, but also an efficient and intuitive way for programmers to access each individual element within the array. Whether the array consists of a handful of elements or spans into the thousands, this indexing system remains consistently effective.

It empowers programmers with the capability to easily iterate over the elements, perform a plethora of operations on them, or retrieve specific data as required. This can range from simple tasks such as sorting or filtering data, to more complex tasks like executing algorithms for data analysis.

By gaining a comprehensive understanding and effectively utilizing the indexing system, programmers can manipulate arrays with ease. This allows them to solve a vast variety of problems, handle data in a highly efficient manner, and ultimately, write code that is both robust and efficient. The knowledge of arrays and their indexing is thus, a crucial tool in the programmer's toolkit, one that greatly enhances their coding prowess.

Example: Accessing Elements

```
let firstFruit = fruits[0];   // Accessing the first element
console.log(firstFruit);   // Outputs: 'Apple'

let secondNumber = numbers[1];   // Accessing the second element
console.log(secondNumber);   // Outputs: 2
```

The example demonstrates how to access elements from arrays. The "fruits" and "numbers" are arrays, and elements in an array are accessed using their index (position in the array). Array indices start at 0, so fruits[0] refers to the first element in the "fruits" array. The code then logs (prints to the console) these accessed elements.

3.1.3 Modifying Arrays

JavaScript arrays are not only dynamic but also flexible, implying that they can comfortably expand and contract in size in accordance with the specific requirements of your program. This is an exceptionally powerful and efficient feature, as it permits us to work with collections of data that are subject to change over time, rather than dealing with data that is static or fixed in size.

This dynamic nature of JavaScript arrays is further facilitated by a variety of methods provided by JavaScript for manipulating these arrays. Some of the most commonly used methods include push(), pop(), shift(), unshift(), and splice().

These methods exhibit incredible versatility, enabling you to add elements to either the end or the beginning of the array (push() and unshift(), respectively), eliminate elements from the end or the beginning of the array (pop() and shift(), respectively), or insert and remove elements from any position within the array (splice()).

These methods empower programmers by providing them with the flexibility to manage and manipulate data according to their specific needs. As a result, JavaScript arrays, through the use of these methods, prove to be an incredibly flexible and powerful tool for managing various collections of data, enhancing the efficiency and functionality of your program.

Example: Modifying Arrays

```
fruits.push('Durian');   // Adds 'Durian' to the end of the array
console.log(fruits);   // Outputs: ['Apple', 'Banana', 'Cherry', 'Durian']

let lastFruit = fruits.pop();   // Removes the last element
console.log(lastFruit);   // Outputs: 'Durian'
console.log(fruits);   // Outputs: ['Apple', 'Banana', 'Cherry']
```

```
fruits.unshift('Strawberry');  // Adds 'Strawberry' to the beginning
console.log(fruits);  // Outputs: ['Strawberry', 'Apple', 'Banana', 'Cherry']

let firstRemoved = fruits.shift();  // Removes the first element
console.log(firstRemoved);  // Outputs: 'Strawberry'
console.log(fruits);  // Outputs: ['Apple', 'Banana', 'Cherry']
```

This JavaScript code demonstrates the use of array manipulation methods.

- .push('Durian'): Adds 'Durian' to the end of the fruits array.
- .pop(): Removes the last element from the fruits array and assigns it to the variable lastFruit.
- .unshift('Strawberry'): Adds 'Strawberry' to the beginning of the fruits array.
- .shift(): Removes the first element from the fruits array and assigns it to the variable firstRemoved.

After each operation, it logs the state of the fruits array or the removed element to the console.

3.1.4 Iterating Over Arrays

When you need to manipulate or interact with every item within an array, numerous tools are at your disposal in JavaScript. Traditional loop structures such as the for loop or the for...of loop can be used. In these loops, you would typically define an index variable and use it to access each element in the array sequentially.

However, JavaScript also provides a number of built-in array methods that can make this process more straightforward and readable. The forEach() method, for instance, executes a provided function once for each array element. The map() method creates a new array populated with the results of calling a provided function on every element in the calling array.

The filter() method creates a new array with all elements that pass a test implemented by the provided function. Lastly, the reduce() method applies a function against an accumulator and each element in the array (from left to right) to reduce it to a single output value. These methods provide a more functional and declarative approach to array iteration.

Example: Iterating Over Arrays

```
// Using forEach to log each fruit
fruits.forEach(function(fruit) {
    console.log(fruit);
```

```
});

// Using map to create a new array of fruit lengths
let fruitLengths = fruits.map(function(fruit) {
    return fruit.length;
});
console.log(fruitLengths);  // Outputs: [5, 6, 6]
```

This JavaScript code snippet demonstrates the use of two powerful array methods: forEach and map.

The forEach method is used to execute a function on each item in an array. In this case, the function is simply logging the name of each fruit in the 'fruits' array. This method is useful when you want to perform the same operation on each item in an array, without altering the array itself or creating a new one. Here, console.log is called for each fruit, which will print the name of each fruit to the console.

The map method, on the other hand, is used to create a new array based on the results of a function that is run on each item in the original array. In this instance, the function is returning the length of each fruit's name, effectively creating a new array that contains the length of each fruit's name. The map method is highly beneficial when you need to transform an array in some way, as it allows you to apply a function to each element and collect the results in a new array.

Finally, the new 'fruitLengths' array is logged to the console. The output of this will be an array of numbers, each representing the number of characters in the corresponding fruit's name from the original 'fruits' array. For example, if the 'fruits' array contained ['Apple', 'Banana', 'Cherry'], the output would be [5, 6, 6] because 'Apple' has 5 characters, 'Banana' has 6 characters, and 'Cherry' also has 6 characters.

By understanding and utilizing these array methods, you can manipulate and transform arrays effectively in JavaScript, which is a fundamental skill in many areas of programming and data handling.

Arrays are a fundamental part of JavaScript and a powerful tool for any developer. By mastering array operations and methods, you can handle collections of data more effectively, making your applications more powerful and responsive.

3.1.5 Multi-dimensional Arrays

JavaScript, a dynamic and versatile programming language, provides support for arrays of arrays, which are commonly known as multi-dimensional arrays. Multi-dimensional arrays are especially useful in certain scenarios due to their ability to represent complex data structures.

For example, they can be employed to depict matrices, an important mathematical concept that is used in various fields ranging from computer graphics to machine learning. Additionally, multi-dimensional arrays can be used for storing data in a tabular form.

This makes them a perfect fit for scenarios where data needs to be organized in rows and columns, like in a relational database or a spreadsheet. Thus, the flexibility and utility of multi-dimensional arrays in JavaScript make them an integral part of the language.

Example: Multi-dimensional Array

```
let matrix = [
    [1, 2, 3],
    [4, 5, 6],
    [7, 8, 9]
];

console.log(matrix[1][2]);   // Accessing the third element in the second array,
Outputs: 6
```

This code is an example of how to create and utilize a two-dimensional array, sometimes referred to as a matrix, in JavaScript.

In this example, 'matrix' is declared as a variable using the 'let' keyword, and it's assigned to a multidimensional array. This array is composed of three smaller arrays, each containing three elements. These sub-arrays represent the rows of the matrix, and the elements within them represent the columns.

In other words, the 'matrix' variable is effectively a grid with three rows and three columns, filled with the numbers 1 through 9. The arrangement of these numbers is significant here, as it's what allows us to retrieve specific elements based on their position within the matrix.

The 'console.log()' function is used to print the result of 'matrix[1][2]' to the console. This expression is accessing the third element (at index 2) of the second sub-array (at index 1) within the matrix, which is the number 6.

Remember that array indices in JavaScript start at 0, so 'matrix[1]' refers to the second sub-array '[4, 5, 6]', and 'matrix[1][2]' refers to the third element of this sub-array, which is the number 6.

This ability to access individual elements within a multidimensional array is crucial when dealing with complex data structures or algorithms in JavaScript, and it's a technique that often comes in handy in various scenarios in programming.

3.1.6 Array Destructuring

The advent of ES6, also known as ECMAScript 2015, ushered in a range of valuable new features designed to enhance JavaScript's functionality and ease of use. Among these key enhancements was the introduction of array destructuring. This powerful feature affords developers a highly convenient method to extract multiple properties from an array or object and assign them to individual variables.

Array destructuring is a game-changer, enabling a more streamlined and simplified code structure. It also dramatically improves readability, a critical factor in any coding project. This is because clear, easy-to-understand code aids both the development process and future maintenance work.

This feature proves its worth especially when navigating complex data structures. Suppose you are dealing with a data structure from which you need to extract multiple values for separate manipulation. In such a scenario, array destructuring can be an invaluable tool. It enables you to extract and work with these values individually in a more effortless and efficient manner.

Overall, the use of array destructuring can enhance the efficiency of your code and its maintainability. It's one of the many features introduced with ES6 that truly elevates the JavaScript coding experience.

Example: Array Destructuring

```
let colors = ['Red', 'Green', 'Blue'];
let [firstColor, , thirdColor] = colors;

console.log(firstColor);  // Outputs: 'Red'
console.log(thirdColor);  // Outputs: 'Blue'
```

This example code demonstrates the use of array destructuring. The array 'colors' is defined with three elements. The line 'let [firstColor, , thirdColor] = colors' uses array destructuring to assign the first and third elements of the 'colors' array to the variables 'firstColor' and 'thirdColor' respectively.

The second element is ignored because of the empty space between the commas. The console.log statements then print the values of 'firstColor' and 'thirdColor', which would be 'Red' and 'Blue', respectively.

3.1.7 Finding Elements in Arrays

When you need to search for a specific element or check if a certain element exists within an array, JavaScript offers a variety of methods that can be utilized. These include the indexOf() method which returns the first index at which a given element can be found in the array, or -1 if it is not present.

Similarly, the find() method returns the value of the first element in an array that satisfies the provided testing function, whereas findIndex() returns the index of the first element that satisfies the same function. Lastly, the includes() method determines whether an array includes a certain value among its entries, returning true or false as appropriate.

Each of these methods provides a unique and efficient way to handle element searches within an array in JavaScript.

Example: Finding Elements

```
let numbers = [1, 2, 3, 4, 5];

console.log(numbers.indexOf(3));        // Outputs: 2
console.log(numbers.includes(4));       // Outputs: true

let result = numbers.find(num => num > 3);
console.log(result);                    // Outputs: 4

let resultIndex = numbers.findIndex(num => num > 3);
console.log(resultIndex);               // Outputs: 3
```

This is a example code snippet demonstrating the use of different array methods.

1. numbers.indexOf(3): This line of code finds the index of the number 3 in the array, which is 2 in this case.
2. numbers.includes(4): This line of code checks if the number 4 is included in the array, which is true in this case.

3. numbers.find(num => num > 3): This line of code uses the find method to return the first number in the array that is greater than 3, which is 4 in this case.
4. numbers.findIndex(num => num > 3): This line of code uses the findIndex method to return the index of the first number in the array that is greater than 3, which is 3 in this case.

3.1.8 Sorting Arrays

In the realm of data manipulation and analysis, one recurrent task that most individuals encounter is sorting of data housed within arrays. This task is essentially concerned with the arrangement of data in a specific order, which can vary from ascending to descending, or even numerical to alphabetical. The ability to properly arrange data in such a manner is of utmost importance, as it plays a vital role in the efficient analysis and manipulation of data.

The sort() method emerges as a powerful tool in this context, offering the much-needed functionality to sort data. It operates by sorting an array in-place. This means that instead of creating and returning a new array that has been sorted, it modifies the original array itself. The implications of this is that the original array gets sorted, and no additional memory is required to store a separate sorted array.

By default, the sort() method is designed to arrange the elements as strings, in an ascending sequence, and in alphabetical order. This implies that if the elements within the array are numbers, the method will start sorting from the smallest numerical value and proceed in an ascending order.

Conversely, if the array consists of words, the sorting will commence from the word that would appear first in an alphabetical listing (starting from A) and will continue in increasing order. This inherent functionality of the sort() method makes it an invaluable tool in any data analyst's toolkit.

Example: Sorting an Array

```
let items = ['Banana', 'Apple', 'Pineapple'];
items.sort();
console.log(items);  // Outputs: ['Apple', 'Banana', 'Pineapple']

let numbers = [10, 1, 5, 2, 9];
numbers.sort((a, b) => a - b);
console.log(numbers);  // Outputs: [1, 2, 5, 9, 10]
```

This example code is creating and sorting two arrays: one consisting of string elements and the other one of numeric elements.

The first part of the code creates an array named 'items' with three string elements - 'Banana', 'Apple', and 'Pineapple'. The sort() method is then called on this array. By default, the sort() method sorts elements as strings in lexicographic (or dictionary) order, which means it sorts them alphabetically in ascending order. When the sorted 'items' array is logged to the console, the output will be ['Apple', 'Banana', 'Pineapple'], which is the result of sorting the original array elements alphabetically.

The second part of the code creates an array called 'numbers', made up of five numeric elements - 10, 1, 5, 2, 9. The sort() method is then called on this array with a comparison function passed as an argument. The comparison function (a, b) => a - b causes the sort() method to sort the numbers in ascending order.

This is because, for any two elements 'a' and 'b', if a - b is less than 0, 'a' will be sorted to an index lower than 'b', i.e., 'a' comes first. If a - b is equal to 0, 'a' and 'b' remain unchanged with respect to each other. If a - b is greater than 0, 'a' is sorted to an index higher than 'b', i.e., 'b' comes first. By doing this, the sort() method sorts the 'numbers' array in ascending numerical order. When we log the sorted 'numbers' array to the console, the output is [1, 2, 5, 9, 10], which has the elements of the original array sorted in increasing order.

This example demonstrates the utility of the sort() method in JavaScript for arranging the elements of an array in a specific order, be it alphabetical for strings or numerical for numbers.

3.1.9 Performance Considerations

When dealing with large arrays in programming, performance can quickly become a significant concern that needs to be addressed carefully. The efficient use of array methods becomes crucial in maintaining the speed and responsiveness of applications. This is particularly true when the goal is to minimize the number of operations, which directly affects the runtime of the code.

For example, chaining methods such as map() and filter() can inadvertently create intermediate arrays. This is not always optimal as it can consume additional memory and slow down the performance. Avoiding such pitfalls can be achieved with more optimal coding strategies. It could involve, for example, using more specialized functions that can handle the tasks in a more efficient manner.

Understanding the implications of your coding choices is key to maintaining high performance when working with large arrays.

Example: Efficient Method Chaining

```
// Less efficient
let processedData = data.map(item => item.value).filter(value => value > 10);

// More efficient
let efficientlyProcessedData = data.reduce((acc, item) => {
    if (item.value > 10) acc.push(item.value);
    return acc;
}, []);
```

This selection shows two ways of processing data in JavaScript.

The first, less efficient way, is to use the 'map' method to create a new array with the 'value' property of each item, and then use 'filter' to select only the values that are greater than 10.

The second, more efficient way, is to use the 'reduce' method. This method traverses through the 'data' array once, and if the 'value' property of an item is greater than 10, it pushes the value into the accumulating array 'acc'. This method is more efficient because it only needs to iterate through the array once, instead of twice.

By diving deeper into these advanced aspects of arrays, you gain the skills to manage and manipulate arrays more effectively, enabling you to handle more complex data structures and improve the performance of your JavaScript applications.

3.2 Objects

In JavaScript programming, objects assume a highly significant role. They are the fundamental building blocks used to store collections of data and even more complex entities. Their importance can't be overstated as they form the backbone of structured data interaction in the language.

Unlike arrays, which are essentially collections indexed by a numeric value, objects introduce a more structured and organized approach to data representation. This structured approach is a key aspect of JavaScript programming, enhancing the readability and maintainability of the code.

Objects operate by using properties that are accessible via specific keys. These keys are used to store and retrieve data, making objects a type of associative array. The utilization of keys in objects provides a clear structure and easy access to the stored data, making them a powerful tool for developers.

This section aims to provide a comprehensive exploration of objects in JavaScript. It delves deeply into the object-oriented nature of JavaScript, covering the creation, manipulation, and practical utilization of objects. This exploration aims to provide a detailed and thorough understanding of objects in JavaScript, their importance, and how they are used.

By gaining a solid understanding of these concepts, you can effectively utilize objects in your JavaScript projects. This leads to the development of more efficient, maintainable, and readable code. Such knowledge is not merely beneficial but crucial for anyone planning to dive deeper into JavaScript and unlock its full potential. By understanding and mastering the use of objects, you're taking a significant step towards becoming a proficient JavaScript developer.

3.2.1 Creating and Accessing Objects

Objects play a pivotal role in shaping the structural design of the language. Objects, in JavaScript, can be conveniently created using a technique known as object literals, which is a straightforward and intuitive method.

When employing this method, the object is instantiated with an array of key-value pairs. These pairs are often referred to as the object's properties, which denote individual characteristics or attributes of the object, thereby providing a detailed description of it.

Each property is composed of two components: a key, which is essentially a unique identifier or the name of the property, and a value, which can be any valid JavaScript value such as strings, numbers, arrays, other objects, and so forth. The key-value pair structure provides a clear and concise way to organize and access data, making it straightforward for developers to work with.

This approach to creating objects is not only incredibly flexible but also extremely powerful. It opens up the possibility of representing complex data structures in a way that is easily comprehensible and manageable, even for developers who are relatively new to the language. This is one of the reasons why JavaScript's object literal notation is so popular and widely used in the programming world.

Example: Creating and Accessing an Object

```
let person = {
    name: "Alice",
    age: 25,
    isStudent: true
```

```
};

console.log(person.name);  // Outputs: Alice
console.log(person['age']);  // Outputs: 25
```

In this example, person is an object with properties name, age, and isStudent. Properties can be accessed using dot notation (person.name) or bracket notation (person['age']).

Here, an object called 'person' is created with the properties 'name', 'age', and 'isStudent'. The 'console.log' statements are used to print the 'name' and 'age' properties of the 'person' object to the console. In the first case, dot notation is used to access the 'name' property, and in the second case, bracket notation is used to access the 'age' property.

3.2.2 Modifying Objects

In programming, JavaScript holds a special place due to its ability to add, modify, and delete properties of objects even after they have been created. This robust feature allows for a high degree of dynamism and flexibility, making object handling exceptionally effective when it comes to managing data. Essentially, this means that you can customise objects to precisely fit your evolving requirements throughout the execution of a program.

Instead of being constrained by the strict parameters of pre-established structures, JavaScript provides the freedom to adapt as you go. This flexibility is immensely valuable as it allows for the addition of new properties as and when required. Simultaneously, it grants the ability to adjust existing properties to better align with your shifting objectives and requirements.

Moreover, the dynamic nature of JavaScript extends to efficiency too. In a scenario where a property becomes redundant or irrelevant, you can simply delete it. This ensures that your objects remain streamlined and efficient, free from unnecessary clutter, which could potentially hamper performance.

This inherent dynamism in the handling of objects is one of the many reasons why JavaScript has proven to be such a versatile programming language. Its popularity among developers is testament to its adaptability and suitability to a wide range of programming needs.

Example: Modifying an Object

```
// Adding a new property
person.email = 'alice@example.com';
console.log(person);
```

```
// Modifying an existing property
person.age = 26;
console.log(person);

// Deleting a property
delete person.isStudent;
console.log(person);
```

This example code snippet that illustrates how to manipulate the properties of an object. Here, we are dealing with an object named person.

In JavaScript, objects are dynamic, which means they can be modified after they have been created. This includes adding new properties, changing the value of existing properties, or even deleting properties. This feature provides a high degree of flexibility and allows us to manage data in a more effective manner.

The first operation in the code is the addition of a new property to the person object. The new property is email and its value is set to 'alice@example.com'. This is done using the dot notation, i.e., person.email = 'alice@example.com';. After this operation, the person object is logged to the console using console.log(person);.

Following this, an existing property of the person object, age, is modified. The new value of the age property is set to 26. This operation is performed using the dot notation again, i.e., person.age = 26;. After this change, the updated person object is logged to the console again.

Finally, a property isStudent is deleted from the person object. This is done using the delete keyword followed by the object and its property, i.e., delete person.isStudent;. After this deletion, the final state of the person object is logged to the console.

The ability to dynamically modify objects is one of the reasons why JavaScript is such a versatile programming language. It allows developers to adapt data structures to fit their evolving requirements throughout the execution of a program. This flexibility is extremely valuable as it lets developers add new properties as needed, adjust existing properties to better align with their objectives, and delete properties that become redundant or irrelevant.

3.2.3 Methods in Objects

In the vast and complex realm of object-oriented programming, one crucial concept stands out: the use of methods. Methods are, in essence, functions that are neatly stored as properties within an object. They are the actions that an object can perform, the tasks that it can carry out. The beauty and key advantage of defining these methods within the objects themselves is that it encapsulates, or bundles together, functionalities that are directly and inherently relevant to the object.

This encapsulation not only neatly packages these functionalities, but it also paves the way for enhanced modularity. This leads to a much more organized, streamlined, and manageable code structure. Instead of having to sift through disjointed pieces of code, developers can easily locate and understand the functionalities thanks to their logical placement within the objects to which they pertain.

Furthermore, this method of organization does more than just improve tidiness - it also significantly increases the reusability of code. By containing these functions within the objects to which they are most relevant, they can easily be called upon and reused as needed. This not only saves time and resources during the code development process, but it also makes debugging a smoother and less arduous process.

The use of methods within objects in object-oriented programming allows for more intuitive understanding of the code, improved efficiency in its development, and easier debugging. It's a method that offers profound benefits, transforming the way that developers approach and handle code.

Example: Methods in Objects

```javascript
let student = {
    name: "Bob",
    courses: ['Mathematics', 'English'],
    greet: function() {
        console.log("Hello, my name is " + this.name);
    },
    addCourse: function(course) {
        this.courses.push(course);
    }
};

student.greet();  // Outputs: Hello, my name is Bob
student.addCourse('History');
console.log(student.courses);  // Outputs: ['Mathematics', 'English', 'History']
```

This is an example of an object in JavaScript, created using the object literal notation. The object, named student, represents a student with specific properties and methods.

The student object has two properties: name and courses. The name property is a string representing the student's name, in this case, "Bob". The courses property is an array that holds the courses the student is currently taking, which are 'Mathematics' and 'English'.

In addition to these properties, the student object also has two methods: greet and addCourse.

The greet method is a function that outputs a greeting to the console when called. It uses the console.log JavaScript function to print a greeting message, which includes the student's name. The this keyword is used to reference the current object, which in this case is student, and access its name property.

The addCourse method is a function that takes a course (represented by a string) as a parameter and adds it to the courses array of the student. It achieves this by using the push method of the array, which adds a new element to the end of the array.

After the student object is defined, the code demonstrates how to use its properties and methods. First, it calls the greet method using the dot notation (student.greet()). Calling this method outputs "Hello, my name is Bob" to the console.

Next, it calls the addCourse method, again using dot notation, and passes 'History' as an argument (student.addCourse('History')). This adds 'History' to the courses array of the student.

Finally, the code prints the courses property of the student to the console (console.log(student.courses)). This outputs the updated courses array which now includes 'History', in addition to 'Mathematics' and 'English'. So, the output would be ['Mathematics', 'English', 'History'].

3.2.4 Iterating Over Objects

When it comes to iterating over objects in JavaScript, several techniques can be implemented. One commonly used technique is the for...in loop, which is specifically designed to enumerate object properties.

This type of loop can be particularly helpful when you have an object with an unknown number of properties and you need to access the keys of these properties. On the other hand, if you prefer a more functional approach to handling data, JavaScript offers methods such as Object.keys(), Object.values(), and Object.entries().

These methods return arrays that contain the object's keys, values, and entries, respectively. This functionality can be incredibly useful when you want to manipulate an object's data in a more declarative manner, or when you need to integrate with other array methods for more complex tasks.

Example: Iterating Over an Object

```
for (let key in student) {
    if (student.hasOwnProperty(key)) {
        console.log(key + ': ' + student[key]);
    }
}

// Using Object.keys() to get an array of keys
console.log(Object.keys(student));     // Outputs: ['name', 'courses', 'greet',
'addCourse']

// Using Object.entries() to get an array of [key, value] pairs
Object.entries(student).forEach(([key, value]) => {
    console.log(`${key}: ${value}`);
});
```

This example code demonstrates various methods for iterating over an object's properties.

The first part of the code uses a 'for in' loop to iterate over each property (or 'key') in the 'student' object. The 'hasOwnProperty' method is used to ensure only the object's own properties are logged to the console, not properties it might have inherited.

The second part uses the 'Object.keys()' method to create an array of the object's keys, then logs this to the console.

The third part uses the 'Object.entries()' method to create an array of [key, value] pairs, then uses a 'forEach' loop to log each key-value pair to the console.

3.2.5 Object Destructuring

The introduction of ECMAScript 6 (ES6) brought with it many new features which significantly improved the JavaScript landscape. One of the most impactful of these is a feature known as object destructuring.

Object destructuring is, in essence, a convenient and efficient method that allows programmers to extract multiple properties from within objects in a single statement. This technique provides an easy way to create new variables by extracting values from an object's properties.

Once these properties have been extracted, they can then be bound to variables. This process helps in simplifying the handling of objects and variables in programming. It eliminates the need for repetitively accessing properties within objects, thereby making code cleaner, easier to understand, and more efficient.

All in all, object destructuring is a highly useful feature for developers. It not only improves code readability but also enhances productivity by reducing the amount of code required for certain tasks. It's one of the many features that make ES6 a powerful tool in the hands of modern JavaScript developers.

Example: Object Destructuring

```
let { name, courses } = student;
console.log(name);  // Outputs: Bob
console.log(courses);  // Outputs: ['Mathematics', 'English', 'History']
```

This example uses destructuring assignment to extract properties from the 'student' object. 'name' and 'courses' are variables that now hold the values of the corresponding properties in the 'student' object. The 'console.log()' statements are used to print these values to the console.

Objects are incredibly powerful and versatile in JavaScript, suitable for representing almost any kind of data structure. By mastering JavaScript objects, you enhance your ability to structure and manage data effectively in your applications, leading to cleaner, more efficient, and scalable code.

3.2.6 Property Attributes and Descriptors

In the world of JavaScript, a key feature that sets it apart is how it deals with the properties of its objects. Each property of an object within JavaScript is uniquely characterized by certain specific attributes. These attributes are not just mere descriptors; they serve a greater purpose by defining the configurability, enumerability, and writability of the properties. These three aspects are crucial as they ultimately determine the way in which the properties of these objects can be interacted with or manipulated, providing a framework for how the objects function within the larger JavaScript environment.

However, JavaScript doesn't stop there. Recognizing that developers need more fine-grained control over how these properties behave to further enhance their coding capabilities, JavaScript offers a built-in function known as Object.defineProperty(). This function is not just powerful, but also a game-changer. It allows for the explicit setting of these attributes, providing developers with a toolkit to define or modify the default behavior of the properties within an object.

What this means in practical terms is that developers can use Object.defineProperty() to tailor their objects to their exact needs, thus enhancing the flexibility and control when coding. This greater level of control can potentially lead to more efficient, effective, and cleaner code, making JavaScript a more powerful tool in the hands of the developer.

Example: Using Property Attributes

```
let person = { name: "Alice" };
Object.defineProperty(person, 'age', {
    value: 25,
    writable: false,   // Makes the 'age' property read-only
    enumerable: true,  // Allows the property to be listed in a for...in loop
    configurable: false  // Prevents the property from being removed or the descriptor
from being changed
});

console.log(person.age);   // Outputs: 25
person.age = 30;
console.log(person.age);   // Still outputs: 25 because 'age' is read-only

for (let key in person) {
    console.log(key);   // Outputs 'name' and 'age'
}
```

This example code creates an object called "person" with a property "name". Then, it uses the Object.defineProperty method to add a new property "age" to the "person" object.

This new property is set with certain attributes:

- Its value is set to 25.
- It's not writable, meaning that attempts to change its value will fail.
- It's enumerable, meaning it will show up in for...in loops.
- It's not configurable, meaning you can't delete this property or change these attributes later on.

The console.log statements demonstrate that the 'age' property cannot be changed due to its 'writable: false' attribute. The final for...in loop demonstrates that 'age' is included in the loop due to its 'enumerable: true' attribute.

3.2.7 Prototypes and Inheritance

JavaScript is a prototype-based language, a type of object-oriented programming language that utilizes a concept known as prototypal inheritance. In this type of language, objects inherit properties and methods from a prototype.

In other words, there is a blueprint, known as a prototype, from which objects are created and derive their characteristics. Each object in JavaScript contains a private property, a unique attribute that holds a link to another object, which is referred to as its prototype.

This prototype serves as the parent, or the base model, from which the object inherits its properties and methods.

Example: Prototypes in Action

```javascript
let animal = {
    type: 'Animal',
    describe: function() {
        return `A ${this.type} named ${this.name}`;
    }
};

let cat = Object.create(animal);
cat.name = 'Whiskers';
cat.type = 'Cat';

console.log(cat.describe());  // Outputs: A Cat named Whiskers
```

In this example, cat inherits the describe method from animal.

This example uses the concept of prototypal inheritance. Here, an object 'animal' is created with properties 'type' and 'describe'. 'describe' is a method that returns a string describing the animal.

Then, a new object 'cat' is created using Object.create() method, which sets the prototype of 'cat' to 'animal', meaning 'cat' inherits properties and methods from 'animal'. The 'name' and 'type' properties of 'cat' are then set to 'Whiskers' and 'Cat' respectively.

Finally, when the 'describe' method is called on 'cat', it uses its own 'name' and 'type' properties due to JavaScript's prototype chain lookup. So it outputs: 'A Cat named Whiskers'.

3.2.8 Object Cloning

In the intricate and complex world of object-oriented programming, there are certain instances when you might find yourself in a situation where there is a need to create an identical copy of an already existing object. This process, known as cloning, can be invaluable in various scenarios.

For instance, let's say you have an object with a specific set of properties or a particular state. Now, you find yourself in a position where you need to create another object that mirrors these exact properties or state. This is where cloning comes into play, allowing you to replicate the original object precisely.

However, the utility of cloning doesn't stop there. One of the crucial aspects of cloning is that you can make modifications to this new, cloned object without having the slightest impact on the original object. This means that the state of the original object remains unaltered, no matter how many changes you make to the cloned one.

In essence, the independent manipulation of two objects, where one is a direct clone of the other, is a significant advantage of the cloning process. It allows for flexibility and freedom in programming, without risking the integrity of the original object. This is what makes cloning a critical tool in the arsenal of every proficient object-oriented programmer.

Example: Cloning an Object

```javascript
let original = { name: "Alice", age: 25 };
let clone = Object.assign({}, original);

clone.name = "Bob";  // Modifying the clone does not affect the original

console.log(original.name);  // Outputs: Alice
console.log(clone.name);     // Outputs: Bob
```

This is an example code snippet demonstrating the concept of object cloning using the Object.assign() method.

In the code, an object named 'original' with properties 'name' and 'age' is created. Then a new object 'clone' is made as a copy of 'original' using Object.assign().

Any modifications made to 'clone' will not affect 'original'. This is shown when 'clone.name' is changed to "Bob", but 'original.name' remains as "Alice".

The console.log() commands at the end are used to verify that the original object remains unchanged when the clone is modified.

3.2.9 Using Object.freeze() and Object.seal()

In JavaScript, there are several methods that can be used to prevent the modification of objects to maintain data integrity and consistency. Among these are the Object.freeze() and Object.seal() methods:

- Object.freeze() is a method that takes an object as an argument and returns an object where changes to existing properties are prevented. This method essentially makes an object immutable by stopping any alterations to the current properties. It also prevents any new properties from being added to the object, ensuring the integrity of the object after it has been defined.
- Another method, Object.seal(), also takes an object as an argument and returns an object that cannot have new properties added to it. This method ensures that the structure of the object remains constant after its definition. In addition to preventing new properties from being added, Object.seal() also makes all existing properties on the object non-configurable. This means that while the values of these properties can be changed, the properties themselves cannot be deleted or reconfigured in any way.

Example: Freezing and Sealing Objects

```
let frozenObject = Object.freeze({ name: "Alice" });
frozenObject.name = "Bob";   // No effect
console.log(frozenObject.name);   // Outputs: Alice

let sealedObject = Object.seal({ name: "Alice" });
sealedObject.name = "Bob";
sealedObject.age = 25;   // No effect
console.log(sealedObject.name);   // Outputs: Bob
console.log(sealedObject.age);    // Outputs: undefined
```

This JavaScript code demonstrates the use of Object.freeze() and Object.seal() methods. The Object.freeze() method makes an object immutable, meaning you can't change, add, or delete its properties. The Object.seal() method prevents new properties from being added and marks all existing properties as non-configurable.

However, properties of a sealed object can still be changed. In the given code, a frozen object and a sealed object are created, both initially having a property name with a value "Alice". Attempting to change the name property of the frozen object has no effect, but the name property of the sealed object can be changed. Attempting to add a new age property to the sealed object also has no effect.

By mastering these advanced features of JavaScript objects, you'll be better equipped to write robust, efficient, and secure JavaScript code. These capabilities enable sophisticated data handling and provide the building blocks for complex and scalable application architectures.

3.3 JSON

JSON, short for JavaScript Object Notation, is a lightweight data-interchange format that stands out due to its simplicity and effectiveness. It's designed to be easily understood and written by humans, while also being easy for machines to parse and generate. This combination of features makes it a valuable tool for transferring data, particularly over the internet.

Over the years, JSON has earned its place as a standard format for structuring data for internet communication. Its widespread use and versatility make it a topic of essential knowledge for any web developer, regardless of their level of experience or the specific nature of their work.

In the following section, we will delve deeper into the world of JSON. We will start by exploring what JSON is in more detail, including its origins, its structure, and the reasons for its popularity. Following that, we will guide you on how to use JSON effectively within JavaScript, one of the most popular programming languages in today's digital world.

Additionally, we will go through some of the most common operations related to JSON. This includes parsing, an essential operation for converting a JSON text into a JavaScript object, and stringifying, the process of converting a JavaScript object into a JSON text. These operations form the backbone of most tasks involving JSON, making their understanding crucial for any aspiring web developer.

3.3.1 What is JSON?

JSON, which stands for JavaScript Object Notation, is a text format that is totally language-agnostic. It uses conventions that are quite familiar to programmers who are well-versed in the C-family of languages. This includes languages such as C, C++, C#, Java, JavaScript, Perl, Python, and a myriad of others. The universal nature of JSON makes it an incredibly useful tool for data interchange.

The JSON structure is built on two fundamental structures, making it simple yet powerful:

- The first is a collection of name/value pairs. This structure is realized in various programming languages in different forms. In some languages, it's known as an object, in others, it's a record. Some languages refer to it as a struct, while others call it a dictionary. You might also hear it referred to as a hash table, a keyed list, or an associative array depending on the language you're using.
- The second structure is an ordered list of values. This, too, is realized differently in most programming languages. It's often known as an array, but can also be referred to as a vector in some languages. Other languages might call this structure a list, while others might refer to it as a sequence.

In essence, JSON's simplicity, versatility, and language-independent nature make it a go-to choice for programmers when it comes to data interchange.

Example: JSON Object

```
{
    "firstName": "John",
    "lastName": "Doe",
    "age": 30,
    "isStudent": false,
    "address": {
        "street": "123 Main St",
        "city": "Anytown",
        "country": "Anycountry"
    },
    "courses": ["Math", "Science", "Art"]
}
```

This example shows a JSON object that describes a person, including their name, age, student status, address, and courses they are taking.

The object contains information about a person named John Doe who is 30 years old, not a student, and lives at 123 Main St, Anytown, Anycountry. He is taking Math, Science, and Art courses.

3.3.2 Parsing JSON

As described before, JavaScript has a unique feature where it handles data received from a server in a format known as JSON, which stands for JavaScript Object Notation. This data, when initially received, is in the form of a JSON string.

A JSON string, while easy to transmit over the internet, is not directly usable for manipulation or retrieval of data within the JavaScript environment. This means that, in its initial state, it can't be used to perform operations or extract specific information.

Consequently, to make this data usable in a JavaScript setting, we need to transform this JSON string into JavaScript objects. These objects can then be easily manipulated and accessed according to the needs of the developer.

This transformation process is done using a specific function provided by JavaScript itself, known as JSON.parse(). It's a powerful function that takes the JSON string as its input and then outputs it as a JavaScript object.

By converting the data into JavaScript objects, developers can easily access specific data points, manipulate the data, and integrate it within their code. Such a feature simplifies the handling of JSON data, making JavaScript a versatile and efficient language for web development.

Example: Parsing JSON

```javascript
let jsonData = '{"firstName":"John","lastName":"Doe","age":30}';
let person = JSON.parse(jsonData);

console.log(person.firstName);  // Outputs: John
console.log(person.age);        // Outputs: 30
```

In this example, JSON.parse() transforms the JSON string into a JavaScript object. It first declares a variable 'jsonData' that holds a string of JSON data. It then uses the 'JSON.parse' function to convert this JSON string into a JavaScript object, which is stored in the variable 'person'. The last two lines use 'console.log' to print out the 'firstName' and 'age' properties of the 'person' object to the console.

3.3.3 Stringifying JSON

On the other hand, there are instances when you need to transport data from a JavaScript application to a server. In such cases, it becomes necessary to alter the format of JavaScript objects into JSON strings.

JSON strings are universally recognized and can be easily handled by servers. The process of converting JavaScript objects into JSON strings is accomplished via a method known as JSON.stringify(). This function allows the data to be sent over the network in a format that can be easily understood and processed by the server.

Example: Stringifying JSON

```
let personObject = {
    firstName: "John",
    lastName: "Doe",
    age: 30
};

let jsonString = JSON.stringify(personObject);
console.log(jsonString);                                    //            Outputs:
'{"firstName":"John","lastName":"Doe","age":30}'
```

Here, JSON.stringify() converts the JavaScript object into a JSON string, which can then be sent to a server. It declares an object named 'personObject' with properties 'firstName', 'lastName', and 'age'. The 'JSON.stringify()' function is then used to convert 'personObject' into a JSON string. This string is stored in the 'jsonString' variable. The last line of code logs the 'jsonString' to the console, outputting the personObject as a JSON string.

3.3.4 Working with Arrays in JSON

JSON, which stands for JavaScript Object Notation, is a widely utilized data format that has the unique capability to incorporate arrays within its structure. This particular feature is incredibly beneficial when you're dealing with the transfer or reception of large quantities of data in list form. With arrays, rather than having to send individual pieces of data one at a time, you can send large sets of data simultaneously.

This bulk transmission of data can be a game-changer for data-driven applications, significantly enhancing their speed and overall efficiency. By enabling the aggregation of data points into organized, easily transmittable packages, arrays within JSON not only simplify data management but also boost the performance of applications handling large volumes of data.

Example: JSON Array

```
let jsonArray = '[{"name":"John"}, {"name":"Jane"}, {"name":"Jim"}]';
let people = JSON.parse(jsonArray);

people.forEach(person => {
    console.log(person.name);
});
```

This example demonstrates parsing a JSON string that contains an array of objects, and then iterating over the resulting array in JavaScript.

It first defines a JSON array of objects, each containing a name attribute. It then parses this JSON array into a JavaScript array of objects using JSON.parse(). Following that, it uses the forEach() method to iterate over each object in the array, and it logs the name of each person to the console.

3.3.5 Handling Dates in JSON

JSON is a popular data-interchange format that has wide-ranging applications. However, one distinctive feature of JSON is that it does not inherently support a date type. This means that any dates that need to be represented within JSON format are typically stored as strings, rather than as actual date objects.

This aspect of JSON can have significant implications when working with dates in your code. Specifically, if you have date strings in JSON and you need to work with them as actual Date objects within your code, you will need to undertake a conversion process. This conversion is not automatically handled by JSON, and thus must be manually implemented by the developer.

This conversion process typically takes place after the JSON data has been parsed. The specific details of this process, including when and how it is performed, will depend on the specific requirements of your application or project. For example, some applications might require immediate conversion of date strings to Date objects upon parsing the JSON data, while others might allow for this conversion to be deferred until a later point in the code execution process.

While JSON is a powerful and versatile data-interchange format, its lack of inherent support for a date type can necessitate additional steps when working with dates in your code. This is an important consideration to keep in mind when planning and implementing your code strategies.

Example: Handling Dates in JSON

```
let eventJson = '{"eventDate":"2022-01-01T12:00:00Z"}';
let event = JSON.parse(eventJson);
event.eventDate = new Date(event.eventDate);

console.log(event.eventDate.toDateString());  // Outputs: Sat Jan 01 2022
```

In this example, the date string from the JSON data is converted into a JavaScript Date object using new Date(). It begins by defining a string eventJson which represents a JSON object with a single property, "eventDate". The JSON.parse() function is used to convert this string into a JavaScript object, event. The "eventDate" property of the event object is then converted from a string into a JavaScript Date object. Finally, the toDateString() method is used to convert the date to a string in the format "Day Month Date Year", and it is logged to the console.

By mastering JSON and its operations in JavaScript, you enhance your ability to handle data in modern web applications efficiently. JSON's universal data format makes it invaluable for data interchange between clients and servers, making it a crucial skill for any web developer.

3.3.6 Handling Complex Nested Structures

JSON, which stands for JavaScript Object Notation, is a popular data format that can sometimes contain deeply nested structures. These structures can be quite intricate, making them challenging to navigate and modify.

This complexity arises from the fact that each level of nesting represents a different object or array, which can contain its own objects or arrays, and so forth. Understanding how to access these nested structures, as well as how to modify the values within them, is an absolutely crucial skill when working with more complex data.

This knowledge will allow you to manipulate the data in ways that suit your specific needs, whether that involves extracting specific information, changing certain values, or structuring the data in a different way.

Example: Accessing Nested JSON

```
{
    "team": "Development",
    "members": [
        {
            "name": "Alice",
            "role": "Frontend",
            "skills": ["HTML", "CSS", "JavaScript"]
        },
        {
            "name": "Bob",
            "role": "Backend",
            "skills": ["Node.js", "Express", "MongoDB"]
        }
    ]
}
```

This example code is a JSON formatted data representing a team and its members. It shows a development team consisting of two members, Alice and Bob. Alice is a frontend developer skilled in HTML, CSS, and JavaScript. Bob is a backend developer skilled in Node.js, Express, and MongoDB.

JavaScript Code:

```
let jsonData = `{
    "team": "Development",
    "members": [
        {"name":   "Alice",   "role":   "Frontend",   "skills":   ["HTML",   "CSS",
"JavaScript"]},
        {"name":   "Bob",   "role":   "Backend",   "skills":   ["Node.js",   "Express",
"MongoDB"]}
    ]
}`;
let teamData = JSON.parse(jsonData);

console.log(teamData.members[1].name);   // Outputs: Bob
teamData.members.forEach(member => {
    console.log(`${member.name} specializes in ${member.skills.join(", ")}`);
});
```

This example demonstrates how to parse JSON containing an array of objects and how to iterate over it to access nested properties.

This JavaScript code declares a variable jsonData which contains a string of JSON data representing a development team and its members. It then parses this JSON data into a JavaScript object teamData using the JSON.parse() method.

Afterward, it prints the name of the second member of the team (Bob) to the console.

Finally, it uses a forEach loop to iterate over each team member and prints a string that includes each member's name and their respective skills.

3.3.7 Safely Parsing JSON

When you are working with JSON data that originates from external sources, there is invariably a risk that the JSON data may not be properly formed or could contain syntax errors. These malformations or errors can result in JSON.parse() throwing a SyntaxError, which can disrupt the flow of your code and potentially cause unwanted behaviors or crashes in your application.

To handle this situation in a more elegant and controlled manner, it is highly recommended to wrap your JSON parsing code within a try-catch block. This way, you can catch the potential SyntaxError and handle it in a way that is most appropriate for your specific application, preventing unexpected crashes and improving the overall robustness of your code.

Example: Safe JSON Parsing

```
let jsonData = '{"name": "Alice", "age": }';  // Malformed JSON

try {
    let user = JSON.parse(jsonData);
    console.log(user.name);
} catch (error) {
    console.error("Failed to parse JSON:", error);
}
```

This approach ensures that your application remains robust and can handle unexpected or incorrect data gracefully. The 'jsonData' string is intended to represent a user object with 'name' and 'age' properties, but it's missing a value for 'age', making it invalid JSON. The 'try-catch' block is used to handle any errors that might occur during JSON parsing. If parsing fails, an error message will be logged to the console.

3.3.8 Using JSON for Deep Copy

A prevalent application of the JavaScript Object Notation (JSON) methods JSON.stringify() and JSON.parse() in tandem is to formulate a deep clone of an object. This approach is particularly efficient and user-friendly for objects that exclusively encompass properties which are compatible with JSON serialization.

This means that these properties can be easily converted into a data format that JSON can read and generate. This pair of methods work harmoniously, with JSON.stringify() transforming the object into a JSON string, and JSON.parse() method converting this string back into a JavaScript object.

This process results in a new object that is a deep copy of the original, allowing for manipulation without altering the initial object.

Example: Deep Copy Using JSON

```
let original = {
    name: "Alice",
```

```
    details: {
        age: 25,
        city: "New York"
    }
};

let copy = JSON.parse(JSON.stringify(original));
copy.details.city = "Los Angeles";

console.log(original.details.city);   // Outputs: New York
console.log(copy.details.city);       // Outputs: Los Angeles
```

This technique ensures that changes to the copied object do not affect the original object, useful for scenarios where immutability is necessary.

This example code creates a deep copy of an object using JSON.parse() and JSON.stringify() methods. It first declares an object named 'original', then creates a deep copy of this object and assigns it to 'copy'. After that, it changes the 'city' property of the 'details' object in the 'copy'. Finally, it logs the 'city' property of the 'details' object in both 'original' and 'copy'. The output shows that changing the 'copy' does not affect the 'original', proving that a deep copy has been made.

3.3.9 Best Practices

1. **Employ the Correct MIME type**: It's important to use the application/json MIME type when you're serving JSON data from a server. This is crucial because it ensures that clients will treat the response as JSON, which helps to avoid any potential issues that could arise from misinterpreting the data type.
2. **Ensure JSON Data Validation**: Particularly when you're dealing with data that comes from external sources, it's absolutely essential to validate your JSON data. By doing this, you can ensure that the data meets the expected structure and types before you begin processing it. This will help to avoid any possible errors or inconsistencies that could occur if the data doesn't match the expected format.
3. **The Importance of Pretty Printing JSON**: When you're debugging or displaying JSON, you can use the JSON.stringify() method with additional parameters to format it in an easy-to-read way. This is known as "pretty printing" and it can make a huge difference when you're trying to understand or debug your JSON data, as it organizes the data in a clean and structured manner.

console.log(JSON.stringify(original, null, 2)); // Indents the output with 2 spaces

This is an example code that uses the console.log function to print out a stringified version of an object called original. The JSON.stringify method is used to convert the original object into a JSON string. The null and 2 parameters indicate that the output JSON string should have no replacements and should be indented with 2 spaces for readability.

By understanding these advanced aspects and best practices of JSON handling, you enhance your capabilities in data management and exchange in web applications. JSON's simplicity and effectiveness in structuring data make it an indispensable tool in the modern developer's toolkit.

3.4 Map and Set

In addition to the already existing data structures like arrays and objects, the JavaScript ES6 update brought with it two powerful, novel data structures. These are Set and Map. They are particularly useful when it comes to handling unique items and key-value pairs in a more efficient manner.

This section of the document is dedicated to exploring these two new structures in detail. We will comprehensively discuss their underlying properties, delve into their typical use cases, and examine how they can be utilized to enhance your JavaScript projects. By integrating Set and Map, you can achieve greater efficiency and simplicity in your JavaScript code, thereby improving the performance of your applications.

3.4.1 Map

A Map in JavaScript is essentially a collection or an aggregate of key-value pairs. This means that you can store data in a way where each value is associated with a unique key. The key aspect that differentiates a Map from an object in JavaScript is that the keys in a Map can be of any type.

This is unlike objects, which only support keys that are of String or Symbol types. Another important distinction to note is that Maps maintain the order of elements as they were inserted, unlike objects where the order is not guaranteed.

This feature of maintaining the order can be beneficial for certain applications where the sequence of data matters. For instance, if you are building a feature where the chronological order of user interactions needs to be preserved, using a Map would be more appropriate than an object.

Creating and Using a Map

Example: Creating a Map and Manipulating Data

```
let map = new Map();

// Setting values
map.set('name', 'Alice');
map.set('age', 30);
map.set({}, 'An object key');

// Getting values
console.log(map.get('name'));   // Outputs: Alice
console.log(map.get('age'));    // Outputs: 30

// Checking for keys
console.log(map.has('age'));    // Outputs: true

// Iterating over a Map
for (let [key, value] of map) {
    console.log(`${key}: ${value}`);
}

// Size of the Map
console.log(map.size);   // Outputs: 3

// Deleting an element
map.delete('name');
console.log(map.has('name'));   // Outputs: false

// Clearing all entries
map.clear();
console.log(map.size);   // Outputs: 0
```

This example demonstrates the basic operations of a Map, including setting and retrieving values, checking for the presence of keys, and iterating over entries.

First, a new Map is created. The set method is used to add key-value pairs to the Map. Here, the keys are 'name', 'age', and an empty object, with corresponding values 'Alice', 30, and 'An object key'.

The get method is used to retrieve values associated with a particular key from the Map.

The has method is used to check if a particular key exists in the Map.

There's a loop that iterates over the Map, logging each key-value pair.

The size property is logged to the console, showing the number of entries in the Map.

Then, the delete method is used to remove the 'name' key and its associated value from the Map.

Finally, the clear method is used to remove all entries from the Map.

3.4.2 Set

In the realm of programming, a Set is a specialized form of data structure. It is specifically designed to house a collection of unique values. These values can be of any type, which imbues the Set with an incredible degree of versatility. This feature makes it the optimal choice for creating a wide variety of lists or collections where the mandate is that each element must be unique and appear only once.

The enforcement of this uniqueness is one of the most important attributes of a Set. It allows for the execution of operations in an efficient manner, as it eliminates the possibility of duplication. This is especially advantageous when a programmer is dealing with large volumes of data. In these instances, duplicate values would not only be superfluous but could also cause significant issues.

Therefore, the Set, with its inbuilt mechanism to prevent duplication, emerges as an ideal solution for handling such scenarios.

Creating and Using a Set

Example: Creating a Set and Manipulating Elements

```
let set = new Set();

// Adding values
set.add('apple');
set.add('banana');
set.add('apple');   // Duplicate, will not be added

// Checking the size
console.log(set.size);   // Outputs: 2

// Checking for presence
console.log(set.has('banana'));   // Outputs: true

// Iterating over a Set
set.forEach(value => {
```

```
    console.log(value);
});

// Deleting an element
set.delete('banana');
console.log(set.has('banana'));  // Outputs: false

// Clearing all elements
set.clear();
console.log(set.size);  // Outputs: 0
```

In this example, you see how to add items to a Set, check for their presence, and iterate through the set. Duplicate entries are automatically rejected, ensuring all elements are unique.

In this code:

1. A new Set is created.
2. 'apple' and 'banana' are added to the Set. The second attempt to add 'apple' is ignored because Sets only store unique values.
3. The size of the Set (the number of elements) is logged to the console.
4. The code checks if 'banana' is in the Set and logs the result to the console.
5. The Set is iterated over using 'forEach', and each value is logged to the console.
6. 'banana' is removed from the Set, and its presence is checked again, logging 'false' to the console.
7. Finally, all elements are removed from the Set with 'clear()', and the size of the Set is logged again, resulting in 0.

3.4.3 Use Cases and Practical Applications

Maps, as a data structure, play a crucial role when there is a requirement for a direct association between keys and values, complemented by the need for efficient insertions and deletions. They become particularly useful in scenarios where the uniqueness of keys is a mandatory condition, and maintaining order is of importance, for instance when caching data derived from a database. This makes them an excellent choice for handling such specific data-related tasks.

On the other hand, **Sets** are the go-to data structure for managing collections of items where duplication is not an option. They are particularly useful in situations such as tracking unique user identifiers or in settings where membership testing is a frequent operation. They provide an efficient way to handle unique items in a collection, thus ensuring data integrity and consistency.

Both Map and Set offer significant performance improvements when dealing with large sets of data. They are especially efficient in operations such as searching for a specific value, providing a clear advantage over other data structures like objects and arrays. Furthermore, they are equipped with a variety of methods that make them particularly user-friendly and efficient when dealing with complex data structures, ensuring they are a valuable tool in handling large and complex data sets.

By integrating Map and Set into your JavaScript toolset, you can handle data more efficiently and elegantly, making your applications faster and more scalable. These structures enhance your ability to deal with data dynamically and can significantly simplify your code when used appropriately.

Practical Exercises

To solidify your understanding of the concepts discussed in Chapter 3: "Working with Data," here are some practical exercises focusing on arrays, objects, JSON, and the new ES6 structures, Map and Set. These exercises will help you apply what you've learned and deepen your knowledge of handling various data structures in JavaScript.

Exercise 1: Manipulating Arrays

Create an array of numbers, reverse it, and then sort it in ascending order.

Solution:

```
let numbers = [3, 1, 4, 1, 5, 9];
numbers.reverse();  // Reverses the array
numbers.sort((a, b) => a - b);  // Sorts the array in ascending order

console.log(numbers);  // Outputs: [1, 1, 3, 4, 5, 9]
```

Exercise 2: Object Operations

Create an object representing a book with properties for title, author, and year of publication. Then, add a method to the object that prints a description of the book.

Solution:

```
let book = {
    title: "JavaScript: The Definitive Guide",
    author: "David Flanagan",
```

```
    year: 2020,
    describe: function() {
        console.log(`${this.title} by ${this.author}, published in ${this.year}`);
    }
};

book.describe();  // Outputs: "JavaScript: The Definitive Guide by David Flanagan,
published in 2020"
```

Exercise 3: JSON Parsing and Stringifying

Convert a JSON string representing a person into a JavaScript object, then modify the age, and convert it back to a JSON string.

Solution:

```
let personJSON = '{"name":"John", "age":28, "city":"New York"}';
let person = JSON.parse(personJSON);

person.age += 1;  // Increment the age

let updatedPersonJSON = JSON.stringify(person);
console.log(updatedPersonJSON);     // Outputs:  '{"name":"John","age":29,"city":"New
York"}'
```

Exercise 4: Using Map

Create a Map to store the names of students and their corresponding grades. Add some entries, modify an entry, and then display all entries.

Solution:

```
let studentGrades = new Map();

studentGrades.set('Alice', 85);
studentGrades.set('Bob', 92);
studentGrades.set('Alice', 88);  // Update Alice's grade

studentGrades.forEach((value, key) => {
    console.log(`${key}: ${value}`);
});
```

Exercise 5: Unique Values with Set

Given an array of numbers with duplicates, use a Set to find and display the unique numbers.

Solution:

```
let numbers = [1, 2, 3, 2, 1, 4, 4, 5];
let uniqueNumbers = new Set(numbers);

console.log(Array.from(uniqueNumbers));  // Outputs: [1, 2, 3, 4, 5]
```

These exercises provide hands-on experience with JavaScript's data structures, enhancing your ability to manipulate and manage data effectively in your coding projects. By completing these tasks, you'll become more adept at recognizing which data structure is most appropriate for a given situation, improving both the performance and readability of your code.

Chapter Summary

In Chapter 3 of "JavaScript from Scratch: Unlock your Web Development Superpowers," we explored various powerful data structures and techniques essential for handling and manipulating data effectively in JavaScript. This chapter provided a comprehensive look at arrays, objects, JSON, Maps, and Sets, each serving unique purposes and offering different benefits in JavaScript programming. Here, we summarize the key concepts discussed in each section to reinforce your understanding and highlight how these components work together to manage data in web applications.

Arrays

We started with arrays, a fundamental data structure for storing ordered collections of items in JavaScript. Arrays are versatile and widely used due to their ability to hold items of any type and offer numerous methods for manipulating these items, including adding, removing, sorting, and searching operations. We discussed how to create, access, and modify arrays and the importance of understanding array methods like map(), filter(), reduce(), and forEach() for effective data manipulation.

Objects

Next, we delved into objects, which are key-value pairs that serve as the backbone of most JavaScript applications. Unlike arrays, objects provide a way to store data in a more structured way, allowing for more flexible and intuitive data access and manipulation. We explored creating, accessing, modifying, and deleting object properties, and we emphasized the role of methods within objects to encapsulate functionality related to the object's data.

JSON

The discussion on JSON (JavaScript Object Notation) highlighted its role as a lightweight data interchange format that is easy for both humans and machines to read and write. We covered how JSON is used to serialize and transmit structured data over a network, particularly between web clients and servers. You learned how to parse JSON into JavaScript objects and how to convert objects back to JSON strings, which is essential for web communications.

Map and Set

Finally, we introduced ES6 enhancements to JavaScript's data handling capabilities with Map and Set. Maps provide an efficient way of storing key-value pairs with any type of key, while Sets allow for the storage of unique values without duplication. Both structures offer methods that enhance performance and usability compared to traditional objects and arrays, especially when dealing with large datasets or when performance is a concern.

Throughout this chapter, we provided practical examples and exercises designed to help you apply these concepts. By mastering the use of these data structures, you enhance your ability to structure, access, and manipulate data efficiently, which is crucial for any web development project.

As we conclude this chapter, remember that the choice of data structure can significantly impact the performance and readability of your application. Understanding the strengths and limitations of each type of data structure allows you to choose the most appropriate one for your specific programming challenges, leading to more robust and maintainable code. Continue to practice these skills as you move forward to ensure that you are prepared to tackle more complex data handling scenarios in your future projects.

Chapter 4: DOM Manipulation

Welcome to Chapter 4, a chapter dedicated to exploring one of the most critical and fundamental areas of web development—DOM manipulation. The Document Object Model (DOM) is not just a programming interface for HTML and XML documents, but it is the backbone that provides a structured representation of the document. This representation comes in the form of a tree of nodes that can be interacted with and modified using programming languages like JavaScript.

Understanding the intricacies of the DOM is not just important, but it is absolutely critical for any web developer as it is the key to dynamically altering the content, structure, and style of web pages. Without a firm grasp of the DOM, a web developer's ability to create dynamic and interactive web pages would be greatly diminished.

This chapter is not just about guiding you through the basics or fundamentals of the DOM. Rather, it delves deeper into the techniques for manipulating the DOM, and it helps develop a strong understanding of the best practices for ensuring your applications are not just efficient, but also responsive. We are going to take a journey together into the heart of web development, starting with a fundamental understanding of the DOM, and emerging with a comprehensive understanding that will enhance your web development skills.

4.1 Understanding the DOM

The Document Object Model (DOM) is essentially a tree-like representation of the contents of a web page. It is a key concept in web development that is crucial to understanding how web pages work. After the HTML document is fully loaded, the browser painstakingly creates this model, turning every HTML element into an object that can be manipulated programmatically using JavaScript or similar languages.

This process allows developers to interact with and modify the contents and structure of a web page in real-time, leading to the dynamic and interactive web experiences we see today. Grasping this structure and how it functions is the first critical step toward mastering the behavior of dynamic web pages and becoming proficient in interactive web development.

4.1.1 What is the DOM?

The Document Object Model, often abbreviated as the DOM, is a crucial concept in web development, although it is not an inherent part of the JavaScript programming language itself. Instead, the DOM is a universally accepted standard established for how to interact with HTML elements. It offers a systematic way to access, modify, add, or delete HTML elements. Essentially, the DOM serves as a bridge or interface that allows JavaScript to communicate and interact with the HTML and CSS of a webpage seamlessly.

This interaction is achieved by treating the various parts of the webpage as objects, which can be manipulated by JavaScript. In essence, the DOM translates the webpage into an object-oriented structure that JavaScript can understand and interact with, thereby enabling the manipulation of webpage elements.

One of the most powerful capabilities of the DOM is its ability to change the structure, style, and content of a webpage dynamically. This dynamic nature of the DOM, combined with JavaScript, creates interactive and robust web experiences that respond to user input and actions. This means that the DOM enables JavaScript to react to user events, change webpage content on the fly, and even alter the appearance and style of the webpage in response to user actions.

Understanding the DOM is essential for any web developer. It provides the necessary tools to make websites more interactive and user-friendly, thereby enhancing the overall user experience.

Example: Viewing the DOM

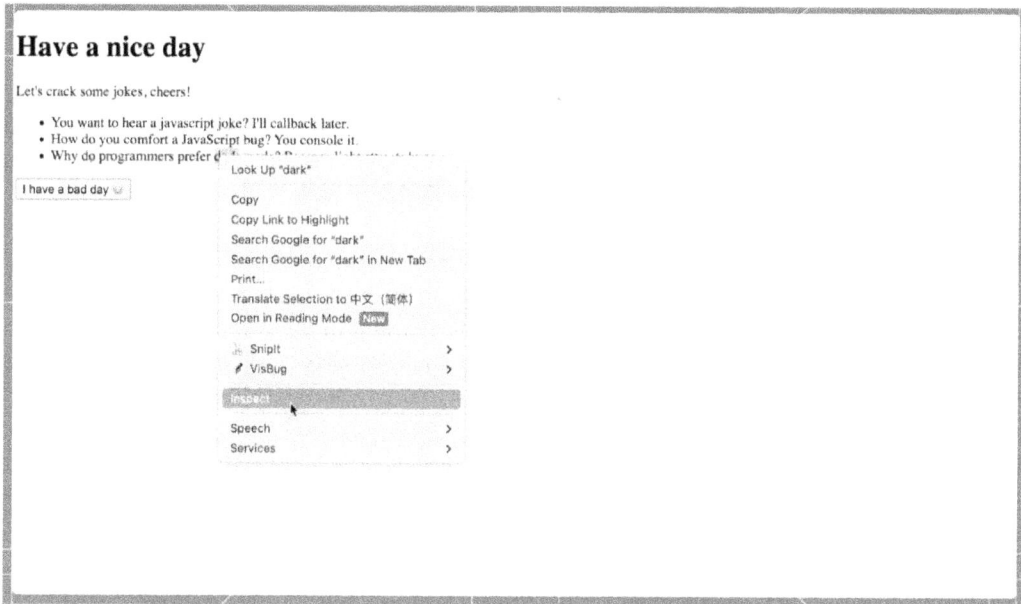

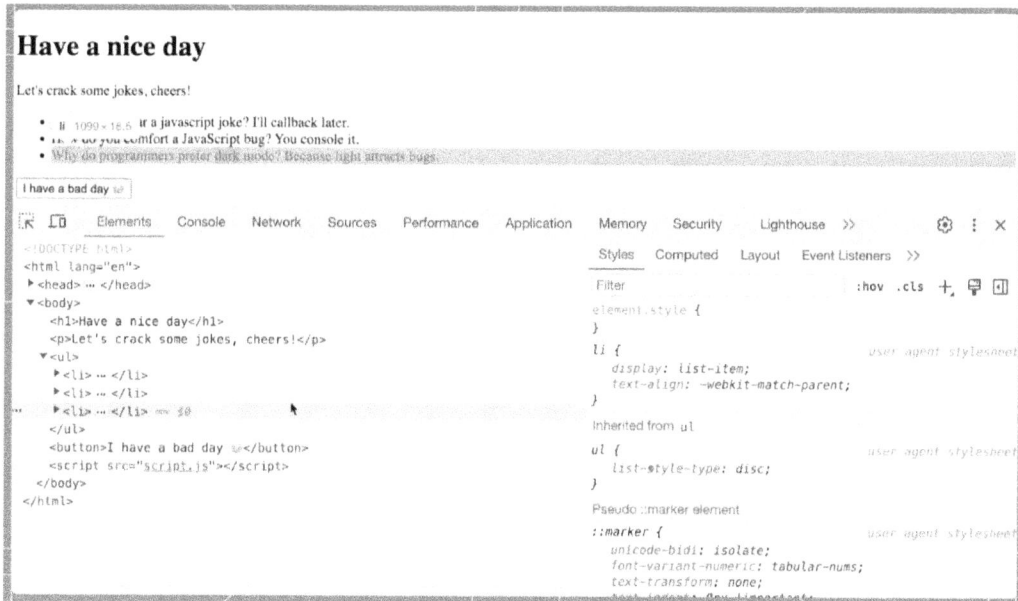

When an HTML document is loaded into a web browser, it is represented internally using a structure known as the Document Object Model, or DOM. This model is a crucial part of web development, as it provides a systematic way for programming languages like JavaScript to interact with the document's elements. The DOM is essentially a tree-like representation of the contents of a webpage.

These developer tools, which are typically accessible via the browser's menu or using a keyboard shortcut, possess various features designed to help developers inspect and debug web pages. One of these features is a visual representation of the DOM.

When you open the developer tools and navigate to the section displaying the DOM (often labeled as 'Elements' or 'Inspector'), you will see a structured, nested list that mirrors the content and organization of the webpage. Each HTML element of the webpage corresponds to a node in the DOM tree. By expanding these nodes (usually by clicking on a small arrow or plus symbol), you can see the child nodes that correspond to the nested HTML elements.

This visual representation of the DOM is interactive. Clicking on a node in the DOM will highlight the corresponding element in the webpage. This is beneficial when trying to understand the layout and structure of complex webpages.

In addition, viewing the DOM allows you to see the current state of the webpage's structure, including any changes that have been made programmatically using JavaScript. You can also edit the DOM directly within the developer tools, allowing you to experiment with changes and immediately see the results.

In essence, the ability to view and interact with the DOM via a browser's developer tools is a powerful asset in a web developer's toolkit. It supports not only understanding the tree structure of a document but also debugging and optimizing web pages.

4.1.2 Structure of the DOM

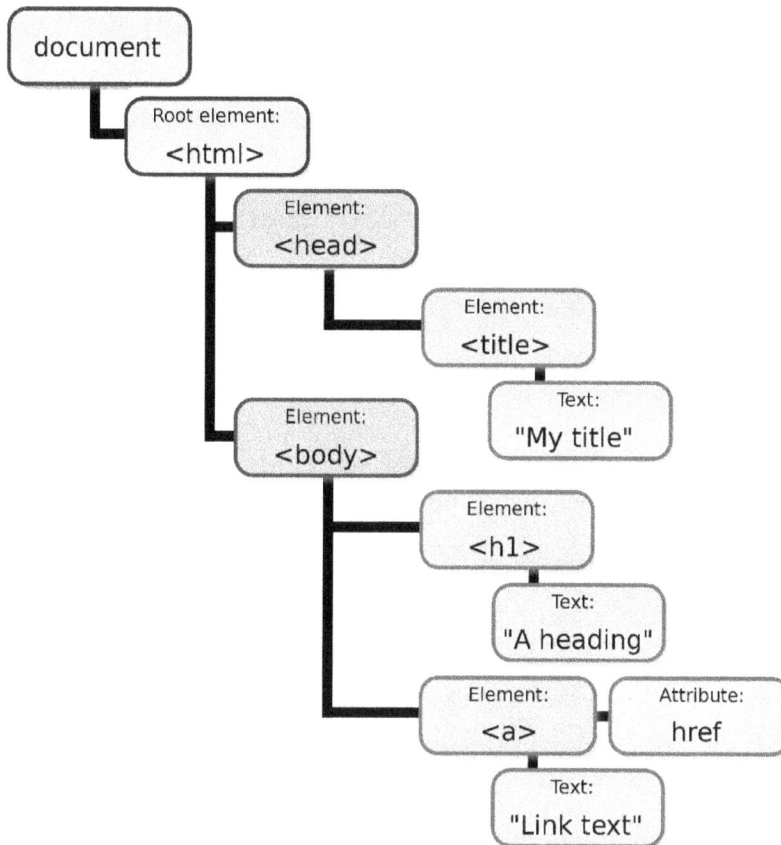

As discussed, the Document Object Model, also known as the DOM, is a structured, tree-like layout that fundamentally represents the contents and structure of a web page. This model is a crucial part of web development as it allows scripts and programming languages to interact with the contents of a webpage dynamically.

The DOM structure is composed of various nodes, with each node representing a different component of the document. These components can be an element, attribute, or even the text within the document. Each of these nodes carries a specific function and plays a crucial role in representing the document's structure and content.

The DOM, with its hierarchical tree-like structure, provides an efficient way to traverse, manipulate and interact with the information on a webpage.

The types of nodes in the DOM structure include:

Nodes

In the context of the Document Object Model, or DOM, every individual component present within the document forms part of the DOM tree and is referred to as a 'node'. This includes every element, attribute, and even piece of text contained within the document.

To provide some specific examples, an <h1> HTML tag would be considered a node within this structure. Similarly, any text that is contained within a paragraph, even if it's a single word or character, is also classified as a separate node.

These nodes are the fundamental building blocks of any webpage and enable the dynamic interaction between the code and the document's content that is so integral to modern web design.

Example:

```
<h1>Welcome to my Website!</h1>
```

In this example, both the <h1> element itself and the text "Welcome to my Website!" are nodes in the DOM tree. The <h1> element is an Element Node, while the text content is a Text Node.

Document Node

The root node is a fundamental component of the document structure. It serves as an essential representation of the entire document, encapsulating its complexity and offering a glimpse into its entirety. Acting as the starting point, the root node holds all other nodes within its structure, creating a hierarchical organization.

The information contained within the root node spreads out to encompass all other elements of the document. It acts like a container, holding within it all the various components that make up the document. This encapsulation is what gives the document its structure and order.

The root node serves as the primary reference point for accessing any part of the document. It is the first point of contact when navigating through the document's structure and provides a roadmap for accessing the various sections of the document. This essential function of the root node makes it a critical component in the overall structure and organization of the document.

Example:

```
<!DOCTYPE html>
<html>
<head>
  <title>My Website</title>
</head>
<body>
  <h1>Welcome to my Website!</h1>
  <p>This is some content.</p>
</body>
</html>
```

This code represents a basic HTML document structure. The entire document, from the <!DOCTYPE html> declaration to the closing </html> tag, along with all elements and text content within, is represented by the Document Node at the root of the DOM tree.

Element Nodes

These are fundamental components that represent HTML elements within a webpage. They are an integral part of the Document Object Model (DOM), which is a structured representation of the HTML elements on the page.

Each HTML tag on the page is represented by an Element Node in the DOM. These nodes can contain other nodes, including text nodes and other element nodes, which allows for the hierarchical structure that HTML documents usually have.

Example:

```
<ul>
  <li>Item 1</li>
  <li>Item 2</li>
  <img src="image.png" alt="My Image">
  <a href="<https://www.example.com>">Visit our website</a>
</ul>
```

This code snippet showcases various HTML elements like <ul>, <li>, <img>, and <a>. Each of these elements (ul, two li, img, and a) is a separate Element Node in the DOM tree.

Text Nodes

These are crucial components of the XML document structure, and they contain the actual text within the elements. They carry the information that the XML document is intended to convey and are enclosed within the start and end tags of an element.

Example:

```
<p>This is a paragraph with some text content.</p>
```

Here, the text "This is a paragraph with some text content." within the <p> tags is a Text Node. Notice that Text Nodes don't have any child nodes of their own.

Attribute Nodes

These are nodes that are associated with the attributes of elements within an XML document. These attribute nodes contain additional information about element nodes and are often accessed directly through these element nodes. They hold data that provide more details about the elements but are not a part of the actual data content.

Example:

<img src="image.png" alt="My Image">

In this example, the <img> element has an attribute named src with a value of "image.png". The src attribute itself is an Attribute Node associated with the <img> Element Node. While Attribute Nodes exist, they are typically accessed through the corresponding Element Node for ease (e.g., element.getAttribute("src")).

4.1.3 Navigating the DOM

In web development, you have the ability to traverse or navigate the Document Object Model (DOM) tree using different properties. These properties provide a roadmap of sorts that allows you to move around and manipulate elements within the DOM tree.

Some examples of these properties include parentNode, childNodes, and firstChild. The parentNode property, for instance, allows you to access the parent node of a specified node in the DOM tree. Similarly, the childNodes property lets you access all child nodes of a specified node, while the firstChild property specifically gives you the first child node of a specified node. By utilizing these properties, developers can efficiently interact with, and modify, elements within the DOM tree.

Example: DOM Navigation in Action

Let's look at a practical example using the provided HTML code:

<!DOCTYPE html>

<html>

<head>

  <title>Sample Page</title>

</head>

<body>

  <div id="content">

    <p>First paragraph</p>

    <p>Second paragraph</p>

  </div>

  <script>

    let contentDiv = document.getElementById('content');

    console.log(contentDiv.childNodes.length); // Outputs: 5 (includes text nodes, like whitespace)

    console.log(contentDiv.firstChild.nextSibling.textContent); // Outputs: 'First paragraph'

  </script>

</body>

</html>

In this example, the childNodes property includes all child nodes, including text nodes (which can be whitespace if the HTML is formatted with indents or spaces). The firstChild.nextSibling navigates to the first <p> element effectively. Now, let's go deeper:

1. Selecting the Element:

The code starts by selecting the element with the ID "content" using document.getElementById('content'). This assigns the <div> element containing the paragraphs to the variable contentDiv.

2. Navigating with childNodes:
    o   The first console.log statement uses contentDiv.childNodes.length. The childNodes property returns a collection (NodeList) of all child nodes of the specified element. In this case, it includes the two <p> elements, any text nodes (like whitespace between elements), and potentially comments (if present). This explains why the output is 5 even though we only see two paragraphs in the HTML.
3. Navigating with firstChild and nextSibling:
    o   The second console.log statement demonstrates navigating to specific child nodes. Here's a step-by-step breakdown:
        ▪   contentDiv.firstChild accesses the first child node of the contentDiv (which is the <div> element). This refers to the first <p> element containing "First paragraph".
        ▪   .nextSibling moves to the next sibling node of the current node. Since the first <p> element has another <p> element as its sibling, this points to the second paragraph element with the text "Second paragraph".
        ▪   Finally, .textContent retrieves the text content within the selected element, which outputs "First paragraph" in this case.

Key Takeaways:

These properties (childNodes, firstChild, nextSibling) allow you to traverse the DOM tree and access specific elements relative to their position within the structure. This is essential for manipulating and interacting with elements in your web pages using JavaScript.

Understanding the DOM's structure and how to navigate it is fundamental for any web developer looking to create dynamic and responsive web applications. By manipulating the DOM, you can update the content, structure, and style of a webpage without needing to send a request to the server for a new page.

4.1.4 DOM as a Live Representation

A critical aspect of the Document Object Model (DOM), which should continually be at the forefront of your considerations, is its unique characteristic of providing what is referred to as a "live" representation of the HTML document.

This essentially means that the DOM is inherently dynamic rather than static. It's an interactive mirror of the current web page which means that any modifications made to the DOM, whether they are additions, deletions, or alterations, are instantaneously reflected on the web page.

This is a two-way street, as any changes to the web page's content or structure, perhaps due to user interaction or an automatic update, are immediately mirrored in the DOM. This continual synchronization between the DOM and the web page ensures that what you manipulate programmatically in the DOM is always consistent with what is visually rendered on the web page.

Example: Live DOM Updates

```javascript
let list = document.createElement('ul');

document.body.appendChild(list);



let newItem = document.createElement('li');

newItem.textContent = 'New item';

list.appendChild(newItem); // Immediately visible on the web page
```

This example demonstrates how dynamically adding an element to the DOM immediately updates the webpage. The ul and li elements are created and added to the document dynamically, and these changes are instantly visible.

4.1.5 Performance Considerations

Working with the Document Object Model (DOM) can sometimes be a resource-heavy operation, particularly when it comes to performance. When there are constant alterations being made to the DOM, it can create a significant strain on the system, especially if the webpage in question is large or complex in its structure.

This can result in obvious performance issues, which can affect the overall user experience. As a result, it becomes absolutely crucial to streamline and optimize the way we interact with the DOM. This optimization is mostly targeted towards minimizing the occurrences of 'reflows' and 'repaints', which are processes that the browser has to undergo in order to recalculate the layout and redraw certain parts of the webpage.

By doing so, we not only make the webpage more efficient but also enhance the overall performance and user experience.

Best Practices for Optimizing DOM Manipulations

- **Minimize DOM changes**: One effective way to improve the performance of your web page is to minimize changes to the Document Object Model (DOM). You can do this by batching updates to the DOM. This means changing properties of elements while they are offscreen, and then appending or updating these elements on the page in one single operation. This reduces the amount of reflow and repaint operations the browser has to perform, leading to smoother rendering and a better user experience.
- **Use document fragments**: Another advantageous technique involves the use of document fragments. Document fragments allow you to hold a part of the DOM off-screen, giving you the freedom to make multiple changes, such as adding, modifying, or deleting nodes, without causing any page reflow. Once all changes are complete, the document fragment can be attached back to the document, causing only one reflow despite the multiple changes.
- **Event delegation**: An important aspect of efficient JavaScript is the appropriate use of event listeners. Instead of attaching event listeners to individual elements, which can be memory-intensive and lead to performance issues, it's often more beneficial to use a single listener on a parent element. This single listener can then handle events from child elements using a process called event propagation. This technique can significantly improve performance, particularly when dealing with a large number of similar elements.

Example: Using Document Fragments

```javascript
let fragment = document.createDocumentFragment();
for (let i = 0; i < 10; i++) {
    let newItem = document.createElement('li');
    newItem.textContent = `Item ${i}`;
    fragment.appendChild(newItem);
}

document.getElementById('myList').appendChild(fragment);
```

This approach ensures that the DOM is updated all at once instead of in ten separate operations, significantly improving performance.

4.1.6 Accessibility and the DOM

When you're working with the Document Object Model (DOM) in web development, it's absolutely critical that you keep accessibility at the forefront of your considerations. The process of adding, removing, or modifying elements on your webpage can have a significant impact on the experience of users who depend on screen readers and other types of assistive technologies to navigate the internet.

This group of users can include those with visual or motor impairments. Therefore, it's important that we, as developers, ensure that our web content is not only dynamic but also fully accessible. One way to achieve this is by using proper ARIA (Accessible Rich Internet Applications) roles and properties.

These are integral tools that make dynamic and complex web content more accessible, especially for people with disabilities. By correctly implementing ARIA roles and properties, you can ensure that the content you add or change on the fly is still accessible to all users, thereby creating a more inclusive digital environment.

Example: Enhancing Accessibility with ARIA

```
let dialog = document.createElement('div');
dialog.setAttribute('role', 'dialog');
dialog.setAttribute('aria-labelledby', 'dialog_label');
dialog.textContent = 'This is an accessible dialog box.';
document.body.appendChild(dialog);
```

This example demonstrates setting ARIA roles and properties to inform assistive technologies about the role of the element and its state, enhancing accessibility for all users.

4.2 Selecting Elements

In the field of web development, the process of manipulating the Document Object Model, or DOM as it is often referred to, typically begins with the act of selecting elements. This elementary operation forms the basis of most interactive functions on a webpage.

The ability to select elements from the DOM, both accurately and efficiently, is of paramount importance for a myriad of tasks including, but not limited to, dynamically updating content, altering styling, and responding to user interactions.

Given the fundamental role that element selection plays in web development, it's crucial to understand the different methods available to us for this purpose. With the advancement of JavaScript, there are now a plethora of methods at our disposal for selecting elements from the DOM. However, each method has its own specific use case, benefits, and downsides.

In this section, we will delve into the exploration of various methods provided by JavaScript for selecting elements from the DOM. This will include a detailed explanation of their usage scenarios and the best practices associated with each method. The goal is to equip you with the knowledge and skills necessary to enhance your scripting effectiveness and overall productivity in the realm of web development.

4.2.1 The document.getElementById() Method

When it comes to selecting an element in the Document Object Model (DOM), the most uncomplicated and direct method is by utilizing its unique identifier, commonly referred to as ID. IDs are designed to be unique within a webpage, which means each ID should correspond to a single element.

The JavaScript function document.getElementById(), therefore, provides an expedient and efficient method of locating a single element within the webpage's structure. By using this function, developers can quickly access and manipulate the properties of the DOM element that corresponds to the specified ID.

Example: Using document.getElementById()

```
<div id="content">This is some content.</div>
<script>
    let contentDiv = document.getElementById('content');
    console.log(contentDiv.textContent);  // Outputs: This is some content.
</script>
```

This method is very fast because the browser can immediately access the element by its unique identifier.

4.2.2 The document.getElementsByTagName() Method

When it comes to selecting elements by their specific tag name in JavaScript, you can use the function document.getElementsByTagName(). What this function does is that it returns a live HTMLCollection of elements that correspond to the given tag name.

This means that the collection automatically updates when the document changes. This functionality is particularly useful for operations that need to be applied to all elements of a specific type. For example, if you wanted to manipulate or perform an action on all div elements in your HTML document, you could use this function.

The live HTMLCollection would contain all div elements, and changes made to these elements in the script would be reflected in the document.

Example: Using document.getElementsByTagName()

```
<ul>
    <li>First item</li>
    <li>Second item</li>
</ul>
<script>
    let listItems = document.getElementsByTagName('li');
    for (let item of listItems) {
        console.log(item.textContent);
    }
</script>
```

This method will log "First item" and "Second item", demonstrating how to iterate over multiple elements.

4.2.3 The document.getElementsByClassName() Method

In the vast realm of web development, it is often necessary to classify elements according to their class attributes. This is where the JavaScript method document.getElementsByClassName() becomes exceptionally useful.

This method serves as a powerful tool, allowing developers to select all elements that bear the same class name. It's important to note that this method doesn't just return a static list of elements, but rather, it returns a live HTMLCollection.

This live HTMLCollection is a dynamic, updated list of all elements adorned with the specified class name, thus providing real-time tracking of all relevant elements within the document.

Example: Using document.getElementsByClassName()

```
<div class="note">Note 1</div>
<div class="note">Note 2</div>
<script>
    let notes = document.getElementsByClassName('note');
    for (let note of notes) {
        console.log(note.textContent);
    }
</script>
```

This example selects all elements with the class "note" and logs their contents.

4.2.4 Query Selectors

When dealing with more intricate or sophisticated selections on a webpage, Cascading Style Sheets (CSS) query selectors prove to be remarkably potent. They offer a methodical and precise way of targeting and manipulating different elements within a webpage's Document Object Model (DOM).

There are primarily two methods used to employ these CSS-style selectors for finding elements within the DOM: document.querySelector() and document.querySelectorAll().

The document.querySelector() function is particularly useful when you are interested in only the first element that matches a specified CSS selector. It will search through the DOM and return the first element it encounters that fits the provided CSS selector. This can be incredibly handy when you need to quickly find and manipulate a specific element.

On the other hand, document.querySelectorAll() is a slightly different yet equally useful tool. Instead of returning the first matching element, it returns a NodeList, essentially a collection, of all elements that correspond to the specified CSS selector.

This method is particularly useful when you need to select multiple elements and perform the same action on all of them, such as adding a specific class or altering the style.

Example: Using Query Selectors

```
<div id="container">
    <div class="item">Item 1</div>
    <div class="item">Item 2</div>
</div>
<script>
    let container = document.querySelector('#container');
    let items = document.querySelectorAll('.item');

    console.log(container);  // Outputs the container div
    items.forEach(item => console.log(item.textContent));  // Outputs: Item 1, Item 2
</script>
```

These methods provide flexibility and power, allowing for complex querying strategies like combining class selectors, id selectors, and pseudo-classes.

4.2.5 Best Practices

When it comes to selecting elements on a webpage, there are a few important considerations to keep in mind:

- **Use IDs for unique elements**: If you have a single, unique element that you find yourself needing to access frequently, the optimal choice is to use an ID. IDs are a powerful tool for pinpointing a specific element and can be leveraged to manipulate that element in various ways.
- **Prefer class names for groups of elements**: If you are dealing with a group of elements that share similar characteristics or need to have similar behavior or styling applied to them, class names are your best bet. They allow you to collectively access and modify a number of related elements in one go.
- **Utilize query selectors for complex selections**: If your selection needs are more complex and cannot be adequately handled with IDs or class names, query selectors can be a useful tool. However, it's important to be aware of the potential performance implications associated with their use. This is especially true when using document.querySelectorAll() on large documents. It can potentially slow down your page load times, so it's essential to use it judiciously.

Understanding these various methods to select elements in the DOM enables you to manipulate web pages more effectively, laying the groundwork for dynamic and interactive user experiences. By mastering element selection, you can efficiently access any part of the DOM to read data, modify attributes, or trigger changes in the document's appearance or behavior.

4.2.6 Caching DOM References

When you're working on a project where you notice you're repeatedly accessing the same element, it becomes beneficial to make use of caching. What this means is that you would store the reference to that particular element in a variable.

This method is often employed to avoid the unnecessary overhead of querying the Document Object Model (DOM) repeatedly. The repeated querying can lead to a reduced performance rate, which is not ideal in any case. However, by using caching, you can significantly enhance the performance of your application.

This is especially pertinent in the case of complex applications where efficiency and quick response times are key. Thus, it's not just about making the code cleaner, but also about improving the overall user experience by speeding up the application.

Example: Caching DOM References

```javascript
const menu = document.getElementById('main-menu');   // Access DOM once and store reference

// Use 'menu' multiple times without re-querying the DOM
menu.classList.add('active');
menu.addEventListener('click', handleMenuClick);
```

Caching is particularly useful in event handlers or any function that is called repeatedly.

This is a example code snippet that accesses the HTML (DOM) element with the ID 'main-menu' and assigns it to the variable 'menu'. The snippet then uses this reference to add the class 'active' to the menu, and to set up an event listener that will call the function 'handleMenuClick' whenever a click event occurs on the menu.

4.2.7 Using data-* Attributes for Selection

HTML5, a major revision of the core language of the World Wide Web, introduced an important feature known as custom data attributes. These attributes provide a means to store extra information directly within standard HTML elements.

The process involves using attributes that are prefixed with data-, which serves as a marker for these user-defined attributes. This new feature is powerful and flexible, allowing developers to enrich elements with custom data, extending the native capabilities of HTML elements.

These custom data attributes can be incredibly useful for a multitude of reasons, some of which include associating data directly with elements without having to resort to non-standard attributes or additional DOM properties.

This not only enhances efficiency but also ensures the integrity of the code. It's a significant step forward in HTML development, offering a more versatile and effective way of managing and manipulating data within HTML documents.

Example: Using data-* Attributes

```
<div id="product-list">
    <div data-product-id="001" data-price="29.99">Product 1</div>
    <div data-product-id="002" data-price="39.99">Product 2</div>
</div>

<script>
    const products = document.querySelectorAll('[data-product-id]');
    products.forEach(product => {
        console.log(`Product ID: ${product.getAttribute('data-product-id')}, Price:
$${product.getAttribute('data-price')}`);
    });
</script>
```

This approach not only keeps your HTML valid but also leverages the dataset for efficient data retrieval and manipulation.

The HTML part of the example code creates a container with the ID of "product-list", which contains two div elements representing two different products. Each product has a unique ID and a price associated with it, set as data attributes.

The JavaScript part of the code selects all elements with the attribute 'data-product-id', which in this case are the div elements representing the products. It then loops over these elements, and for each product, it logs the product ID and the price to the console.

4.2.8 Considerations When Using NodeList and HTMLCollection

In web development, it's crucial to grasp the distinctions between NodeList and HTMLCollection. These are two different types of collections of DOM nodes, and they vary significantly in their behaviors, especially in terms of their "live" versus "static" nature.

When you use document.getElementsByClassName(), it returns what is known as a live HTMLCollection. The term "live" means that this HTMLCollection is dynamically updated to reflect any changes that occur in the DOM. For example, if elements that match the class name specified are added or removed from the document after the call to getElementsByClassName(), the HTMLCollection will automatically update to include or exclude these elements.

On the other hand, document.querySelectorAll() returns a NodeList that is static, not live. This means that, unlike an HTMLCollection, the NodeList returned by querySelectorAll() does not automatically update to reflect changes in the DOM. If elements that match the selectors passed to querySelectorAll() are added or removed from the document after the call to querySelectorAll(), these changes will not be reflected in the NodeList.

Understanding this difference is fundamental to ensure the correct manipulation of the DOM in your JavaScript code.

Example: Static vs. Live Collections

```javascript
const liveCollection = document.getElementsByClassName('item');
const staticList = document.querySelectorAll('.item');

// Adding a new element with class 'item'
const newItem = document.createElement('div');
newItem.className = 'item';
document.body.appendChild(newItem);

console.log(liveCollection.length);  // Includes the newly added element
console.log(staticList.length);      // Does not include the newly added element
```

Understanding the behavior of these collections is crucial for correctly managing DOM elements in dynamic applications.

This example code snippet demonstrates the difference between getElementsByClassName() and querySelectorAll(). Both functions are used to select HTML elements with the class 'item'. When a new element with class 'item' is added to the document, getElementsByClassName() reflects this change immediately and includes the new element in its collection, this is because it returns a live collection of elements. On the other hand, querySelectorAll() does not include the new element, as it returns a static NodeList that does not update to reflect changes in the DOM.

4.2.9 Efficient Querying and Scope Limitation

By carefully limiting the scope of your queries, you can drastically enhance the performance of your code, especially when dealing with expansive DOM structures. Rather than indiscriminately querying the entirety of the document, a more efficient approach would be to confine your query within a specific subtree of the DOM.

This approach ensures that the search operation is conducted within a reduced set of elements, thus reducing the time and resources required to execute the query. This technique is particularly beneficial when dealing with large-scale, complex DOM structures where unnecessary querying can result in significant performance degradation.

Example: Scope Limitation

```html
<div id="sidebar">
    <!-- Sidebar content -->
</div>

<script>
    const sidebar = document.getElementById('sidebar');
    const links = sidebar.querySelectorAll('a');  // Only search within 'sidebar'
</script>
```

This method is more efficient than document.querySelectorAll() when the target elements are known to reside within a specific part of the DOM.

The HTML part creates a division (div) with the id 'sidebar' to hold the sidebar content. The JavaScript part is used to select the 'sidebar' div and all the anchor ('a') elements within it. As a result, this script is used to gather all the links present in the 'sidebar' section of the webpage.

4.3 Modifying Content

JavaScript is renowned for its robust and versatile capabilities, with one of its most potent features being the ability to dynamically modify the content of a webpage. This capability is not merely a neat trick; it's a vital component when it comes to creating web applications that are interactive, responsive, and capable of adapting in real-time to user input and other external stimuli.

In this comprehensive section, we will delve into the many techniques available for content modification. This encompasses a wide range of methods, from altering the textual content to changing the underlying HTML, and even tweaking the attributes of DOM elements. Each method we'll explore comes with its unique advantages and potential use cases, thereby augmenting your toolkit for dynamically customizing web page behavior and visual presentation.

By mastering these techniques, you'll be better equipped to create engaging, interactive web experiences that not only respond to user input but also adapt their behavior and appearance to better suit the needs and expectations of the end-user.

4.3.1 Changing Text Content

When it comes to modifying the content of an element in programming, particularly in JavaScript, there are a couple of straightforward ways to achieve this by changing its text. JavaScript offers two primary properties that are instrumental for this purpose, namely textContent and innerText.

The first property, **textContent**, provides an unembellished way to both retrieve and alter the text content of an element along with all its descendant elements. An interesting aspect of textContent is that it does not take into account any styles that may be applied to hide the text. As a result, it returns the content in its raw form, without any alterations.

On the other hand, the second property, **innerText**, operates a bit differently. It is cognizant of any styles that have been applied to the text, and therefore, it will not return the text of elements that have been "hidden" using certain styles, such as display: none. This is in stark contrast to textContent. Furthermore, innerText respects the visual presentation of the text, meaning it takes into consideration how the text is visually formatted and displayed on the webpage.

Example: Using textContent and innerText

```
<div id="message">Hello <span style="display: none;">hidden</span> World!</div>
<script>
    const element = document.getElementById('message');
    console.log(element.textContent);  // Outputs: "Hello hidden World!"
    console.log(element.innerText);    // Outputs: "Hello World!"
</script>
```

This example illustrates the difference between textContent and innerText. While textContent retrieves all text regardless of CSS styles, innerText provides a representation closer to what is visible to a user.

4.3.2 Modifying HTML Content

The innerHTML property is a powerful tool when it comes to manipulating the HTML content of an element. This property gives you the capability to either set or retrieve the HTML content (i.e., markup) that is contained within the element.

One of the key features of the innerHTML property is that it encompasses not just the text within the element, but also any HTML tags that might be included within it. This means that you can use innerHTML to insert complex HTML structures directly into an element, or extract such structures for use elsewhere.

Hence, the innerHTML property provides a highly efficient and versatile method for dynamically manipulating the content of a webpage.

Example: Using innerHTML

```
<div id="content">Original Content</div>
<script>
    const contentDiv = document.getElementById('content');
    contentDiv.innerHTML = '<strong>Updated Content</strong>';

    console.log(contentDiv.innerHTML);          //    Outputs:    "<strong>Updated
Content</strong>"
</script>
```

This method is powerful for adding complex HTML structures within an element but should be used carefully to avoid cross-site scripting (XSS) vulnerabilities.

The HTML defines a div element with the id of "content" that contains the text "Original Content". The JavaScript code then selects this div using its id, and changes its innerHTML to "Updated Content", which makes the text bold and changes it to "Updated Content". The last line of the JavaScript code outputs the current innerHTML of the div, which will be "Updated Content", to the console.

4.3.3 Updating Attributes

One of the common requirements in web development is the manipulation of the Document Object Model, or DOM, elements' attributes. This is often necessary for dynamic and interactive web applications where elements' properties need to be adjusted based on user interaction or other factors.

JavaScript, being the language of the web, provides several methods to dynamically manage these attributes. Among these methods are setAttribute, getAttribute, and removeAttribute. The setAttribute method allows us to assign a specific value to an attribute, getAttribute allows us to retrieve the current value of an attribute, and removeAttribute allows us to remove an attribute altogether.

These methods offer powerful ways to manipulate the properties of DOM elements, thus enabling more dynamic and interactive user experiences.

Example: Modifying Attributes

```
<a id="link" href="<http://example.com>">Visit Example</a>
<script>
    const link = document.getElementById('link');
    console.log(link.getAttribute('href'));  // Outputs: "<http://example.com>"

    link.setAttribute('href', '<https://www.changedexample.com>');
    link.textContent = 'Visit Changed Example';

    console.log(link.getAttribute('href'));                    //          Outputs:
"<https://www.changedexample.com>"
</script>
```

This example demonstrates how to change the href attribute of an anchor tag, effectively redirecting users to a different URL.

Initially, the example sets up a hyperlink (anchor tag) with the id "link" that points to "http://example.com" with the link text "Visit Example". Then, a script is run.

The script gets the element with the id "link" and logs its href attribute to the console, which is "http://example.com".

Then, it changes the href attribute of the link to "https://www.changedexample.com" and also changes the text of the link to "Visit Changed Example".

Finally, it logs the new href attribute to the console, which is "https://www.changedexample.com".

4.3.4 Handling Classes

In web development, managing CSS classes is a frequent requirement, especially when you need to dynamically change the content. This is particularly important when you want to alter the appearance of elements based on user interactions.

For instance, you might want to change the color of a button when a user hovers over it or change the layout of a page based on user preferences. To facilitate this, JavaScript provides a property called classList.

The classList property gives you access to several useful methods that make managing CSS classes a breeze. These methods include add, remove, toggle, and contains. The add method lets you add a new class to an element, the remove method allows you to delete a class, the toggle method enables you to switch a class on and off, and the contains method checks if a specific class is assigned to an element.

Example: Using classList

```
<div id="toggleElement">Toggle My Style</div>
<script>
    const element = document.getElementById('toggleElement');

    // Toggle a class
    element.classList.toggle('highlight');
    console.log(element.classList.contains('highlight'));  // Outputs: true

    // Remove a class
    element.classList.remove('highlight');
    console.log(element.classList.contains('highlight'));  // Outputs: false
</script>
```

This example shows how to toggle a class to visually highlight an element, and then remove the class to revert to its original style.

The example code includes a div element with the ID "toggleElement". The JavaScript code accesses this div by its ID and toggles the 'highlight' class. If the 'highlight' class is present, it's removed; if it's absent, it's added. Following each operation, the code checks for the presence of the 'highlight' class on the div and logs the outcome to the console.

4.3.5 Efficient Batch Updates

When developing web applications, it's important to understand that making changes directly to the Document Object Model, also known as the DOM, can be quite costly in terms of performance.

This is particularly the case when such modifications are done repeatedly within a loop or during a complex sequence of operations. The reason behind this is that every time you make a change to the DOM, the browser needs to recalculate the layout, repaint the screen, and perform other tasks that can slow down your application.

To optimize performance and ensure that your application runs smoothly, it's advisable to minimize direct interactions with the DOM. Instead, consider using a technique known as batching updates.

This approach involves making multiple changes to the DOM in a single operation, which can significantly reduce the amount of work the browser needs to do and thus improve the speed of your application. Always remember that efficient DOM manipulation is key to a performant web application.

Example: Efficient Batch Update

```
<div id="listContainer"></div>
<script>
    const listContainer = document.getElementById('listContainer');
    let htmlString = '';

    for (let i = 0; i < 100; i++) {
        htmlString += `<li>Item ${i}</li>`;
    }

    listContainer.innerHTML = htmlString;   // Updates the DOM once, rather than in
each iteration
</script>
```

This example demonstrates creating a string of HTML and updating the DOM once, rather than updating the DOM in every loop iteration, which would be significantly less efficient.

4.3.6 Working with Document Fragments

A DocumentFragment is a minimal, lightweight document object that has the unique characteristic of storing a portion of a document's structure, but it does not possess a parent node. Its primary function is to hold nodes just like any other document, but there's a key distinction - it exists outside of the main DOM tree.

This means that changes made to a DocumentFragment do not affect the document, trigger reflow, or incur any performance impact. The benefit of this becomes clear when you need to append multiple elements to the DOM.

Rather than individually appending each node, which could result in multiple reflows and consequent performance hits, you can instead append these nodes to a DocumentFragment. Then, you append this fragment to the DOM. By doing this, you only trigger a single reflow, thereby optimising performance.

Example: Using Document Fragments

```
<ul id="myList"></ul>
<script>
    const myList = document.getElementById('myList');
    const fragment = document.createDocumentFragment();

    for (let i = 0; i < 5; i++) {
        let li = document.createElement('li');
        li.appendChild(document.createTextNode(`Item ${i}`));
        fragment.appendChild(li);
    }

    myList.appendChild(fragment);  // Appends all items in a single DOM update
</script>
```

This approach is particularly effective when constructing complex or large-scale DOM structures dynamically.

The example code creates an unordered list in HTML with the id "myList". Then, using JavaScript, it creates a DocumentFragment (a minimal document object that can hold nodes). It loops 5 times creating a list item (li) in each iteration. Each of these items gets appended to the DocumentFragment. Finally, this fragment is appended to the unordered list "myList". The advantage of this approach is that appending the fragment triggers only one reflow, making the operation more performance efficient.

4.3.7 Modifying Styles

One of the crucial aspects of creating dynamic web content is the ability to manipulate the style of elements. This is what allows you to create a visually engaging and interactive user experience. The className and classList attributes are particularly useful tools in this regard, as they allow for the efficient management of CSS classes.

These can be used to alter the appearance of HTML elements in response to user interactions, or to dynamically adjust the layout of a webpage. However, there are instances where direct inline style changes are necessary. These situations typically arise when you need to make specific adjustments to the style of an element on the fly, without affecting the overall class properties.

In such cases, being able to edit the inline styles directly gives you a greater degree of control over the precise look and feel of your web content.

Example: Changing Styles Dynamically

```html
<div id="dynamicDiv">Dynamic Style</div>
<script>
    const dynamicDiv = document.getElementById('dynamicDiv');
    dynamicDiv.style.backgroundColor = 'lightblue';
    dynamicDiv.style.padding = '10px';
    dynamicDiv.style.border = '1px solid navy';
</script>
```

This technique is useful for quick, one-off modifications and animations but should be used judiciously as it can override CSS stylesheets.

This code snippet first defines a div element with the id "dynamicDiv". Then, using JavaScript, it selects that div element and applies several CSS styles to it dynamically: changing the background color to light blue, adding padding of 10 pixels, and setting a navy-colored border of 1 pixel width.

4.3.8 Conditionally Modifying Content

There may be occasions when you find it necessary to alter the content of a webpage or application in response to specific conditions or parameters. This task is where the intersection of Document Object Model (DOM) manipulation and JavaScript's robust control structures really shines and proves to be incredibly powerful.

By utilizing JavaScript's control structures such as loops and conditional statements, you can dynamically change the DOM, or the structure of the webpage, based on the user's interaction or other specific conditions. This combination allows for a more interactive and responsive user experience.

Example: Conditional Content Modification

```
<div id="message">Welcome, guest!</div>
<script>
    const user = { name: 'Alice', loggedIn: true };
    const messageDiv = document.getElementById('message');

    if (user.loggedIn) {
        messageDiv.textContent = `Welcome, ${user.name}!`;
        messageDiv.classList.add('loggedIn');
    }
</script>
```

In this example, the message and style are changed based on the user's login status, demonstrating how JavaScript's logical capabilities integrate with DOM manipulation.

The HTML creates a div element with the id "message" and the text "Welcome, guest!". The JavaScript code creates a user object with properties name and loggedIn. It then selects the div element with the id "message". If the user is logged in (i.e., user.loggedIn is true), the text content of the div is changed to "Welcome, Alice!" (or whatever the user's name is) and the class 'loggedIn' is added to the div.

4.4 Creating and Removing Elements

The ability to dynamically create and remove elements is a crucial aspect of web development. These techniques give developers the power to modify the document structure in real time, making it responsive to user interactions, data alterations, or varying other conditions. This can significantly enhance the interactivity and responsiveness of a web application, making it more engaging and user-friendly.

This section will walk you through the intricate process of adding new elements to the Document Object Model (DOM) and removing existing ones. The DOM is a programming interface for web documents. It represents the structure of a document and allows programs to manipulate the document's structure, style, and content. Adding and removing elements are fundamental operations in DOM manipulation, and mastering them can tremendously enhance your web development skills.

However, it's not just about adding or removing elements at will. There are practical considerations to bear in mind when manipulating the DOM. One of the key aspects to remember is to ensure that your manipulations improve the user experience and do not introduce performance issues or erratic behavior. Performance bottlenecks can occur if DOM manipulations are not handled correctly, leading to a sluggish user experience. Similarly, improper manipulations could lead to unexpected behavior, confusing the user, and potentially causing them to abandon your application.

Therefore, this section will not only teach you how to add and remove elements in the DOM but also how to do so correctly and effectively, keeping in mind the best practices and potential pitfalls. By the end of this guide, you should be well-equipped to manipulate the DOM dynamically, improving the responsiveness, performance, and user experience of your web applications.

4.4.1 Creating Elements

JavaScript, offers a method named document.createElement(). This method is specifically designed to create a new element node within the document. Once this new element node has been generated using this method, it can then be configured as needed.

The configuration can include defining the type of the element, setting its attributes, or even specifying its content. After it has been fully configured, the new element can then be seamlessly inserted into the current document.This process allows for dynamic modification of the document structure, providing a high degree of flexibility and interactivity.

Example: Creating and Inserting an Element

```
<div id="container"></div>
<script>
    const container = document.getElementById('container');

    // Create a new paragraph element
    const newParagraph = document.createElement('p');
    newParagraph.textContent = 'This is a new paragraph.';

    // Append the new element to the container
    container.appendChild(newParagraph);
</script>
```

In this example, a new paragraph element is created, text is added to it, and it is appended to a div container in the DOM.

This code first selects a HTML element with the id 'container' using the document.getElementById method. Then, it creates a new paragraph (<p>) element, sets its text content to 'This is a new paragraph.', and appends this new paragraph to the 'container' element. The result of this code would be adding a paragraph saying 'This is a new paragraph.' inside the 'container' element on the web page.

4.4.2 Removing Elements

When it comes to removing an element from the DOM (Document Object Model), there are a couple of methods that you can resort to. The first method is the removeChild() method. This method allows you to target a specific child element and remove it from the DOM. The other method, if it is supported by your environment, is the remove() method.

This method is directly applied on the element that you want to remove. Both methods are effective, and your choice largely depends on the specific requirements of your project and the compatibility of the method with the browsers you are targeting.

Example: Removing an Element

```
<div id="container">
    <p id="oldParagraph">This paragraph will be removed.</p>
</div>
<script>
    const container = document.getElementById('container');
    const oldParagraph = document.getElementById('oldParagraph');

    // Remove the old paragraph using removeChild
    container.removeChild(oldParagraph);

    // Alternatively, use the remove method if you don't need a reference to the parent
    // oldParagraph.remove();
</script>
```

This demonstrates two methods to remove an element. The choice depends on whether you need to perform actions on the parent node or not.

In the HTML part, there is a 'div' element with an ID of 'container', containing a 'p' (paragraph) element with an ID of 'oldParagraph'. The JavaScript part first accesses the 'div' and the 'p' element through their respective IDs.

Then, it removes the 'p' element from the 'div' using the 'removeChild' method. There is also a commented-out code suggesting an alternative way of removing the 'p' element directly using the 'remove' method, which doesn't require a reference to the parent 'div'.

4.4.3 Using Document Fragments for Batch Operations

When you have the task of creating a multitude of elements, an efficient approach would be to utilize a feature known as DocumentFragment. This powerful tool allows you to assemble all the elements together in one cohesive unit.

Once you have structured your elements within the DocumentFragment, you can then append them to the Document Object Model (DOM) in a single operation. This method is particularly beneficial as it significantly reduces the amount of page reflow.

Page reflow is a process that can impact the performance of your page negatively as it involves the calculation of layout changes and re-rendering in response to alterations in elements. By using DocumentFragment, you can minimize this reflow, thereby enhancing the performance and responsiveness of your page.

Example: Using Document Fragments

```html
<ul id="list"></ul>
<script>
    const list = document.getElementById('list');
    const fragment = document.createDocumentFragment();

    for (let i = 0; i < 5; i++) {
        let listItem = document.createElement('li');
        listItem.textContent = `Item ${i + 1}`;
        fragment.appendChild(listItem);
    }

    // Append all items at once
    list.appendChild(fragment);
</script>
```

This method is particularly useful when you need to add a large number of elements to the DOM.

This is a script written to dynamically create a list of 5 items in HTML. It first selects an unordered list element with the id "list". Then it creates a document fragment, which is a lightweight container for storing temporary elements.

It then creates a loop that runs five times, each time creating a new list item ('li'), setting its text content to "Item" followed by the current loop index plus one. These items are then appended to the document fragment.

After the loop completes, all the list items are appended to the 'list' element in the HTML document in one operation. This approach is efficient because it minimizes changes to the actual DOM.

4.4.4 Cloning Elements

When working with web development or any task that requires manipulation of Document Object Model (DOM) elements, there may be instances when you need to create a duplicate of an existing element. This could be for a variety of reasons, such as wanting to replicate the element with or without its child elements, or perhaps you want to introduce some modifications to the element without influencing the original. In such scenarios, the cloneNode() method proves to be extremely useful.

The cloneNode() method, as the name suggests, helps in creating a copy or clone of the node on which it is invoked. The method works by creating and returning a new node that is an identical copy of the node you wish to clone. The beauty of this method is the added control it provides. When you use the cloneNode() method, you're given the option to specify whether you want to clone the node's entire subtree (which is referred to as a 'deep clone') or if you just want to clone the node itself without its child elements.

This level of flexibility makes the cloneNode() method an indispensable tool when handling DOM elements, allowing developers to maintain the integrity of the original element while still being able to work with its copy.

Example: Cloning Elements

```
<div id="original" class="sample">Original Element</div>
<script>
    const original = document.getElementById('original');
    const clone = original.cloneNode(true); // true means clone all child nodes
    clone.id = 'clone';
    clone.textContent = 'Cloned Element';
    original.parentNode.insertBefore(clone, original.nextSibling);
</script>
```

This example shows how to clone an element and modify its ID and text before inserting it back into the DOM.

This code example identifies an HTML element using its id "original", creates a duplicate of it, alters the id and text content of the duplicate, and finally adds the duplicate to the DOM, immediately following the original element.

4.4.5 Practical Considerations

When it comes to the process of creating and removing elements within any given framework or programming language, there are two key areas of concern that must be addressed with utmost care and attention:

Management of Memory and Resources

One of the most significant concerns during this process is the efficient and effective management of memory and resources. It's vital to be extremely cautious of potential memory leaks, especially when it comes to the removal of elements that have event listeners attached to them.

These event listeners, if not properly managed, can lead to memory leaks, which can severely impact the performance of your application. Therefore, it's critically important to always remove event listeners if and when they're no longer needed in order to prevent such issues.

Maintaining Accessibility Standards

The other crucial area to focus on is maintaining accessibility standards. It's essential to ensure that all content that is dynamically added to your application is fully accessible to all users. This includes managing focus for elements that are either added or removed and updating aria attributes as and when it's necessary.

These steps are crucial in ensuring that your application is inclusive and accessible to all users, regardless of any potential disabilities or limitations they may have.

4.4.6 Efficiently Managing Element IDs

When working with dynamic element creation in your web development process, it becomes crucially important to manage your element IDs with care and precision. The reason for this is that you want to avoid creating duplicate IDs, which can introduce problems into your website's operation.

Duplicates can lead to unpredictable behavior in your site's interface, confusing your users and potentially leading to loss of data or incorrect operation. Furthermore, these duplicates can cause errors in your JavaScript logic, leading to failure in executing the intended functions and operations.

This could significantly disrupt the user experience and complicate debugging processes. Therefore, careful management of element IDs when dynamically creating elements is not just good practice, but a necessary aspect of robust, reliable web development.

Example: Managing Dynamic IDs

```
function createUniqueElement(tag, idBase) {
    let uniqueId = idBase + '_' + Math.random().toString(36).substr(2, 9);
    let element = document.createElement(tag);
    element.id = uniqueId;
    return element;
}

const newDiv = createUniqueElement('div', 'uniqueDiv');
document.body.appendChild(newDiv);
console.log(newDiv.id);   // Outputs a unique ID like 'uniqueDiv_15gs6kd1i'
```

This approach ensures that each element has a unique ID, preventing conflicts and enhancing the stability of your DOM manipulations.

This example code snippet includes a function named 'createUniqueElement'. This function takes two parameters: 'tag' (the type of HTML element to create) and 'idBase' (the base string for creating a unique ID). It generates a unique ID by appending a random string to the 'idBase', creates a new HTML element of the type specified by 'tag', assigns the unique ID to this element, and then returns the element.

The code then uses this function to create a new 'div' element with a unique ID starting with 'uniqueDiv', appends this new 'div' to the body of the document, and logs its unique ID to the console.

4.4.7 Handling Memory Leaks

In web development, when elements are removed from the Document Object Model, or DOM, it is of paramount importance to ensure that any associated resources are also cleaned up. This cleanup operation is necessary to prevent memory leaks that can lead to performance issues over time.

Memory leaks happen when memory resources allocated to tasks are not released back to the system after the tasks are completed. In the case of DOM elements, these resources can include event listeners or external resources like images or custom data. Event listeners, in particular, can cause significant memory leaks if not properly managed.

This is because they hold onto memory in the DOM even after the element they were attached to has been removed. The same can be said for external resources like images or custom data. Hence, a thorough cleanup is crucial for maintaining optimal performance in any web application.

Example: Preventing Memory Leaks

```javascript
const button = document.getElementById('myButton');
button.addEventListener('click', function handleClick() {
    console.log('Button clicked!');
});

// Before removing the button, remove its event listener
button.removeEventListener('click', handleClick);
button.parentNode.removeChild(button);
```

Always clean up after your elements, especially in single-page applications where long-term performance is critical.

This code is using the DOM to manipulate a button on a webpage. First, it gets a reference to a button element using its 'id' attribute ('myButton'). Then, it adds an event listener to the button that will log 'Button clicked!' to the console every time the button is clicked. Finally, before removing the button from the webpage, it removes the event listener from the button to prevent memory leaks.

4.4.8 Using Custom Data Attributes

HTML5 data attributes, often referred to as data-* attributes, represent a valuable feature that can significantly streamline the process of interacting with elements that are created dynamically within a web page. These attributes provide a convenient method of storing necessary data directly within the DOM (Document Object Model) element.

This approach offers distinct advantages as it eliminates the need for extra code or separate storage to handle this data. Thus, it helps in keeping the code clean and manageable. Moreover, one of the major benefits of using data-* attributes is that they can be easily and directly accessed via JavaScript.

This ease of access simplifies the process of data manipulation and retrieval, making the overall coding experience more efficient and less error-prone.

Example: Using Data Attributes

```
<div id="userContainer"></div>
<script>
    for (let i = 0; i < 5; i++) {
        let userDiv = document.createElement('div');
        userDiv.setAttribute('data-user-id', i);
        userDiv.textContent = 'User ' + i;
        userDiv.onclick = function() {
            console.log('Selected user ID:', this.getAttribute('data-user-id'));
        };
        document.getElementById('userContainer').appendChild(userDiv);
    }
</script>
```

This method provides an elegant way to associate data with elements without complicating your JavaScript logic.

This code creates a 'div' container with the id 'userContainer'. Within this container, it generates five 'div' elements using a for loop, each representing a different user. These 'div' elements are assigned an id (from 0 to 4), and when clicked, the id of the selected user is printed to the console.

4.4.9 Optimizing for Accessibility

When you are looking to dynamically add or remove elements in your digital interface, it is crucial to take into account how these changes might impact users who are dependent on assistive technologies. These users might include those with visual or auditory impairments who use tools like screen readers or captioning.

By managing focus in an appropriate manner and updating ARIA (Accessible Rich Internet Applications) attributes as and when needed, you can help ensure a seamless and inclusive user experience. This not only enhances accessibility but also promotes a more universal design that can be beneficial to all users, regardless of their individual needs or abilities.

Example: Managing Accessibility

```
let modal = document.createElement('div');
modal.setAttribute('role', 'dialog');
```

```
modal.setAttribute('aria-modal', 'true');
modal.setAttribute('tabindex', '-1'); // Make it focusable
document.body.appendChild(modal);
modal.focus();  // Set focus to the new modal for accessibility

// When removing
modal.parentNode.removeChild(modal);
document.body.focus(); // Return focus safely
```

This ensures that the application remains accessible, particularly during dynamic content updates, which might otherwise disrupt the user experience for those using screen readers or other accessibility tools.

This code is creating an accessible modal dialog box. First, it creates a new 'div' element. Then, it sets several attributes to make it behave as a modal dialog. 'role' is set to 'dialog' to inform assistive technologies that this is a dialog box. 'aria-modal' is set to 'true' to indicate that it's a modal, and 'tabindex' is set to '-1' to allow focus.

The modal is then added to the document and given focus. When it's time to remove the modal, the code removes it from the document and returns focus to the body of the document.

4.5 Event Handling in the DOM

Event handling serves as a fundamental aspect of interactive web development, playing a critical role in transforming static web pages into dynamic, interactive platforms. It is through event handling that web pages can react and respond to a variety of user actions, such as clicks, key presses, and mouse movements, thereby making the user's web experience more dynamic, engaging, and personalized.

In this comprehensive section, we will delve deeper into the intricate world of event handling within the Document Object Model (DOM), the programming interface for web documents. We will explore and discuss the different methods to attach event listeners to web elements, allowing us to detect and respond to user actions in real-time.

Moreover, we will also outline some of the best practices for managing and handling events in an efficient and effective manner, ensuring that your web pages remain responsive and user-friendly. We will introduce techniques to optimize your event handling, minimizing unnecessary processing and keeping your web pages running smoothly.

By the end of this section, you will have a thorough understanding of event handling in web development, empowering you to create more interactive and user-friendly web experiences.

4.5.1 Basics of Event Handling

In order to effectively respond to user actions within a web application or website, it is fundamental to first establish a mechanism for listening for events. Events can be any type of interaction from the user, such as clicks or key presses. JavaScript, as one of the cornerstone technologies of the web, offers a multitude of ways to attach these event listeners to HTML elements within your code.

By doing so, you enable your code to react and respond to any events triggered by the user, making your application interactive and responsive. This is a crucial aspect of creating a dynamic and engaging user experience.

Attaching Event Listeners

In modern JavaScript, the primary technique for listening to events is through the use of the addEventListener method. This method is characterized by its power and versatility when it comes to event handling.

One of its main features is its ability to attach multiple event handlers to a single event on a single element. This means you can have several different actions or reactions triggered by one event on the same element, which can significantly enhance the interactivity of your application.

Furthermore, the addEventListener method provides options for controlling how events are captured and bubbled. This allows developers to finely tune the behavior of events in their applications, offering more control over the user experience and interaction flow.

Understanding and effectively utilizing the addEventListener method is a crucial skill for any JavaScript developer aiming to create dynamic and responsive web applications.

Example: Using addEventListener

```
<button id="clickButton">Click Me!</button>
<script>
    document.getElementById('clickButton').addEventListener('click', function() {
        alert('Button was clicked!');
    });
</script>
```

This example adds an event listener to a button that triggers an alert when clicked.

This is an example code that demonstrates how to create interactivity on a webpage using the concept of DOM manipulation and event handling.

The HTML part of the code creates a button element on the webpage with an ID of "clickButton" and a label that reads "Click Me!". The ID is a unique identifier that allows the JavaScript code to locate this specific button on the webpage.

The JavaScript code adds an event listener to the button using the addEventListener method. This method takes two arguments: the type of event to listen for and the function to execute when the event occurs. Here, the event type is 'click', which means the function will be executed when the button is clicked.

The function defined here is an anonymous function, which is a function without a name that is defined right where it's used. This function uses the JavaScript alert function to display a pop-up message on the webpage. The message says "Button was clicked!", indicating that the button was indeed clicked by the user.

This simple piece of code effectively demonstrates how HTML and JavaScript can be combined to create interactive elements on a webpage. By using JavaScript to listen for and respond to user events, developers can create dynamic, engaging webpages that respond to user input.

4.5.2 Event Propagation: Capturing and Bubbling

Grasping the concept of event propagation is key for executing efficient and effective event handling within the Document Object Model (DOM). This is especially important for developers working with interactive web interfaces. Events in the DOM have a unique flow that consists of two distinct phases, known as capturing and bubbling.

Capturing phase

The capturing phase is the initial step in the event propagation process. Designed like a descending hierarchy, this phase begins at the highest level of the document structure and systematically works its way down towards the element where the event actually occurred.

It's akin to a ripple effect that is initiated at the outermost part of the web page; this ripple then moves inward, gradually getting closer to the event target. This process ensures that the event is acknowledged and registered at each level of the document's structure, facilitating a robust and comprehensive event handling mechanism.

Bubbling phase

The journey of an event in the web world does not end once it reaches its intended target element. In fact, reaching the target is just half the journey. What follows next is known as the bubbling phase. During this crucial second part of its journey, the event bubbles up from the target element and moves gradually towards the top of the document.

This interesting phenomenon can be visualized much like a bubble in a liquid. When a bubble is formed underwater, it doesn't stay where it was formed. Instead, it rises up towards the surface of the liquid in a path that can be tracked.

Similarly, during the bubbling phase, the event moves in an upward direction, from the depths of the target element towards the surface of the document. This is why it is referred to as the 'bubbling phase', as it mirrors the movement of bubbles in a liquid.

As a developer, you have the power to control whether an event listener is invoked during the capture phase (the descending ripple) or the bubble phase (the ascending bubble). This can be accomplished by setting the useCapture parameter in the addEventListener method. By understanding and controlling this propagation, you can create more robust and interactive web experiences.

Example: Capturing vs. Bubbling

```
<div id="parent">
    <button id="child">Click Me!</button>
</div>
<script>
    // Capturing
    document.getElementById('parent').addEventListener('click', function() {
        console.log('Captured on parent');
    }, true);

    // Bubbling
    document.getElementById('child').addEventListener('click', function() {
        console.log('Bubbled to child');
    });

    // This will log "Captured on parent" first, then "Bubbled to child"
</script>
```

The code example demonstrates event capturing and event bubbling.

Event capturing is where an event starts at the outermost element (the parent) and then fires on each descendant (child) in nesting order. It's set by the third parameter in addEventListener as 'true'.

Event bubbling, on the other hand, is the opposite: the event starts at the innermost element (the child) and then fires on each ancestor (parent) in nesting order.

In this example, when the 'child' button is clicked, the browser first runs the capturing event listener on the 'parent' (logs 'Captured on parent'), then the bubbling event listener on the 'child' (logs 'Bubbled to child').

4.5.3 Removing Event Listeners

In the development of software applications, especially those of larger scale, it is essential to manage event listeners in an effective manner. Both the addition and removal of these listeners are equally significant, particularly in order to avoid potential memory leaks that can impact the application's performance.

Event listeners are added to elements to listen for certain types of events like clicks or presses. However, when these listeners are no longer needed, or when the element associated with them is being removed from the Document Object Model (DOM), it becomes necessary to remove these event listeners.

This can be accomplished by using the removeEventListener method. By properly managing event listeners, we can ensure that our applications run smoothly and efficiently, without unnecessary consumption of resources.

Example: Removing an Event Listener

```
<script>
    const button = document.getElementById('clickButton');
    const handleClick = function() {
        console.log('Clicked!');
        // Remove listener after handling click
        button.removeEventListener('click', handleClick);
    };

    button.addEventListener('click', handleClick);
</script>
```

This snippet is a practical example of how you can add and remove an event listener to an HTML button element using JavaScript.

We begin by defining a constant named 'button' that uses the document.getElementById function to return the element in the document with the id 'clickButton'. This is the button we will be working with throughout this code snippet.

Next, we define a function named 'handleClick'. This function contains a console.log command to output the text 'Clicked!' to the web console every time it's called.

The button.addEventListener line is where we attach the 'handleClick' function to the 'click' event on the button. The 'click' event is triggered every time a user clicks on the button with their mouse. When the 'click' event is fired, the 'handleClick' function is called, and 'Clicked!' is logged to the console.

Inside the 'handleClick' function, we also have a line of code button.removeEventListener('click', handleClick); that removes the event listener from the button immediately after the button has been clicked and 'Clicked!' has been logged to the console.

This means that the 'click' event will only fire once for the button. After the first click, the event listener is removed, so clicking the button additional times will not output 'Clicked!' to the console.

This is a simple yet practical example of how you can manipulate DOM elements using JavaScript, adding and removing event listeners as needed. This can be a powerful tool in enhancing the interactivity and user experience of your web applications.

4.5.4 Event Delegation

Event delegation is a highly efficient technique in JavaScript, which involves assigning a single event listener to a parent element in order to manage events originating from its multiple child elements.

This technique takes advantage of the 'event bubbling' phase - a concept in JavaScript where an event starts at the most deeply nested element, and then 'bubbles up' through its ancestors. Instead of attaching individual event listeners to each child element, which can lead to decreased performance and increased memory usage, the event delegation technique allows for the handling of these events at a higher, more general level.

This method not only optimizes memory usage but also simplifies the code, making it easier to manage and debug.

Example: Event Delegation

```
<ul id="menu">
    <li>Home</li>
    <li>About</li>
    <li>Contact</li>
</ul>
<script>
    document.getElementById('menu').addEventListener('click', function(event) {
        if (event.target.tagName === 'LI') {
            console.log('You clicked on', event.target.textContent);
        }
    });
</script>
```

This is particularly useful for handling events on elements that are dynamically added to the document, as the listener does not need to be reattached every time an element is added.

The example illustrates the use of HTML and JavaScript to create an interactive webpage element. In particular, it presents an unordered list that serves as a navigation menu, and JavaScript code to handle click events on the menu items.

The HTML part of the code defines an unordered list (<ul>) with the ID "menu". This list contains three list items (<li>), each representing a different section of the website: Home, About, and Contact. The ID "menu" serves as a unique identifier for the unordered list, allowing the JavaScript code to easily find and interact with it.

The JavaScript part of the code adds an event listener to the unordered list. This event listener listens for click events that occur within the list. The addEventListener function is used to attach this event listener to the list. This function takes two parameters: the type of event to listen for ('click' in this case) and a function to execute when the event occurs.

The function that's executed on a click event receives an event object as a parameter. This object contains information about the event, including the target element that the event occurred on (event.target). In this case, the function checks whether the clicked element is a list item by comparing the target element's tag name (event.target.tagName) to the string 'LI'. If the clicked element is a list item, the function logs the text content of the clicked item (event.target.textContent) to the console.

This mechanism allows the webpage to respond to user interactions in a dynamic fashion. When a user clicks on different items in the navigation menu, the webpage can identify which section the user is interested in, and respond accordingly. This could be by highlighting the selected menu item, loading the appropriate section of the website, or any other interaction defined by the developer.

4.5.5 Using Custom Events

In today's digital landscape, modern web applications frequently necessitate intricate interactions that extend beyond the scope of standard Document Object Model (DOM) events. These complex interactions often demand a more tailored approach, which is where custom events come into play.

Custom events provide developers with the ability to define and trigger their very own events. This level of customization offers a highly flexible platform for managing behaviors specific to their application. Moreover, this is done in a decoupled way, ensuring that these specific behaviors do not interfere with or depend on other parts of the application.

This method of managing application-specific behaviors allows for greater control, versatility, and adaptability in developing modern web applications.

Example: Creating and Dispatching Custom Events

```
<script>
    // Create a custom event
    const loginEvent = new CustomEvent('login', {
        detail: { username: 'user123' }
    });

    // Listen for the custom event
    document.addEventListener('login', function(event) {
        console.log('Login event triggered by', event.detail.username);
    });

    // Dispatch the custom event
    document.dispatchEvent(loginEvent);
</script>
```

This example demonstrates how to create a custom event with additional data (username) and how to listen and respond to it, which can be particularly useful for more complex application states or interactions that are not covered by native DOM events.

This code creates a custom event named 'login', listens for it, and dispatches it. The event carries data in its 'detail' property, specifically a username 'user123'. When the 'login' event is triggered, an event listener activates a function that logs a message to the console, indicating the username involved in the login event.

4.5.6 Throttling and Debouncing Event Handlers

In web development, handling events is a fundamental aspect. Events such as resize, scroll, or mousemove can fire frequently. When this happens, it becomes crucial to optimize the event handlers to prevent potential performance issues, which could negatively impact the user experience. Throttling and debouncing are two commonly used techniques that serve to limit the rate at which an event handler function is invoked.

Throttling, as a technique, ensures that the event handler function gets called at most once every certain number of milliseconds. It's like setting a fixed pace at which the event handler gets to run. This ensures a steady stream of function invocations, thereby helping to manage the frequency and prevent overloading.

On the other hand, **Debouncing** is a technique that ensures the event handler function is invoked only after the event has stopped firing for a certain number of milliseconds. This helps prevent the handler from being called too often within a very short amount of time.

Debouncing can be particularly useful in scenarios where you want to ensure that the function is fired only after a user has stopped performing a certain action, such as typing in a search box.

Example: Throttling an Event Handler

```
<script>
    let lastCall = 0;
    const throttleTime = 100; // milliseconds

    window.addEventListener('resize', function() {
        const now = new Date().getTime();
        if (now - lastCall < throttleTime) {
            return;
        }
        lastCall = now;
        console.log('Window resized');
    });
</script>
```

This script throttles the resize event to prevent the handler from executing too frequently, which helps maintain performance even when the event fires rapidly, such as during window resizing.

This is an example code snippet that implements a throttling mechanism. It's used to prevent the 'resize' event from firing too frequently, which can cause performance issues. The event will only fire if 100 milliseconds have passed since the last time it was called. When the 'resize' event is triggered, it logs 'Window resized' to the console.

4.5.7 Ensuring Accessibility in Dynamic Content

When it comes to adding, removing, or modifying elements in response to specific events, it becomes crucial to maintain accessibility for all users. This involves a number of key steps.

Firstly, managing focus is necessary to ensure that users can navigate efficiently through the site or application. Secondly, changes should be announced to assistive technologies, which is a vital step in supporting users with varying abilities and ensuring they can access all available information and functions.

Lastly, ensuring keyboard navigability is essential, particularly for users who may rely on keyboard input over mouse navigation. By taking these steps, we can ensure our content remains accessible to all, regardless of their mode of interaction with the site or application.

Example: Managing Focus and Accessibility

```html
<div id="modal" tabindex="-1" aria-hidden="true">
    <p>Modal content...</p>
    <button id="closeButton">Close</button>
</div>
<script>
    document.getElementById('toggleButton').addEventListener('click', function() {
        const modal = document.getElementById('modal');
        modal.style.display = 'block';
        modal.setAttribute('aria-hidden', 'false');
        modal.focus();
    });

    document.getElementById('closeButton').addEventListener('click', function() {
        const modal = document.getElementById('modal');
        modal.style.display = 'none';
        modal.setAttribute('aria-hidden', 'true');
        document.getElementById('toggleButton').focus();
    });
</script>
```

In this example, focus management and ARIA attributes are used to enhance the accessibility of dynamically shown and hidden modal content.

This example explores various aspects of the DOM manipulation and event handling in JavaScript. It highlights how these skills are fundamental for creating dynamic and interactive web experiences.

The example starts with a brief introduction on DOM manipulation and how it contributes to creating an interactive web experience. It describes how a modal dialogue box is added to the document and given focus, and how it's removed when no longer needed.

The code then delves into event handling, which allows web pages to react and respond to a variety of user actions such as clicks, key presses, and mouse movements. This makes the user's web experience more dynamic and personalized. The document discusses different methods to attach event listeners to web elements, allowing real-time detection and response to user actions.

An example of event handling is provided using the addEventListener method in JavaScript. This method is versatile, allowing the attachment of multiple event handlers to a single event on a single element.

The concept of event propagation, consisting of two phases - capturing and bubbling, is also discussed. The capturing phase begins at the highest level of the document structure and works its way down towards the element where the event occurred. On the other hand, the bubbling phase starts from the target element and moves gradually towards the top of the document.

The example also highlights the importance of managing event listeners effectively to avoid potential memory leaks that can impact application performance. An example of how to add and remove event listeners is provided.

Event delegation, a technique of assigning a single event listener to a parent element to manage events from its multiple child elements, is discussed. It is a method that optimizes memory usage and simplifies code management.

The example further explores the use of custom events that provide developers with the ability to define and trigger their own events. This offers a highly flexible platform for managing behaviors specific to their application.

Next, the concept of throttling and debouncing event handlers is introduced. These techniques limit the rate at which an event handler function is invoked, ensuring efficient performance.

Lastly, the code example emphasizes the importance of maintaining accessibility when adding, removing, or modifying elements in response to specific events. It discusses managing focus, announcing changes to assistive technologies, and ensuring keyboard navigability.

A practical example of a modal dialog box is given. The HTML code creates the modal and the JavaScript code manages its display and focus. When the modal is activated, it is displayed and focus is given to it. The 'aria-hidden' attribute is set to false, making it accessible to screen readers. When the 'closeButton' is clicked, the modal is hidden, focus is returned to the 'toggleButton', and 'aria-hidden' is set to true, making it inaccessible to screen readers.

This example presents a comprehensive understanding of how to create more interactive and user-friendly web experiences using DOM manipulation and event handling.

Practical Exercises

To reinforce the concepts covered in Chapter 4 on DOM Manipulation, here are some practical exercises designed to test and enhance your understanding of selecting elements, modifying content, creating and removing elements, and handling events. These exercises will provide hands-on experience and prepare you to apply these techniques in real-world scenarios.

Exercise 1: Select and Style Elements

Select all paragraph elements on a page and change their text color to blue.

Solution:

```
<p>Paragraph one</p>
<p>Paragraph two</p>
<script>
    const paragraphs = document.querySelectorAll('p');
    paragraphs.forEach(p => {
        p.style.color = 'blue';
    });
</script>
```

Exercise 2: Create and Append Elements

Create a list of items dynamically from an array of strings and append it to a div element.

Solution:

```
<div id="listContainer"></div>
<script>
    const items = ['Item 1', 'Item 2', 'Item 3'];
    const list = document.createElement('ul');
```

```
    items.forEach(item => {
        let listItem = document.createElement('li');
        listItem.textContent = item;
        list.appendChild(listItem);
    });

    document.getElementById('listContainer').appendChild(list);
</script>
```

Exercise 3: Event Handling

Attach an event listener to a button that logs a message to the console when clicked. Ensure the button is removed after being clicked once.

Solution:

```
<button id="myButton">Click me</button>
<script>
    const button = document.getElementById('myButton');
    button.addEventListener('click', function handleClick() {
        console.log('Button was clicked!');
        button.removeEventListener('click', handleClick);
        button.remove(); // Removes the button after clicking
    });
</script>
```

Exercise 4: Modify Attributes Dynamically

Create a function that changes the src attribute of an image element and logs the old and new src to the console.

Solution:

```
<img id="myImage" src="original.jpg" alt="Original Image">
<script>
    function changeImageSrc(newSrc) {
        const image = document.getElementById('myImage');
        console.log('Old src:', image.src);
        image.src = newSrc;
        console.log('New src:', image.src);
    }

    changeImageSrc('updated.jpg');
</script>
```

Exercise 5: Custom Event Creation and Handling

Define a custom event called 'userLoggedIn' and dispatch it after setting a listener that updates the content of a div to show a welcome message when the event is triggered.

Solution:

```
<div id="welcomeMessage"></div>
<script>
    // Listener for the custom event
    document.addEventListener('userLoggedIn', function(event) {
        document.getElementById('welcomeMessage').textContent        =        `Welcome,
${event.detail.username}!`;
    });

    // Create and dispatch the custom event
    const loggedInEvent = new CustomEvent('userLoggedIn', { detail: { username:
'Alice' } });
    document.dispatchEvent(loggedInEvent);
</script>
```

These exercises provide practical applications for the DOM manipulation techniques discussed in the chapter, allowing you to build proficiency in creating dynamic and interactive web pages. By completing these tasks, you'll deepen your understanding of how JavaScript can manipulate the DOM in response to user inputs and other events, a critical skill for any web developer.

Chapter Summary

Chapter 4 of "JavaScript from Scratch: Unlock your Web Development Superpowers" provided a comprehensive exploration of DOM manipulation, an essential skill set for any web developer aiming to create dynamic, interactive, and user-friendly web applications. This chapter covered a range of topics, from selecting and modifying elements to creating, removing, and handling events in the DOM. Let's summarize the key points and insights from each section to consolidate your understanding and highlight the practical applications of these skills.

Selecting Elements

We started with various methods to select elements within the DOM, which is fundamental for any interaction or manipulation. Methods such as document.getElementById(), document.getElementsByTagName(), document.getElementsByClassName(), and the more powerful query selectors document.querySelector() and document.querySelectorAll() were discussed. Each method serves different needs, from selecting single elements to retrieving lists of elements based on complex criteria. Mastering these selectors ensures that you can efficiently find and interact with any part of the web page.

Modifying Content

Modifying the content, style, and attributes of DOM elements allows you to dynamically change web pages in response to user interactions or programmatic conditions. We explored how to use properties like textContent, innerHTML, and style, along with methods to manipulate CSS classes such as classList.add(), remove(), toggle(), and more. These capabilities are crucial for tasks like updating UI elements, showing or hiding content, and applying new styles dynamically.

Creating and Removing Elements

The ability to dynamically add and remove elements from the DOM enables developers to build highly interactive and responsive interfaces. We covered how to create new elements using document.createElement() and insert them into the DOM using methods like appendChild() and insertBefore(). Similarly, removing elements using removeChild() or the simpler remove() method was discussed, emphasizing the importance of managing DOM elements efficiently to ensure performance and prevent resource leaks.

Event Handling

Effective event handling is critical for interactive applications. We delved into adding event listeners with addEventListener(), which provides robust control over how events are handled, including options for capturing and bubbling phases. Techniques for removing event listeners to avoid memory leaks were also discussed, as well as advanced strategies like event delegation, which allows for more efficient management of events, especially in dynamic applications with numerous elements.

Practical Applications and Best Practices

Throughout the chapter, emphasis was placed on best practices such as minimizing direct DOM interactions to enhance performance, using document fragments for batch updates, and ensuring accessibility through proper management of focus and ARIA attributes. We also explored custom events for handling complex application-specific interactions and the importance of managing events responsibly to create seamless user experiences.

By the end of this chapter, you should have a solid foundation in DOM manipulation techniques, equipped with the knowledge to select, modify, and manage elements and their interactions effectively. These skills are vital for developing modern web applications that are not only functional but also engaging and accessible. As you continue to practice and apply these techniques, you'll be able to tackle more complex development challenges, enhancing both the user experience and the capabilities of your web projects.

Quiz for Part I: Getting Started with JavaScript

Test your understanding of the fundamental concepts covered in the first part of "JavaScript from Scratch: Unlock your Web Development Superpowers" with this quiz. Each question is designed to reinforce the key points from each chapter, ensuring you have a solid grasp of the basics of JavaScript and DOM manipulation.

Question 1: Basic JavaScript

What is the output of the following JavaScript code?

```
console.log(typeof (typeof 1));
```

A) "string"

B) "number"

C) "object"

D) "boolean"

Question 2: Data Structures

Which method would you use to add an element to the beginning of an array?

A) push()

B) pop()

C) shift()

D) unshift()

Question 3: JSON Handling

Which statement about JSON is correct?

A) JSON is a programming language.

B) JSON strings must use single quotes.

C) JSON can include functions as values.

D) JSON is commonly used for data interchange between a server and web applications.

Question 4: DOM Manipulation

Which method is used to select an element by its ID?

A) document.getElementByClassName()

B) document.getElementById()

C) document.querySelectorAll()

D) document.getElementsByTagName()

Question 5: Creating and Removing DOM Elements

What is the correct way to remove an element from the DOM?

A) element.delete()

B) element.removeChild()

C) element.remove()

D) element.erase()

Question 6: Event Handling

What is the correct syntax to add an event listener that executes when a user clicks a button with the ID "submitBtn"?

```
document.querySelector('???').addEventListener('???', function() {
    alert('Button clicked!');
});
```

Fill in the '???' to correctly set up the event listener.

Question 7: Modifying Element Content

How do you change the text content of an element with the ID "header" to "Welcome to JavaScript"?

A) document.getElementById('header').innerHTML = 'Welcome to JavaScript';

B) document.getElementById('header').textContent = 'Welcome to JavaScript';

C) document.getElementById('header').innerText = 'Welcome to JavaScript';

D) Both B and C are correct.

Question 8: Custom Events

True or False: Custom events can be used to trigger specific functionality that is not covered by native DOM events.

A) True

B) False

Answers

1. A) "string"
2. D) unshift()
3. D) JSON is commonly used for data interchange between a server and web applications.
4. B) document.getElementById()
5. C) element.remove()
6. #submitBtn, 'click'
7. D) Both B and C are correct.
8. A) True

This quiz covers the basic to intermediate concepts introduced in the first part of the book and will help solidify your understanding of JavaScript's core features and DOM manipulation techniques.

Project 1: Building a Simple Interactive Website

1. Project Overview

1.1 Objective

The primary goal of this project is to build a simple interactive website that utilizes core JavaScript skills and DOM manipulation techniques. This site will serve as a practical application of the concepts learned in the first part of this book. By the end of this project, you will have created a dynamic web page that responds to user inputs and changes state accordingly.

1.2 Key Features

The interactive website will feature several key components that will allow users to engage with the content dynamically:

- **Dynamic Content Loader**: A section of the website will dynamically update content based on user selections from a dropdown menu or buttons. This might include displaying text, images, or other media relevant to the user's choice.
- **Interactive Form**: Incorporate a form with fields such as name, email, and a message. The form will include live validation to provide immediate feedback on the input provided by the user, ensuring all required fields are filled out correctly before submission.
- **Theme Toggler**: A button or switch that allows users to change the theme of the website from light to dark mode (and vice versa). This will demonstrate real-time style manipulation using JavaScript.
- **To-Do List**: Users can add, remove, and mark tasks as completed. This feature will utilize DOM manipulation to dynamically update the list as well as demonstrate how to handle user events.
- **Local Storage Integration**: To enhance the user experience by making the website's state persistent across sessions, local storage will be used to save and retrieve key user data or preferences.

These features are designed to provide hands-on practice with a variety of JavaScript functionalities, including event handling, working with the DOM, and local storage, thereby reinforcing your learning and increasing your confidence in using JavaScript for web development.

2. Setup and Initial Configuration

To get started with building our simple interactive website, we first need to establish a solid foundation by setting up our development environment and defining the project structure. This section will guide you through selecting the necessary tools and organizing your files for efficient development.

2.1 Tools and Environment

Before diving into coding, ensure you have the following essential tools set up on your computer:

- **Text Editor**: A text editor is crucial for writing your code. Some popular choices for web development include Visual Studio Code, Sublime Text, and Atom. These editors offer features such as syntax highlighting, code completion, and extensions that can enhance your coding experience.
- **Web Browser**: You'll need a modern web browser to test and view your web application. Google Chrome, Mozilla Firefox, or Microsoft Edge are recommended because of their developer-friendly tools, such as the Developer Console and live DOM inspector.
- **Local Server (Optional)**: While not strictly necessary for this project, running a local server can be beneficial, especially as you expand into more complex projects. Tools like XAMPP, MAMP, or even simple server setups using Node.js or Python can serve your files more reliably than opening HTML files directly in a browser.

2.2 Project Structure

Organizing your project files from the beginning can help manage the development process more smoothly. Here's a basic structure to start with:

```
simple-interactive-website/

├── index.html         # The main HTML document
├── css/               # Folder for CSS stylesheets
│   └── styles.css     # Main stylesheet for the website
├── js/                # Folder for JavaScript files
│   └── script.js      # Main JavaScript file for handling logic
└── assets/            # Folder for images and other assets (if needed)
```

- **HTML File**: Your index.html will be the entry point of your site. This file will contain the basic HTML structure and links to your CSS and JavaScript files.
- **CSS Folder**: This folder will store your stylesheets. Starting with styles.css, you can add more files as needed if your project grows or requires more complex styling.
- **JavaScript Folder**: All your JavaScript files will go in this folder. While you might only need script.js initially, organizing your JavaScript code into modules or separate files can help maintain cleaner code.
- **Assets Folder**: If your project includes images, fonts, or other multimedia files, keeping them in an assets folder makes managing these resources easier.

2.3 Initializing Your Project

1. **Create the Project Directory**: Make a new folder on your computer or development environment named simple-interactive-website.
2. **Set Up the Files**: Inside the project directory, create the file and folder structure as outlined above. You can create the files and folders manually or via command line.
3. **Prepare the HTML Template**: Open index.html and set up a basic HTML5 template. Here's a simple example to get you started:

```
<!DOCTYPE html>
<html lang="en">
<head>
    <meta charset="UTF-8">
    <meta name="viewport" content="width=device-width, initial-scale=1.0">
    <title>Simple Interactive Website</title>
    <link rel="stylesheet" href="css/styles.css">
</head>
<body>
    <header>
        <h1>Welcome to Our Interactive Website</h1>
    </header>
    <main>
        <!-- Content will be dynamically added here -->
    </main>
    <script src="js/script.js"></script>
</body>
</html>
```

1. **Write Placeholder Content**: Add some basic styles in styles.css and a few lines of JavaScript in script.js to ensure everything is connected properly.

This setup provides a solid starting point for developing your interactive website. With this structure, you can begin implementing the interactive features discussed in the project overview, ensuring that each component is manageable and efficiently developed.

3. Designing the User Interface

A well-designed user interface (UI) is crucial for ensuring a pleasant user experience. It should be intuitive, accessible, and visually appealing to engage users effectively. In this section, we'll outline the design considerations for our simple interactive website, including the layout of HTML elements and the styling with CSS.

3.1 HTML Structure

Start by laying out the basic structure of your website using HTML. This will form the skeleton of your project, upon which all dynamic functionalities will be built. Here's how you might structure your HTML to accommodate the features outlined in the project overview:

```
<!DOCTYPE html>
<html lang="en">
<head>
    <meta charset="UTF-8">
    <meta name="viewport" content="width=device-width, initial-scale=1.0">
    <title>Simple Interactive Website</title>
    <link rel="stylesheet" href="css/styles.css">
</head>
<body>
    <header>
        <h1>Interactive Features Website</h1>
        <button id="theme-toggler">Toggle Theme</button>
    </header>
    <nav>
        <ul id="menu">
            <li>Home</li>
            <li>About</li>
            <li>Contact</li>
        </ul>
    </nav>
    <main>
        <section id="dynamic-content">
            <h2>Dynamic Content Area</h2>
            <p>Select an option to change this content.</p>
        </section>
        <section id="todo-list">
            <h2>To-Do List</h2>
            <ul id="tasks"></ul>
            <input type="text" id="new-task" placeholder="Add a new task">
```

```html
                <button id="add-task">Add Task</button>
        </section>
        <section id="form-section">
            <h2>Contact Us</h2>
            <form id="contact-form">
                <input type="text" id="name" name="name" placeholder="Your name"
required>
                <input type="email" id="email" name="email" placeholder="Your email"
required>
                <textarea id="message" name="message" placeholder="Your message"
required></textarea>
                <button type="submit">Send</button>
            </form>
        </section>
    </main>
    <footer>
        <p>© 2024 Interactive Website Project</p>
    </footer>
    <script src="js/script.js"></script>
</body>
</html>
```

3.2 CSS Styling

Next, add CSS to enhance the visual appearance of your site. Start with some basic styles to improve layout and typography, then add more specific styles for interactive elements like buttons and forms.

Example CSS (styles.css):

```css
/* Basic layout and typography styles */
body {
    font-family: Arial, sans-serif;
    line-height: 1.6;
    margin: 0;
    padding: 0;
    background: #f4f4f4;
    color: #333;
}

header, nav, main, footer {
    padding: 20px;
    text-align: center;
}

/* Styling for the navigation menu */
nav ul {
    list-style: none;
```

```css
    padding: 0;
}
nav ul li {
    display: inline;
    margin-right: 10px;
}

/* Theme toggler button */
#theme-toggler {
    position: absolute;
    top: 20px;
    right: 20px;
}

/* Form styling */
input, textarea {
    width: 90%;
    margin-bottom: 10px;
    padding: 10px;
    box-sizing: border-box;
}

button {
    padding: 10px 20px;
    cursor: pointer;
}

/* Dynamic content and To-Do List styles */
#dynamic-content, #todo-list {
    margin-top: 20px;
    padding: 20px;
    background: white;
    box-shadow: 0 0 5px #ccc;
}

/* Footer styling */
footer {
    margin-top: 20px;
    color: #666;
}
```

These styles provide a clean and modern look, but you can certainly expand on them based on your aesthetic preferences or additional functional requirements.

3.3 Responsive Design Considerations

Finally, ensure that your website is accessible and looks great on all devices:

- **Responsive Layout**: Use CSS media queries to adjust styles based on device screen size.
- **Accessibility**: Ensure that all interactive elements are accessible, including providing proper ARIA roles and ensuring that all form elements have associated labels for screen readers.

By thoughtfully designing your user interface with attention to layout, styling, and accessibility, you set the stage for a positive user experience. This foundational work supports the interactive features that will be implemented in the following sections, creating a cohesive and engaging web application.

4. Core Functionality Implementation

Now that we have established a strong foundation with a well-designed user interface, we'll focus on implementing the core functionalities of our simple interactive website. This section delves into adding interactive elements and dynamic content manipulation using JavaScript. We'll cover how to implement the Dynamic Content Loader, Interactive Form, Theme Toggler, and To-Do List, ensuring each feature not only functions correctly but also enhances the overall user experience.

4.1 Implementing Interactivity

1. Dynamic Content Loader

We'll start by allowing users to change the content displayed in a section of the page based on their selection from a menu or set of buttons.

HTML Setup:

```
<select id="contentSelector">
    <option value="content1">Content 1</option>
    <option value="content2">Content 2</option>
    <option value="content3">Content 3</option>
</select>
<div id="contentDisplay">Select an option to see the content.</div>
```

JavaScript:

```
document.getElementById('contentSelector').addEventListener('change', function() {
    const selectedValue = this.value;
    const displayDiv = document.getElementById('contentDisplay');
```

```
    displayDiv.innerHTML = `<p>You selected ${selectedValue}. Here is some more
information about it.</p>`;
});
```

This script listens for changes on the contentSelector dropdown and updates the contentDisplay div with a message related to the selected option.

2. Interactive Form

Next, add validation to ensure all fields are filled out before the form can be submitted.

HTML Setup:

```
<!-- Already defined in the HTML structure -->
```

JavaScript:

```
document.getElementById('contact-form').addEventListener('submit', function(event) {
    event.preventDefault();  // Prevent the form from submitting traditionally
    const name = document.getElementById('name').value;
    const email = document.getElementById('email').value;
    const message = document.getElementById('message').value;

    if (!name || !email || !message) {
        alert('Please fill in all fields.');
    } else {
        alert('Thank you for your message!');
        // Here you could also add an AJAX request to send the data to a server
    }
});
```

3. Theme Toggler

Create a simple toggle button to switch between light and dark themes.

JavaScript:

```
document.getElementById('theme-toggler').addEventListener('click', function() {
    document.body.classList.toggle('dark-theme');
    const isDark = document.body.classList.contains('dark-theme');
    this.textContent = isDark ? 'Switch to Light Mode' : 'Switch to Dark Mode';
});
```

CSS:

```
/* Add to styles.css */
.dark-theme {
    background: #333;
    color: #fff;
}
```

4. To-Do List

Allow users to add, mark as complete, or remove tasks from a list.

HTML Setup:

```
<!-- Already defined in the HTML structure -->
```

JavaScript:

```
document.getElementById('add-task').addEventListener('click', function() {
    const newTask = document.getElementById('new-task').value;
    if (newTask) {
        const listItem = document.createElement('li');
        listItem.textContent = newTask;
        listItem.addEventListener('click', function() {
            this.classList.toggle('completed');
        });
        document.getElementById('tasks').appendChild(listItem);
        document.getElementById('new-task').value = '';    // Clear the input after
adding
    } else {
        alert('Please enter a task.');
    }
});

/* CSS for completed tasks */
.completed {
    text-decoration: line-through;
}
```

These examples illustrate how to dynamically interact with the DOM to create a responsive, user-engaged web application. By implementing these features, you not only apply the core JavaScript and DOM manipulation concepts discussed in the book but also create a practical, functional project that solidifies your understanding of these crucial web development skills.

5. Adding Advanced Features

To elevate our simple interactive website, we will integrate some advanced features that enhance functionality, improve user interaction, and demonstrate more complex JavaScript concepts. In this section, we'll cover how to implement form validation, use local storage to maintain state between sessions, and introduce some additional JavaScript techniques that can provide a richer user experience.

5.1 Form Validation

Enhancing the form with JavaScript-powered validation allows for more interactive feedback. This will ensure users input valid information before submitting.

JavaScript for Enhanced Validation:

```javascript
document.getElementById('contact-form').addEventListener('submit', function(event) {
    event.preventDefault();
    let isValid = true;

    const name = document.getElementById('name');
    const email = document.getElementById('email');
    const message = document.getElementById('message');

    // Simple validation checks
    [name, email, message].forEach(input => {
        if (!input.value) {
            input.style.border = '2px solid red';
            isValid = false;
        } else {
            input.style.border = '';
        }
    });

    if (isValid) {
        alert('Thank you for your message!');
        // Optionally clear the form or handle the data further
        this.reset();
    } else {
        alert('Please fill in all fields correctly.');
    }
});
```

5.2 Local Storage Integration

Using local storage allows the website to remember certain user settings or data across sessions. For example, we can remember the user's theme preference or save the to-do list items.

JavaScript for Local Storage:

```javascript
// Save theme preference
document.getElementById('theme-toggler').addEventListener('click', function() {
    const body = document.body;
    body.classList.toggle('dark-theme');
    const isDark = body.classList.contains('dark-theme');
    localStorage.setItem('theme', isDark ? 'dark' : 'light');
    this.textContent = isDark ? 'Switch to Light Mode' : 'Switch to Dark Mode';
});

// Load theme preference on page load
document.addEventListener('DOMContentLoaded', () => {
    const savedTheme = localStorage.getItem('theme');
    if (savedTheme === 'dark') {
        document.body.classList.add('dark-theme');
        document.getElementById('theme-toggler').textContent = 'Switch to Light Mode';
    }
});

// Save and load tasks
document.getElementById('add-task').addEventListener('click', function() {
    const taskList = document.getElementById('tasks');
    const newTaskValue = document.getElementById('new-task').value;
    if (newTaskValue) {
        const listItem = document.createElement('li');
        listItem.textContent = newTaskValue;
        taskList.appendChild(listItem);
        updateTasksLocalStorage();
        document.getElementById('new-task').value = '';  // Clear the input
    }
});

function updateTasksLocalStorage() {
    const tasks = [];
    document.querySelectorAll('#tasks li').forEach(task => {
        tasks.push(task.textContent);
    });
    localStorage.setItem('tasks', JSON.stringify(tasks));
}

document.addEventListener('DOMContentLoaded', () => {
    const savedTasks = JSON.parse(localStorage.getItem('tasks'));
    if (savedTasks) {
```

```
        savedTasks.forEach(taskText => {
            const listItem = document.createElement('li');
            listItem.textContent = taskText;
            document.getElementById('tasks').appendChild(listItem);
        });
    }
});
```

5.3 Additional JavaScript Techniques

To further enhance the site, consider adding animations or transitions with JavaScript for a smoother user experience, especially when elements enter or leave the DOM, or when the user interacts with UI elements.

Example: Adding Simple Animation

```
const buttons = document.querySelectorAll('button');
buttons.forEach(button => {
    button.addEventListener('mouseover', () => {
        button.style.transition = 'all 0.3s';
        button.style.transform = 'scale(1.1)';
    });
    button.addEventListener('mouseout', () => {
        button.style.transform = 'scale(1)';
    });
});
```

By integrating these advanced features, our simple interactive website not only becomes more functional and personalized but also demonstrates the application of powerful JavaScript techniques in real-world scenarios. These enhancements will help ensure the site is engaging and maintains user data effectively, providing a richer and more interactive user experience.

6. Testing and Debugging

A crucial step in any web development project is testing and debugging. This process ensures that your website functions correctly across different environments and that any issues are identified and resolved. In this section, we will discuss strategies for testing the interactive website and provide tips for effective debugging.

6.1 Testing Strategies

1. Functional Testing:

- **Manual Testing**: Go through all the features of your website manually to ensure they work as expected. Test each button, form submission, and interaction.

- **Browser Developer Tools**: Utilize the console and network tabs in browser developer tools to inspect elements, monitor network requests, and check console outputs for errors.
- **Responsive Testing**: Resize your browser window to test different screen sizes or use device simulation tools available in browser developer tools to ensure the website is responsive and functional on all devices.

2. Cross-Browser Testing:

- **Different Browsers**: Test your website on different browsers (like Chrome, Firefox, Safari, and Edge) to ensure consistent behavior and appearance.
- **Compatibility Tools**: Use tools like BrowserStack or Caniuse for checking compatibility and performing cross-browser tests.

3. Usability Testing:

- **User Feedback**: Gather feedback from users to understand their experience with the website. This can highlight unexpected issues or potential improvements.
- **Accessibility Compliance**: Check that your website is accessible by testing with screen readers, keyboard-only navigation, and using accessibility testing tools like the WAVE tool.

6.2 Debugging Tips

1. Console Logs:

- Use console.log(), console.warn(), and console.error() strategically to output debugging information in the JavaScript console. This can help you understand what part of your code is executing and pinpoint where things may be going wrong.

2. Breakpoints:

- Set breakpoints in your JavaScript code using browser developer tools to pause execution and inspect the values of variables at various points. This can be crucial for understanding the flow of logic and state changes.

3. Step Through Code:

- Utilize the step-through feature in browser developer tools to execute your JavaScript line by line. This allows you to see exactly how your code runs and how data is being manipulated.

4. Network Monitoring:

- Monitor and analyze network requests that your website makes, especially when interacting with APIs or server endpoints. Ensure that requests are successful and that data returned is as expected.

5. Error Handling:

- Implement robust error handling in your code. Catch errors and display useful information to the user or log it for further analysis. Use try...catch blocks in JavaScript to handle exceptions gracefully.

6. Validation Checks:

- Regularly validate both client-side and server-side (if applicable) inputs to ensure data integrity and to prevent common security vulnerabilities like XSS (Cross-Site Scripting).

6.3 Common Issues and Their Solutions

- **Caching Issues**: Sometimes, changes to CSS or JavaScript might not appear due to browser caching. Clear the browser cache or use cache-busting techniques during development, such as appending a version query string to your CSS and JavaScript file references.
- **Asynchronous Bugs**: Issues related to asynchronous code can be tricky. Use async and await or promises effectively, and ensure that operations dependent on the completion of asynchronous calls are placed correctly in the logic flow.
- **Responsive Design Problems**: Use CSS media queries to handle different screen sizes and ensure that all elements resize and reposition correctly. Test responsiveness regularly during development.

By following these testing and debugging strategies, you can ensure that your interactive website not only functions flawlessly across different platforms and devices but also provides a robust user experience. Debugging is an ongoing process, and continual testing and refinement are key to developing a reliable and professional website.

7. Deployment

Once your interactive website is tested and all features are functioning as expected, the next step is to deploy it to a live server so that it can be accessed by users on the internet. Deployment involves several key steps, from choosing the right hosting provider to ensuring your site remains secure and performs well under traffic. This section will guide you through the process of getting your website online.

7.1 Choosing a Hosting Provider

The first step in deploying your website is to select a hosting provider. There are many options available, each with different features and price points. Here are a few popular choices:

- **Shared Hosting**: Providers like Bluehost and HostGator offer cost-effective plans suitable for small to medium websites. Shared hosting means your site will be hosted on a server shared with other websites.
- **VPS Hosting**: A Virtual Private Server (VPS) provides a dedicated slice of a server, which means more control and better performance than shared hosting. Providers like DigitalOcean and Linode offer VPS hosting with scalable options.
- **Cloud Hosting**: Services like AWS, Google Cloud, and Microsoft Azure offer robust cloud hosting solutions that scale automatically to handle traffic spikes and provide extensive geographic coverage.

7.2 Setting Up Your Domain

Your website will need a domain name, which is the web address users will type into their browsers to find your site. You can purchase a domain from registrars such as GoDaddy, Namecheap, or Google Domains. Once purchased, configure the domain's DNS settings to point to your hosting server, a process that typically involves setting A records or CNAME records based on your host's instructions.

7.3 Preparing for Deployment

Before uploading your website, make sure everything is optimized for the best performance:

- **Minimize CSS and JavaScript**: Use tools to minify your CSS and JavaScript files, reducing their size to improve load times.
- **Optimize Images**: Ensure that images are not larger than necessary, use appropriate file formats, and consider using compression tools to reduce file sizes without losing quality.

- **SSL/TLS Certificate**: Secure your website by obtaining an SSL/TLS certificate, which encrypts data sent to and from your site. Many hosts offer free SSL certificates through Let's Encrypt.

7.4 Uploading Your Website

To deploy your website, you will need to upload your files to your hosting provider. This can generally be done using one of the following methods:

- **FTP/SFTP**: Use an FTP client like FileZilla to transfer your website files from your local computer to your hosting server. SFTP is a secure version of FTP that encrypts file transfers.
- **Hosting Control Panel**: Many hosting providers offer a control panel (such as cPanel) that includes a file manager. You can use this tool to upload your files directly through your web browser.
- **Version Control Systems**: If you are using a version control system like Git, some hosts allow you to deploy directly from your repository. This method is particularly effective for managing updates and rollbacks.

7.5 Post-Deployment Checks

Once your website is live, perform the following checks to ensure everything is functioning as expected:

- **Test all features**: Ensure all interactive elements and functionalities work as they did in your development environment.
- **Monitor performance**: Use tools like Google PageSpeed Insights to analyze your site's performance and get suggestions for improvements.
- **Set up analytics**: Implement tracking with Google Analytics or similar to monitor visitor behavior and gather insights that could guide future enhancements.

7.6 Continuous Monitoring and Updates

Deployment is not the end of the website development process. It's essential to continuously monitor your site for any issues and regularly update your content, technologies, and security measures to ensure optimal performance and protection against threats.

By following these deployment steps, your interactive website will be live and accessible to users worldwide, representing a significant milestone in your journey as a web developer. Keep learning and iterating based on user feedback and analytics to maintain and improve your site over time.

8. Challenges and Extensions

After successfully deploying your simple interactive website, you may wish to further enhance your skills and the functionality of your project. This section provides challenges and extension ideas that not only increase the complexity and usability of your website but also encourage continued learning and improvement.

8.1 Further Challenges

1. **Implement Server-Side Interaction**:
    - o Integrate a backend framework such as Node.js with Express or another server technology to handle form submissions more dynamically and securely. Explore creating RESTful APIs to interact with your front-end more efficiently.
2. **Add More Complex User Interactions**:
    - o Incorporate drag-and-drop functionality within the to-do list or other parts of your site to enhance user experience.
    - o Develop a live chat feature using WebSockets for real-time communication between users.
3. **Incorporate Third-Party APIs**:
    - o Integrate APIs such as Google Maps for location-based services or payment gateways like Stripe or PayPal for processing transactions.
    - o Fetch external data, such as weather information or news articles, and display it dynamically on your site.
4. **Create a Progressive Web App (PWA)**:
    - o Convert your website into a PWA, which allows for offline functionality, push notifications, and a native app-like experience on mobile devices.

8.2 Learning Extensions

1. **Explore Modern JavaScript Frameworks**:
    - o Learn and apply frameworks like React, Angular, or Vue.js to your project. These tools can help manage more complex state and user interfaces, offering powerful patterns and optimizations out of the box.
2. **Advanced CSS Techniques**:
    - o Dive deeper into advanced CSS concepts such as Flexbox, Grid, animations, and transitions to improve the layout and visual dynamics of your website.
3. **Optimization and Best Practices**:

- o Study website optimization techniques, focusing on improving the load times and performance of your website. Learn about lazy loading, service workers, and efficient asset management.

4. **Security Enhancements**:
   - o Enhance the security of your website by implementing features such as HTTPS, content security policies, and other modern security practices to protect user data and interactions.

5. **Accessibility Improvements**:
   - o Make your website more accessible by adhering to WCAG (Web Content Accessibility Guidelines). Implement features that aid navigation for all users, including those with disabilities.

6. **Automated Testing and Continuous Integration (CI)**:
   - o Set up automated testing frameworks such as Jest or Mocha for JavaScript testing. Integrate your project with CI tools like Jenkins or GitHub Actions to automate testing and deployment processes.

7. **Explore Database Integration**:
   - o Add a database solution like MongoDB, PostgreSQL, or Firebase to store user data persistently. This will allow you to handle user accounts, store application data, and retrieve it dynamically.

Each of these challenges and extensions provides a pathway to not only enhance your project but also deepen your understanding of web development. Tackling these will help you build more robust, efficient, and user-friendly web applications while preparing you for advanced roles in web development. As you progress, continue to seek out new technologies and methodologies to stay current in the fast-evolving field of web development.

9. Conclusion

Congratulations on completing Part I of "JavaScript from Scratch: Unlock your Web Development Superpowers" and successfully building and deploying your simple interactive website. Through this journey, you've laid a solid foundation in JavaScript, learning to manipulate the DOM, handle events, and dynamically update content based on user interactions. This final section serves to recap the skills you've developed, reflect on what you've learned, and consider how these skills can be applied to future projects or professional tasks.

9.1 Recap of Skills Learned

Throughout this part of the book, you've acquired and applied a range of essential web development skills:

- **JavaScript Fundamentals**: You've gained proficiency in JavaScript basics, including variables, data types, control structures, and functions.

- **DOM Manipulation**: You've learned how to select, modify, create, and remove HTML elements dynamically, which is crucial for building interactive web pages.
- **Event Handling**: You've mastered setting up event listeners and responding to user actions, allowing for interactive and responsive design.
- **Form Validation and Handling**: You've implemented client-side form validation and learned basic techniques to enhance user input handling.
- **Local Storage**: You've used local storage to maintain state across user sessions, enhancing the user experience by remembering user preferences and data.
- **Deployment**: You've successfully deployed your website, making it accessible to users worldwide and gaining experience in the final crucial steps of web development.

9.2 Reflecting on the Project Process

This project has not only helped solidify your understanding of technical concepts but also taught you about the project lifecycle from conception through to deployment. Reflect on the following aspects:

- **Problem Solving**: How did you approach troubleshooting and solving issues that arose during development? What strategies did you find most effective?
- **Design Thinking**: Consider how design decisions impact user experience. How did the design choices you made influence how users interacted with your website?
- **Performance Optimization**: Reflect on the importance of website performance. What steps did you take to ensure your site loaded quickly and ran smoothly?

9.3 Future Directions

With these foundational skills in hand, you're well-prepared to tackle more complex projects. Consider exploring the following paths:

- **Advanced JavaScript and Frameworks**: Dive deeper into JavaScript and explore frameworks like React, Angular, or Vue.js, which can help manage more complex application states and user interfaces.
- **Full-Stack Development**: Expand your knowledge to include server-side programming. Learning Node.js, Python, or Ruby can help you build complete applications from front to back.
- **Specialization**: Consider specializing in areas such as front-end performance optimization, accessibility, or user experience design. Each of these areas offers deep, rewarding career paths.

9.4 Continued Learning

Web development is a rapidly evolving field. Staying informed about new technologies, best practices, and emerging standards is crucial. Engage with the development community through forums, social media, and conferences. Continue learning through online courses, tutorials, and by building projects that challenge you to apply and expand your skills.

In closing, remember that the journey of learning and development is ongoing. Each project builds on the last, and every challenge offers a new opportunity to grow. Keep coding, keep experimenting, and most importantly, keep enjoying the process of creating something new. Your journey as a web developer is just beginning, and the possibilities are limitless.

Part II: Intermediate JavaScript

Chapter 5: Advanced Functions

Welcome to Chapter 5 of "**JavaScript from Zero to Superhero: Unlock your web development superpowers**," where we embark on a journey into the complex and rewarding world of advanced functions. This chapter is meticulously designed to broaden your understanding of JavaScript functions. It introduces you to a variety of more sophisticated concepts that have the potential to significantly enhance the readability, efficiency, and functionality of your code. As we delve into the heart of JavaScript programming, you'll discover that functions are the basic building blocks, and mastering them is absolutely crucial for developing complex applications.

In this enlightening chapter, we'll explore the many faces of JavaScript functions. We'll navigate through various types of functions, uncover their subtle nuances, and learn how to leverage them effectively in real-world scenarios. This section is your guide to mastering the art of JavaScript functions. We'll cover topics ranging from the simplicity and elegance of arrow functions, to the versatility of higher-order functions, the utility of callbacks, the cleverness of closures, and the power of async functions.

Each section is thoughtfully crafted to not only provide you with a deep theoretical understanding but also practical knowledge that you can apply immediately in your projects. By the end of this chapter, you'll have a comprehensive understanding of advanced functions and be ready to tackle any challenges that come your way in your JavaScript programming journey.

5.1 Arrow Functions

Arrow functions, a novel feature introduced in the sixth edition of ECMAScript (ES6), provide a more streamlined syntax for writing functions, thus simplifying the code and improving its readability. These functions are particularly beneficial for cases where functions are primarily used to return a computed value, thereby reducing the need for excess coding.

The primary advantage of arrow functions is their concise syntax which enables developers to write less verbose function expressions. They eliminate the need for the 'function' keyword, curly brackets, and the 'return' keyword when there is only a single statement in the function. This makes the code cleaner and easier to understand at a glance.

In addition, arrow functions share the same lexical 'this' as their surrounding code. This is a significant feature, as it eliminates the common confusion around the behavior of 'this' in JavaScript. 'This' inside an arrow function always represents the object that defined the arrow function. This makes them ideal for use in contexts where you might otherwise need to bind a function to the current object, resulting in a more intuitive behavior.

Therefore, the introduction of arrow functions in ES6 has enhanced the brevity and readability of JavaScript, making it a powerful tool for modern web development.

5.1.1 Syntax and Basic Use

The arrow function, which is a fundamental feature of modern JavaScript, has a basic syntax that is relatively straightforward and easy to understand. It's a more concise way of creating functions in JavaScript, and it also has some differences in behavior compared to traditional function expressions. This makes it an essential tool for JavaScript programmers to understand and utilize effectively.

The basic syntax of an arrow function is straightforward:

```
const functionName = (parameters) => expression;
```

This syntax is a more concise version of a function expression. The return statement is implied and omitted for single-expression functions. Here's a simple comparison:

This is a simple example of JavaScript arrow function syntax. It defines a function named "functionName" that takes arguments specified in "parameters" and returns the result of the "expression". This syntax is widely used in JavaScript, especially in the React.js library.

Traditional Function Expression:

```
const add = function(a, b) {
    return a + b;
};
```

The example code is a simple example of a function in JavaScript, one of the most widely used programming languages in web development. The function is named 'add', which implies its purpose - to add two numbers together.

The function is declared using the 'const' keyword, which means it's a constant variable. This essentially means that once the function is defined, you cannot redeclare it later in the code.

Now, let's break down the code:

Here, 'add' is the name of the function. The 'function' keyword is used to declare a function. Following the function keyword, in parentheses, we have 'a' and 'b'. These are the parameters or inputs to the function. This particular function expects two inputs.

Inside the function body, enclosed in curly brackets {}, there is a single line of code: return a + b;. The return keyword is used in functions to specify the output of the function. In this case, the function returns the sum of 'a' and 'b'.

So, to sum it up, this function named 'add' takes two parameters, 'a' and 'b', adds them together and returns the result.

Arrow Function:

```
const add = (a, b) => a + b;
```

This is a simple example. It defines a function named "add" that takes two parameters "a" and "b". The function returns the sum of "a" and "b". The syntax is in ES6 format, which uses the arrow function.

No Parameters and Multiple Expressions

If there are no parameters, you must include an empty pair of parentheses in the definition:

```
const sayHello = () => console.log("Hello!");
```

This example illustrates that in JavaScript, when defining a function using arrow function syntax, you must include an empty set of parentheses even if the function doesn't require any parameters. The provided code snippet defines a function named "sayHello" that logs the string "Hello!" to the console.

When the function contains more than one expression, you need to wrap the body in curly braces and use a return statement (if returning a value):

```
const multiply = (a, b) => {
    let result = a * b;
    return result;
};
```

5.1.2 Use Cases for Arrow Functions

Arrow functions are a more streamlined, concise way of writing functions in JavaScript, significantly simplifying the code and improving its readability. They are especially beneficial in cases where functions are primarily used to return a computed value, thereby reducing the need for extra coding.

In this section, you will gain a deeper understanding of arrow functions in programming. We will explore various scenarios, demonstrating when and how to effectively use these arrow functions in real-world programming situations.

The aim is to equip you with the necessary knowledge and skills to seamlessly integrate the use of arrow functions into your coding practices, thereby enhancing your programming efficiency and proficiency.

Concise Iteration Operations: Arrow functions are particularly useful for carrying out array manipulations such as mapping, filtering, or reducing. These operations are made more succinct and easier to read with arrow functions, saving developers time and effort, while also making the code cleaner and easier to understand:

```javascript
const numbers = [1, 2, 3, 4, 5];
const squared = numbers.map(number => number * number);
console.log(squared);  // Output: [1, 4, 9, 16, 25]
```

The code declares an array 'numbers' consisting of five elements. It then generates a new array 'squared' by squaring each item in the 'numbers' array using the 'map' function. This function works by accepting a function as an argument and applying it to each array element.

The arrow function 'number => number * number' is used to square each number. Finally, the 'squared' array is logged to the console, resulting in a new array of the original numbers squared: [1, 4, 9, 16, 25].

As Callbacks: In the realm of JavaScript programming, the lexical binding of this that arrow functions provide is particularly useful, especially when dealing with callbacks. Traditional function expressions might inadvertently bind the this keyword to a different context, which could potentially lead to unexpected results or bugs in the code.

This is a pitfall that could be easily avoided through the use of arrow functions, which automatically bind this to the context of the enclosing scope where the function was defined, thus maintaining the intended context and contributing to cleaner, more predictable code.

Example:

```javascript
document.getElementById("myButton").addEventListener('click', () => {
    console.log(this);  // 'this' refers to the context where the function was defined,
not to the element.
});
```

The code adds an event listener to the HTML element with the ID "myButton". When this button is clicked, a function is triggered which outputs the current context, which is where the function was defined, to the console. However, in this case 'this' doesn't refer to the clicked HTML element because an arrow function was used, and arrow functions do not have their own 'this' context.

Functional Programming: Arrow functions play a significant role in functional programming. By enabling a more concise and readable syntax, they contribute to the creation of cleaner, more maintainable code. This is particularly beneficial in larger projects, where code clarity and readability are paramount.

The adoption of a functional programming style can lead to improved debugging and testing processes, ultimately resulting in higher quality software. It is important to note that while arrow functions can greatly enhance the functional programming experience, understanding their usage and implications is crucial for effective development.

Limitations of Arrow Functions

- **No Binding of this:** One of the characteristics of arrow functions is that they do not bind this. This can often be beneficial, providing more flexibility in certain scenarios. However, it can also be limiting in situations where you need the function to bind to a different context.
- **No arguments Object:** Another important feature of arrow functions is that they do not have their own arguments object. This means that if you need to access an array-like object of arguments, you are required to use rest parameters instead.
- **Not Suitable for Methods:** While arrow functions can be very useful, they are not suitable for defining object methods. This is because when you define an object method with an arrow function, you cannot access the object via this, which can limit functionality.
- **No Constructor Use:** Lastly, it's essential to note that arrow functions cannot be used as constructors. Attempting to use an arrow function with the new keyword will result in an error being thrown, as this is not a supported use case for this type of function.

Arrow functions are a powerful addition to JavaScript's function syntax arsenal. They provide syntactical benefits and handle this differently, which can lead to more predictable behavior when used appropriately.

5.1.3 Returning Objects Literally

Arrow functions provide a shorter and more concise way to declare functions. However, they have certain nuances that might trip up developers who are not aware of them. One such common pitfall that developers often encounter while working with arrow functions is when they want to return an object literal directly from the function.

The unique syntax of arrow functions can lead to a misunderstanding by the JavaScript interpreter, especially when it comes to curly braces. In JavaScript, curly braces are used to denote the body of a function, and they are also used to define object literals. This can lead to confusion. When you use curly braces after the arrow in an arrow function, the JavaScript interpreter thinks you are starting the function's body rather than declaring an object literal.

This might cause unexpected results or errors if not handled properly. To mitigate this issue and to return an object literal directly from an arrow function, you need to take a simple but important step. You must wrap the object literal in parentheses. By doing this, you're giving a clear signal to the JavaScript interpreter that the curly braces are not denoting the function body but instead are used for the object literal.

This is a crucial point to remember when writing arrow functions and it's an example of how understanding the nuances of syntax can prevent bugs and errors in your code.

Example: Returning an Object Literal

```
const createPerson = (name, age) => ({name: name, age: age});
console.log(createPerson("John Doe", 30));  // Outputs: { name: 'John Doe', age: 30 }
```

This technique ensures that the object literal is not confused with the function's body, allowing you to return objects succinctly.

The code defines a function named "createPerson". This function takes two parameters, "name" and "age", and returns an object with these two properties. Then, the function is called with arguments "John Doe" and "30", and the output is logged to the console. The output is an object with the properties name being 'John Doe' and age being 30.

5.1.4 Arrow Functions in Array Methods

Arrow functions, have become a game-changer in JavaScript programming, especially when working with array methods that expect callback functions. This includes methods such as map(), filter(), reduce(), and others. The real power of arrow functions shines through their concise syntax which eliminates the need for verbose function and return keywords.

map(), filter(), and reduce() are powerful array methods in JavaScript that are often used in the context of functional programming.

The map() function is used to create a new array by applying a specific function to all the elements of an existing array. In simple terms, it "maps" each element of the existing array to a new element in the resulting array, based on the transforming function provided. This is particularly useful when you want to apply a transformation or operation to every element in an array, without altering the original array.

```
const numbers = [1, 2, 3, 4, 5];
const squared = numbers.map(number => number * number);
console.log(squared);  // Output: [1, 4, 9, 16, 25]
```

The filter() function, on the other hand, is used to create a new array from a given array including only those elements that satisfy a condition specified by a provided function. It essentially "filters" out the elements that don't meet the condition. This can be useful when you want to select certain elements from an array based on specific criteria.

```
const numbers = [1, 2, 3, 4, 5];
const evenNumbers = numbers.filter(n => n % 2 === 0);
console.log(evenNumbers);  // Outputs: [2, 4]
```

Lastly, the reduce() function is used to apply a function to each element in the array (from left to right) so as to reduce the array to a single value - hence the name "reduce". This function takes two arguments: an accumulator and the current value. The accumulator accumulates the callback's return values, while the current value represents the current element being processed in the array.

```
const numbers = [1, 2, 3, 4, 5];
const sum = numbers.reduce((total, current) => total + current, 0);
console.log(sum);  // Outputs: 15
```

In this example, the reduce() method is used to sum all the numbers in the array. The 'total' parameter is the accumulator that stores the ongoing total, and 'current' is the current number in the array as reduce() works its way from left to right through the array. The 0 after the callback function is the initial value of 'total'.

These functions offer a robust and efficient way of handling arrays and can significantly enhance your ability to write concise, understandable, and maintainable code.

5.1.5 Handling this in Event Listeners

While arrow functions share the same lexical this as their surrounding code, which is usually beneficial, this can lead to unexpected behavior when adding event listeners that rely on this referring to the element that fired the event:

Example: Arrow Function in Event Listeners

```
document.getElementById('myButton').addEventListener('click', () => {
    console.log(this.innerHTML);   // Does not work as expected; 'this' is not the
button
});
```

To handle such cases, you can either use a traditional function expression or use the event object, which is passed to the event handler:

Corrected Example Using Event Object

```
document.getElementById('myButton').addEventListener('click', event => {
    console.log(event.target.innerHTML);   // Correctly logs the button's innerHTML
});
```

5.1.6 No Duplicate Named Parameters

In contrast to regular function expressions, arrow functions present a unique characteristic in that they strictly prohibit the use of duplicate named parameters, regardless of whether they're operating in strict or non-strict mode. This particular enforcement mechanism plays an instrumental role in identifying potential errors at an early stage of development.

As a result, it significantly enhances the reliability and cleanliness of the code. By ensuring that each parameter holds a unique name, developers can avoid confusion and potential malfunctions down the line, leading to a more robust and fault-tolerant software product.

Example: Arrow Function Syntax Error

```
const add = (a, a) => a + a;   // Syntax error in strict and non-strict mode
```

5.1.7 Debugging Arrow Functions

Debugging arrow functions in JavaScript can at times pose a challenge, particularly due to their concise and compact syntax, which often leaves little room for detailed debugging. Arrow functions, while incredibly useful for writing clean and efficient code, can become a tricky area when things go wrong and the need for debugging arises.

When complex logic is involved in these functions, a good practice is to consider transforming the concise body syntax into a more verbose block body syntax. This allows for the inclusion of proper debugging statements, such as the use of console.log().

Such a strategy can provide a clearer picture of what the function is doing at each step, thereby making the debugging process more manageable and effective.

Example: Debugging Within an Arrow Function

```javascript
const complexCalculation = (a, b) => {
    console.log("Input values:", a, b);
    const result = a * b + 100;
    console.log("Calculation result:", result);
    return result;
};
```

5.2 Callbacks and Promises

In the expansive realm of JavaScript, a language that is highly utilized in a variety of applications and scenarios, managing operations that are asynchronous in nature such as network requests, file operations, or timers is of crucial importance.

These operations are a critical part of most JavaScript applications, and their proper handling can greatly affect the performance and user experience of your application. In this comprehensive section, we delve deeply into two fundamental concepts that are widely used to handle such complex tasks: callbacks and promises.

These two concepts are cornerstones of asynchronous programming in JavaScript, and they provide different ways to organize and structure your asynchronous code. By gaining a thorough understanding of these concepts, you will significantly enhance your ability to write clean, effective, and maintainable asynchronous code.

This will, in turn, allow you to develop more robust and efficient applications, and also make your code easier to read and debug, thus enhancing your overall productivity as a JavaScript developer.

5.2.1 Understanding Callbacks

A **callback** is a specialized type of function that is designed to be passed into another function as an argument. The purpose of a callback is to defer a certain computation or action until later. In other words, the callback function is "called back" at some specified point in the future.

This arrangement is ideally suited to asynchronous programming paradigms, where we want to start a long-running task (such as a network request) and then move on to other tasks.

The callback function allows us to specify what should happen when the long-running task completes. This allows us to ensure that the right code will be executed at the right time.

Basic Example of a Callback

```
function greeting(name, callback) {
    console.log('Hello ' + name);
    callback();
}

greeting('Alice', function() {
    console.log('This is executed after the greeting function.');
});
```

In this example, the greeting function takes a name and a callback function as arguments. The callback function is called right after the greeting message is printed to the console.

This example defines a function called "greeting" that takes two parameters: a name (as a string) and a callback function. The function prints a greeting message ('Hello ' + name) to the console and then executes the callback function.

After the greeting function is defined, it is called with the name 'Alice' and an anonymous function as parameters. This anonymous function will be executed after the greeting function, printing 'This is executed after the greeting function.' to the console.

Callbacks in Asynchronous Operations

In programming, callbacks serve a critical role, especially when dealing with asynchronous operations. Asynchronous operations are those that allow other processes to continue before they complete.

For instance, suppose you're fetching data from a server, which is a common operation in web development. This process can take an unspecified amount of time. To keep your application responsive and efficient, you don't want to halt the entire program while waiting for the data. Here's where callbacks come into play.

Using a callback function, you can effectively say, "Continue running the rest of the program. Once the data arrives from the server, execute this function to handle it." This way, the callback function acts as a practical means to manage the data once it becomes available.

Example: Using Callbacks with Asynchronous Operations

```
function fetchData(callback) {
    setTimeout(() => {
        callback('Data retrieved');
    }, 2000);  // Simulates a network request
}

fetchData(data => {
    console.log(data);  // Outputs: 'Data retrieved'
});
```

While callbacks are simple and effective for handling asynchronous results, they can lead to issues like "callback hell" or "pyramid of doom," where callbacks are nested within callbacks, leading to deeply indented and hard-to-read code.

This code defines a function named fetchData. The function takes a callback function as its argument, simulates a network request through a delay of 2000 milliseconds (2 seconds) using the setTimeout method, and then calls the callback function with the argument 'Data retrieved'. Below the function definition, the fetchData function is called with a callback function that logs the data (in this case 'Data retrieved') to the console.

5.2.2 Promises: A Cleaner Alternative

In response to the intricate challenges and complexities that are often associated with the use of callbacks in JavaScript, the ECMAScript 6 (ES6) version brought about a significant improvement in the form of **Promises**.

Promises are not just ordinary objects; they hold a special meaning in the context of asynchronous operations. They signify the eventual conclusion of these operations, be it a successful completion or an unfortunate failure.

Additionally, these Promises are not just about the completion or failure of operations, they also carry the resulting value of the operations. This feature provides a more streamlined and efficient way of handling asynchronous tasks in JavaScript.

Creating a Promise

```javascript
const promise = new Promise((resolve, reject) => {
    setTimeout(() => {
        resolve('Data loaded successfully');
        // reject('Error loading data');  // Uncomment to simulate an error
    }, 2000);
});

promise.then(data => {
    console.log(data);
}).catch(error => {
    console.error(error);
});
```

In this example code a new Promise is created. As discussed, a Promise is an object representing the eventual completion or failure of an asynchronous operation.

In this case, the Promise is resolved with the message 'Data loaded successfully' after a delay of 2 seconds. If there's an error, it's rejected with the message 'Error loading data'.

The then method is used to schedule code to run when the Promise resolves, logging the data. If the Promise is rejected, the catch method catches the error and logs it.

Key Points About Promises:

- A promise in JavaScript programming has three states, which are: pending, where the outcome is not yet determined; fulfilled, where the operation completed successfully; and rejected, where the operation failed.
- The then() method, which is an integral part of working with promises in JavaScript, is used to schedule a callback function that will be executed as soon as the promise is resolved, meaning it has fulfilled.
- Lastly, the catch() method is used to handle any errors or exceptions that may occur during the promise's execution. It acts as a safety net, ensuring that any failure conditions or errors are properly dealt with and not left unhandled.

Chaining Promises A fundamental strength inherent in Promises is their inherent ability to be chained or linked together. This feature is made possible because each invocation of the then() method on a Promise returns a completely new Promise object.

This new Promise can then be used as the base for another then() method, creating a chain. This chain of promises makes it possible for asynchronous methods to be called in a specific order, ensuring that each operation is executed sequentially, one after the other.

This is a critical aspect of Promises that allow for structured and predictable handling of asynchronous operations.

Example: Promise Chaining

```
function fetchUser() {
    return new Promise(resolve => {
        setTimeout(() => resolve({ name: 'Alice' }), 1000);
    });
}

function fetchPosts(userId) {
    return new Promise(resolve => {
        setTimeout(() => resolve(['Post 1', 'Post 2']), 1000);
    });
}

fetchUser().then(user => {
    console.log('User fetched:', user.name);
    return fetchPosts(user.name);
}).then(posts => {
    console.log('Posts fetched:', posts);
}).catch(error => {
    console.error(error);
});
```

This example demonstrates how you can perform multiple asynchronous operations in sequence, where each step depends on the outcome of the previous one.

This example illustrates the concept of Promises and asynchronous programming. The code initially fetches a user's data (simulated using setTimeout to resolve a promise after 1 second with a user object). After fetching the user, it logs the user's name and fetches the user's posts (another simulated fetch using setTimeout). The posts are then displayed in the console. Any errors encountered during the process are caught and logged to the console.

Understanding and properly utilizing callbacks and promises are fundamental to effective JavaScript programming, especially in scenarios involving asynchronous operations. Promises, in particular, provide a cleaner, more manageable approach to asynchronous code than traditional callbacks, reducing complexity and improving readability.

5.2.3 Error Handling in Promises

In JavaScript, every Promise is in one of three states: pending (the operation is ongoing), fulfilled (the operation completed successfully), or rejected (the operation failed). Error handling in Promises primarily deals with the rejected state. When a Promise is rejected, this usually means an error occurred.

For example, a Promise might be used to request data from a server. If the server responds with the data, the Promise is fulfilled. But if the server doesn't respond or sends an error response, the Promise would be rejected.

To handle these rejections, you can use the .catch() method on the Promise object. This method schedules a function to be run if the Promise is rejected. The function can take the error as a parameter, allowing you to handle the error appropriately, for example by logging the error message to the console or displaying an error message to the user.

Additionally, the .finally() method can be used to schedule code to run after the Promise is either fulfilled or rejected, which is useful for cleanup tasks like closing a database connection.

By implementing effective error handling in Promises, you can ensure that your application remains robust and reliable, even when dealing with the inherent uncertainties of asynchronous operations.

When dealing with promises in programming, proper error handling becomes an aspect of paramount importance. This is because it serves as an efficient safeguard, ensuring that any errors, whether they are minor or major, do not go undetected or unnoticed. Instead, they are caught early and dealt with effectively.

Moreover, incorporating efficient error handling in your application equips it with the ability to handle unexpected situations in a graceful manner. This means your application will not crash or behave unpredictably when an error occurs. Instead, it will continue to function as smoothly as possible while also providing meaningful feedback about the error, thus allowing for timely and effective troubleshooting.

Example: Comprehensive Error Handling

```
const fetchUserData = () => {
    return new Promise((resolve, reject) => {
        setTimeout(() => {
            if (Math.random() > 0.5) {
                resolve({ name: "Alice", age: 25 });
            } else {
                reject(new Error("Failed to fetch user data"));
            }
        }, 1000);
    });
};

fetchUserData()
    .then(data => {
        console.log("User data retrieved:", data);
    })
    .catch(error => {
        console.error("An error occurred:", error.message);
    });
```

In this example, the catch() method is used to handle any errors that occur during the promise's execution, ensuring that all possible failures are managed.

The code defines a function called fetchUserData that returns a Promise. This Promise simulates the process of fetching user data: after a delay of 1 second (1000 milliseconds), it either resolves with an object containing user data (name and age), or rejects with an error. The outcome is randomly determined with a 50% chance for each.

After the fetchUserData function is defined, it is immediately called. It uses the .then method to handle the case where the Promise is resolved, logging the user data to the console. It also uses the .catch method to handle the case where the Promise is rejected, logging the error message to the console.

5.2.4 Promise.all

In scenarios where you're dealing with multiple asynchronous operations that need to be executed simultaneously, and it's essential to wait for all these operations to complete before proceeding, the Promise.all method becomes an invaluable tool in JavaScript.

The Promise.all method works by accepting an array of promises as its input parameter. In response, it gives back a new promise. In terms of this new promise's behavior, it's designed to resolve only when all the promises in the input array have successfully resolved. This means it waits for each asynchronous operation to complete successfully.

On the other hand, if any one of the promises in the input array fails or rejects, the new promise returned by Promise.all will immediately reject as well. This means it doesn't wait for all operations to complete if any one operation fails, thus allowing you to handle errors promptly.

Example: Using Promise.all

```
const promise1 = Promise.resolve(3);
const promise2 = 42;
const promise3 = new Promise((resolve, reject) => {
    setTimeout(resolve, 100, 'foo');
});

Promise.all([promise1, promise2, promise3]).then(values => {
    console.log(values);   // Output: [3, 42, "foo"]
}).catch(error => {
    console.error("Error:", error);
});
```

This is particularly useful for aggregating results of multiple promises and ensures that your code only proceeds when all operations are complete.

In this code, there are three promises: promise1 is a promise that resolves with a value of 3, promise2 is a direct value of 42, and promise3 is a promise that resolves with a value of 'foo' after 100 milliseconds.

The Promise.all() method is used to handle these promises. It takes an array of promises and returns a single promise that resolves when all of the input promises have resolved. In this case, it resolves with an array of resolved values from the input promises, in the same order as the input promises: [3, 42, 'foo'].

If any of the input promises are rejected, the Promise.all() promise is also rejected, and the .catch() method is used to handle the error.

5.2.5 Handling Promises with finally()

The finally() method, which is an important aspect of JavaScript Promise, returns a promise. This happens when the promise has been settled, which means it has been either fulfilled or rejected. At this point, the callback function that you have specified is executed.

This functionality is particularly useful because it allows you to run specific types of code, commonly referred to as cleanup code. Examples of cleanup code could include closing any open database connections or clearing out resources that are no longer in use.

What's especially useful about this is that you can run this cleanup code regardless of the outcome of the promise chain. This means whether the promise was fulfilled or rejected, your cleanup code will still run, ensuring a tidy and efficient execution.

Example: Using finally with Promises

```
fetch('<https://api.example.com/data>')
    .then(data => data.json())
    .then(json => console.log(json))
    .catch(error => console.error('Error fetching data:', error))
    .finally(() => console.log('Operation complete.'));
```

This is an example code using the Fetch API to retrieve data from a specified URL (https://api.example.com/data). The fetch() function returns a Promise that resolves to the Response of the request. This response is then converted to JSON format with data.json(). The JSON data is then logged in the console. If there's any error during the fetch operation, it's caught with the catch() method and logged in the console as an error. Finally, regardless of the outcome (success or error), 'Operation complete.' is logged in the console, thanks to the finally() method.

Mastering callbacks, promises, and async/await is essential for effective JavaScript programming, particularly in dealing with asynchronous operations. By understanding these patterns and how to handle errors properly, you can write cleaner, more efficient, and more robust asynchronous code. This knowledge is invaluable as you build more complex applications that require interacting with APIs, performing long-running operations, or handling multiple asynchronous tasks simultaneously.

5.3 Async/Await

In modern JavaScript development, handling asynchronous operations elegantly is crucial. Introduced in ES2017, the async and await syntax provides a cleaner, more readable way to work with promises, making asynchronous code easier to write and understand. This section dives deep into the async/await syntax, demonstrating how to effectively integrate it into your JavaScript projects.

Async/await is a powerful tool that provides a more comfortable and readable way to work with promises, significantly simplifying asynchronous code. An async function is a function explicitly defined as asynchronous and contains one or more await expressions.

These expressions literally pause the execution of the function, making it appear as though it's synchronous, but without blocking the thread. When you use await, it pauses the specific part of your function until a promise is resolved or rejected, allowing other tasks to continue their execution in the meantime.

This way, async/await enables us to write promise-based asynchronous code as if it were synchronous, but without blocking the main thread.

Example: Using async/await

```javascript
async function loadUserData() {
    try {
        const response = await fetch('<https://api.mydomain.com/user>');
        const userData = await response.json();
        console.log("User data loaded:", userData);
    } catch (error) {
        console.error("Failed to load user data:", error);
    }
}

loadUserData();
```

Async/await makes your asynchronous code look and behave a little more like synchronous code, which can make it easier to understand and maintain.

This 'async' function, named 'loadUserData', works by trying to fetch user data from a specific URL (https://api.mydomain.com/user) and then logging the data to the console. If it fails to fetch the data for any reason (e.g., server issues, network problems), it will catch the error and log a failure message to the console. The 'await' keyword is used to pause and wait for the Promise returned by 'fetch' and 'response.json()' to resolve or reject, before moving on to the next line of code.

5.3.1 Understanding Async/Await

In JavaScript programming, the async keyword plays a crucial role in declaring a function to be asynchronous. By using the async keyword before a function, you are essentially instructing JavaScript to automatically encapsulate the function's return value inside a promise. This promise, a key concept in asynchronous programming, represents a value that may not be available yet but is expected to be available in the future, or it may never be available due to an error.

Moving on to the await keyword, its primary function is to pause the ongoing execution of the async function. It's important to note that the await keyword can only be utilized within the context of async functions. When await is used, it halts the function until a Promise is either fulfilled (resolved) or failed (rejected). This allows the function to asynchronously wait for the promise's resolution, enabling the function to proceed only when it has the necessary data or a confirmed failure.

Thus, the async and await keywords together provide a powerful tool for handling asynchronous operations in JavaScript, making the code easier to write and understand.

Basic Syntax

```
async function fetchData() {
    return "Data fetched";
}

fetchData().then(console.log); // Outputs: Data fetched
```

In this example, fetchData is an async function that returns a string. Despite not explicitly returning a promise, the function's return value is wrapped in a promise.

The code defines an asynchronous function named 'fetchData' that returns a promise which resolves to "Data fetched". The function is then called with its result being handled by a promise handler (.then), which logs the result to the console. As a result, "Data fetched" is printed to the console.

5.3.2 Using Await with Promises

In the world of programming, the true strength and potential of the async/await functions become dramatically evident when performing operations that are asynchronous in nature. Asynchronous tasks are those that run separately from the main thread and notify the calling thread of its completion, error, or progress updates. These tasks include, but are not limited to, executing operations such as communicating with APIs, handling file operations, or any other tasks that are promise-based.

The async/await functions provide a far more elegant and readable syntax for managing these asynchronous operations than traditional callback-based approaches. What this means is that instead of nesting callbacks within callbacks, leading to the infamous "callback hell", you can write code that looks like it's synchronous, but actually operates asynchronously. This makes it exponentially easier to understand and maintain the code, especially for those who are new to asynchronous programming in JavaScript.

This is where async/await truly shines, and its power is fully realized. It allows for code that is not only more readable and easier to understand, but also easier to debug and maintain. This is a significant advantage in today's complex and fast-paced development environments where readability and maintainability are as important as functionality.

Example: Fetching Data with Async/Await

```javascript
async function getUser() {
    let response = await fetch('<https://api.example.com/user>');
    let data = await response.json();
    return data;
}

getUser().then(user => console.log(user));
```

Here, await is used to pause the function execution until the promise returned by fetch() is resolved. The subsequent await pauses until the conversion of the response to JSON is completed. This approach avoids the complexity of chaining promises and makes the asynchronous code look and feel synchronous.

This code uses the Fetch API to retrieve user data from a certain API endpoint ('https://api.example.com/user'). It is an asynchronous function, meaning it operates in a non-blocking manner. The 'getUser()' function first sends a request to the given URL, waits for the response, then processes the response as JSON. The processed data is then returned. The last line of the code calls this function and logs the returned user data to the console.

5.3.3 Error Handling

Managing errors in asynchronous code through the use of async/await is a process that is relatively simple and straightforward. This is usually accomplished by the implementation of try...catch blocks. If you have experience in working with synchronous code, this approach should feel familiar and intuitive.

The basic principle involves placing the segment of asynchronous code that you anticipate might cause an error within the try block. If an error does indeed occur while the try block is executing, the flow of the code is immediately shifted over to the catch block.

The catch block serves as a designated area where you can dictate how to handle the error. This could involve simply logging the error for debugging purposes, or it could involve more complex operations designed to recover from the error and ensure the continued running of the remaining code.

This approach of using try...catch blocks with async/await provides a clean, efficient, and systematic way to manage any errors that may occur throughout the execution of asynchronous code. By handling errors effectively, you can greatly increase the stability and robustness of your JavaScript applications, leading to better performance and enhanced user experience.

Example: Error Handling with Async/Await

```javascript
async function loadData() {
    try {
        let response = await fetch('<https://api.example.com/data>');
        let data = await response.json();
        console.log(data);
    } catch (error) {
        console.error('Failed to fetch data:', error);
    }
}

loadData();
```

In this example, any error that occurs during the fetch operation or while converting the response to JSON is caught in the catch block, allowing for clean error management.

The function's main task is to fetch data from a specified URL (https://api.example.com/data) using the 'fetch' API. The 'fetch' function is a web API provided by modern browsers for retrieving resources across the network. It returns a Promise that resolves to the Response object representing the response to the request. This Promise is handled using the 'await' keyword, which makes the function wait until the Promise resolves before proceeding to the next line of code.

After the 'fetch' Promise resolves, the function attempts to parse the response body as JSON using the 'response.json()' method. This operation also returns a Promise, which is again handled using 'await'. Once this Promise resolves, the resulting JSON data is logged to the console using 'console.log(data)'.

All of these operations are enclosed in a 'try...catch' block. This is a common error handling pattern in many programming languages. The 'try' block contains the code that might throw an exception, and the 'catch' block contains the code to execute if an exception is thrown.

In this case, if an error occurs during the fetch operation or the JSON conversion — for example, if the network request fails due to connectivity issues, or if the response data cannot be parsed as JSON — the error will be caught and passed to the 'catch' block. The 'catch' block then logs the error message to the console using 'console.error('Failed to fetch data:', error)', providing a useful debug message that indicates what went wrong.

Finally, the 'loadData()' function is called to execute the data fetching operation.

This code demonstrates how to perform asynchronous operations in JavaScript using the 'async/await' syntax and error handling techniques. It is a fundamental pattern in modern JavaScript programming, especially in scenarios that involve networking or any other operations that might take some time to complete.

5.3.4 Handling Multiple Asynchronous Operations

In situations where there are numerous independent asynchronous operations that need to be executed, a highly effective way to manage these operations is to run them concurrently. This can be achieved by using Promise.all() in combination with async/await. By doing so, you're able to significantly enhance the performance of your code.

This is because Promise.all() allows multiple promises to be handled at the same time, rather than sequentially, and when used with async/await, it ensures that your code will wait until all promises have either resolved or rejected, before moving on. This way, you're making the most out of your resources and improving the responsiveness of your application.

Example: Concurrent Asynchronous Operations

```
async function fetchResources() {
    try {
        let [userData, postsData] = await Promise.all([
            fetch('<https://api.example.com/users>'),
            fetch('<https://api.example.com/posts>')
        ]);
        const user = await userData.json();
        const posts = await postsData.json();
        console.log(user, posts);
    } catch (error) {
        console.error('Error fetching resources:', error);
    }
}

fetchResources();
```

This pattern is particularly useful when the asynchronous operations are not dependent on each other, allowing them to be initiated simultaneously rather than sequentially.

The function fetchResources is a JavaScript function that is defined as asynchronous, indicated by the async keyword before the function declaration. This keyword informs JavaScript that the function will return a Promise and that it can contain an await expression, which pauses the execution of the function until a Promise is resolved or rejected.

Inside this function, we have a try...catch block, which is used for error handling. The code inside the try block is executed and if any error occurs during the execution, instead of failing and potentially crashing the program, the error is caught and passed to the catch block.

Within the try block, we see an await expression along with Promise.all(). Promise.all() is a method that takes an array of promises and returns a new promise that only resolves when all the promises in the array have been resolved. If any promise in the array is rejected, the promise returned by Promise.all() is immediately rejected with the reason of the first promise that was rejected.

In this case, Promise.all() is used to send two fetch requests simultaneously. Fetch is an API for making network requests similar to XMLHttpRequest. Fetch returns a promise that resolves to the Response object representing the response to the request.

The await keyword is used to pause the execution of the function until Promise.all() has resolved. This means the function will not proceed until both fetch requests have completed. The responses from the fetch requests are then destructured into userData and postsData.

Next, we see two more await expressions - await userData.json() and await postsData.json(). These are used to parse the response body to JSON. The json() method also returns a promise, so we need to use await to pause the function until this promise is resolved. The resulting data is then logged to the console.

In the catch block, if any error occurred during the fetch requests or while converting the response body to JSON, it is caught and logged to the console with console.error('Error fetching resources:', error).

Finally, the fetchResources() function is called to execute the data fetching operation. This function demonstrates how async/await is used with Promise.all() to handle multiple simultaneous fetch requests in JavaScript.

This pattern can significantly improve performance when dealing with multiple independent asynchronous operations, as it allows them to be initiated simultaneously rather than sequentially.

5.3.5 Detailed Practical Considerations

- **Performance and Efficiency**: One of the key benefits of the async/await syntax is that it simplifies the process of writing asynchronous code. However, it's important to be mindful of how and where you use the await keyword. Despite its convenience, unnecessary or improper use of await can lead to performance bottlenecks. This potential issue highlights the importance of understanding the underlying principles and mechanics of asynchronous programming and the await keyword.
- **Debugging and Error Handling**: Debugging asynchronous code written with async/await can be more intuitive compared to code written using promises. This is primarily due to the fact that error stack traces in async/await are generally clearer and provide more informative data. This enhanced clarity can significantly improve the debugging process and expedite the identification and resolution of bugs or issues within the code.
- **Usage in Loop Constructs**: Special attention must be given when using await within loop constructs. Asynchronous operations inside a loop are inherently sequential and should be managed correctly. Mismanagement or incorrect usage can lead to performance issues, making the code less efficient. It's important to understand the intricacies of using async/await within loops and ensure that the asynchronous operations are handled optimally to prevent unnecessary performance degradation.

5.3.6 Combining Async/Await with Other Asynchronous Patterns

The async/await syntax in JavaScript is a powerful tool that can be utilized in an elegant manner to handle asynchronous code, thereby improving readability and maintainability. This feature allows us to write asynchronous code as if it were synchronous.

This can significantly simplify the logic behind handling promises or callbacks, making your code easier to understand. In addition to this, async/await can be seamlessly integrated with other JavaScript features such as generators or event-driven code.

When used together, they can effectively tackle complex problems and make the development process much more efficient and enjoyable. It's a combination that is potent and can help in building robust, efficient, and scalable applications.

Example: Using Async/Await with Generators

```
async function* asyncGenerator() {
    const data = await fetchData();
    yield data;
    const moreData = await fetchMoreData(data);
```

```
    yield moreData;
}

async function consume() {
    for await (const value of asyncGenerator()) {
        console.log(value);
    }
}

consume();
```

This example demonstrates how async/await can be used with asynchronous generators to handle streaming data or progressive loading scenarios.

The function asyncGenerator() is an asynchronous generator function, which is a special kind of function that can yield multiple values over time. It first fetches data asynchronously using fetchData(), yields the fetched data, then fetches more data asynchronously using fetchMoreData(data) and yields the additional fetched data.

The function consume() is an asynchronous function that iterates over the values yielded by the asyncGenerator() using a for-await-of loop. This loop waits for each promise to resolve before moving on to the next iteration. It logs each yielded value to the console.

Finally, the consume() function is called to start the process.

5.3.7 Best Practices for Code Structure

Avoiding Await in Loops

It is important to note that directly incorporating await within loops can result in a decrease in performance. This is because each loop iteration is forced to wait until the one before it has fully completed. To circumvent this potential issue, a useful strategy is to collect all of the promises that are generated by the loop.

Once these promises have been collected, the Promise.all function can be used to await all of them in a concurrent manner, rather than sequentially. This optimizes the code by allowing multiple operations to run simultaneously, thereby improving the overall speed and efficiency of the code.

Top-Level Await

For those utilizing modules in their code, the use of top-level await can be an effective way to simplify the initialization of asynchronous modules. However, it is imperative to use this feature judiciously.

Overuse or inappropriate use of top-level await can result in the blocking of the module graph, which can lead to performance issues. Proper use of this feature can simplify your code and make asynchronous operations easier to handle, but it is always important to consider the potential implications on the rest of your module graph.

Example: Optimizing Await in Loops

```
async function processItems(items) {
    const promises = items.map(async item => {
        const processedItem = await processItem(item);
        return processedItem;
    });
    return Promise.all(promises);
}

async function processItem(item) {
    // processing logic
}
```

This code defines two asynchronous functions. The first one, named processItems, takes an array of items as an argument. It creates an array of promises by mapping each item to an asynchronous operation. This operation involves calling the second function, processItem, which also takes an item as an argument and processes it.

Once all the promises are resolved, processItems returns an array of processed items. The second function, processItem, is where the logic for processing an individual item would be written. This code is written using JavaScript's async/await syntax, which allows for writing asynchronous code in a more synchronous, readable manner.

5.4 Closures

Closures represent a fundamental and extraordinarily powerful feature of JavaScript, one which can empower developers to write more complex and efficient code. They allow functions to remember and maintain access to variables from an outer function, even after the outer function has completed its execution. This feature is not just a byproduct of the language, but rather an intentional and integral part of JavaScript's design.

This section delves deep into the concept of closures, aiming to demystify their workings and to provide a comprehensive understanding of their functionality. We will explore how they operate, the mechanics behind their implementation, and the scope of variables within them. Additionally, we will examine their practical applications in real-world coding scenarios.

By mastering closures, you can significantly enhance your ability to write efficient and modular JavaScript code. You can utilize closures to control access to variables, thus promoting encapsulation and modularity, crucial principles in software development. Understanding closures could potentially open up new avenues for your JavaScript development and propel your code to the next level.

5.4.1 Understanding Closures

A closure is a concept that is characterized by the declaration of a function within another function. This structure inherently allows the inner function to have access to the variables of its outer function. The ability to do so is not just a random occurrence, but rather a feature that is pivotal for achieving certain goals in programming.

Specifically, this capability plays a significant role in the creation of private variables. By leveraging closures, we can create variables that are only visible and accessible within the scope of the function in which they are declared, thereby effectively creating a shield against unwanted access from the outside.

Furthermore, closures also play an indispensable role in encapsulating functionality in JavaScript. By using closures, we can group related functionality together and make it a self-contained, reusable unit, thus promoting modularity and maintainability in our code.

Basic Example of a Closure

```javascript
function outerFunction() {
    let count = 0;  // A count variable that is local to the outer function

    function innerFunction() {
        count++;
        console.log('Current count is:', count);
    }

    return innerFunction;
}

const myCounter = outerFunction(); // myCounter is now a reference to innerFunction
myCounter(); // Outputs: Current count is: 1
myCounter(); // Outputs: Current count is: 2
```

In this example, innerFunction is a closure that retains access to the count variable of outerFunction even after outerFunction has executed. Each call to myCounter increments and logs the current count, demonstrating how closures can maintain state.

The outerFunction declares a local variable count and defines an innerFunction that increments count each time it's called and logs its current value.

When outerFunction is called, it returns a reference to innerFunction. In this case, myCounter holds that reference.

When myCounter (which is effectively innerFunction) is called, it continues to have access to count from its parent scope (the outerFunction), even after outerFunction has finished executing.

So, when you call myCounter() multiple times, it increments count and logs its value, preserving the changes to count across invocations because of closure.

5.4.2 Practical Applications of Closures

In the realm of programming, closures are not merely theoretical constructs or concepts that are confined to the academic sphere. In fact, they have practical, everyday applications in coding tasks, serving as an essential tool in the toolkit of any proficient programmer.

1. Data Encapsulation and Privacy

The first, and arguably most important, practical use of closures is in the realm of **Data Encapsulation and Privacy**.

In programming, the concept of encapsulation refers to the bundling of related data and methods into a single unit, while hiding the specifics of the class's implementation from the user. This is where closures come into play.

They provide a method to create private variables. This can be of paramount importance when it comes to hiding intricate implementation details. Moreover, closures help in preserving state securely, which is a crucial aspect of any application that deals with sensitive or confidential data. In essence, closures play an indispensable role in maintaining the integrity and security of an application.

Example: Encapsulating Data

```
function createBankAccount(initialBalance) {
    let balance = initialBalance; // balance is private

    return {
```

```
        deposit: function(amount) {
            balance += amount;
            console.log(`Deposit ${amount}, new balance: ${balance}`);
        },
        withdraw: function(amount) {
            if (amount > balance) {
                console.log('Insufficient funds');
                return;
            }
            balance -= amount;
            console.log(`Withdraw ${amount}, new balance: ${balance}`);
        }
    };
}

const account = createBankAccount(100);
account.deposit(50);    // Outputs: Deposit 50, new balance: 150
account.withdraw(20);   // Outputs: Withdraw 20, new balance: 130
```

This is a code snippet that defines a function createBankAccount. This function takes an initialBalance as an argument and creates a bank account with a private variable balance. The function returns an object with two methods: deposit and withdraw.

The deposit method takes an amount as an argument, adds it to the balance, and prints out the new balance. The withdraw method also takes an amount as an argument, checks if the amount is greater than the balance (in which case it prints 'Insufficient funds' and returns early), otherwise it deducts the amount from the balance and prints out the new balance.

Finally, the code creates a new account with an initial balance of 100, deposits 50 into it, and then withdraws 20 from it.

2. Creating Function Factories

Closures represent a powerful concept in programming which permit the creation of function factories. These factories, in turn, have the ability to generate new, distinct functions based on the unique arguments passed to the factory.

This allows for increased modularity and customization in code, making closures an invaluable tool in the toolbox of any skilled programmer.

Example: Function Factory

```
function makeMultiplier(x) {
```

```
      return function(y) {
          return x * y;
      };
}

const double = makeMultiplier(2);
const triple = makeMultiplier(3);

console.log(double(4));   // Outputs: 8
console.log(triple(4));   // Outputs: 12
```

This example code snippet demonstrates the concept of closures and function factories. A closure is a function that has access to its own scope, the scope of the outer function, and the global scope. A function factory is a function that returns another function.

The function makeMultiplier is a function factory. It accepts a single argument x and returns a new function. This returned function is a closure because it has access to its own scope and the scope of makeMultiplier.

The returned function takes a single argument y and returns the result of multiplying x by y. This works because x is available in the scope of the returned function due to the closure.

The makeMultiplier function is used to create two new functions double and triple which are stored in constants. This is done by calling makeMultiplier with arguments 2 and 3 respectively.

The double function is a closure that multiplies its input by 2, and triple multiplies its input by 3. This is because they have 'remembered' the x value that was passed to makeMultiplier when they were created.

The console.log statements at the end of the code are examples of how to use these new functions. double(4) executes the double function with the argument 4, and because double multiplies its input by 2, it returns 8. Similarly, triple(4) returns 12.

This is a powerful pattern which allows you to create specialized versions of a function without having to manually rewrite or copy the function. It can make code more modular, easier to understand, and reduce redundancy.

3. Managing Event Handlers

Closures play a particularly pivotal role when it comes to event handling. They enable programmers to attach specific data to an event handler effectively, thereby allowing for a more controlled use of that data.

What makes closures so beneficial in these scenarios is that they provide a way to associate this data with an event handler without the need to expose the data globally. This leads to a much more contained and safer utilization of data, ensuring that it is only accessible where it is needed, and not available for potential misuse elsewhere in the code.

Example: Event Handlers with Closures

```javascript
function setupHandler(element, text) {
    element.addEventListener('click', function() {
        console.log(text);
    });
}

const button = document.createElement('button');
document.body.appendChild(button);
setupHandler(button, 'Button clicked!');
```

The example code snippet illustrates how to handle click events on an HTML element using the Event Listener.

The code begins by declaring a function named setupHandler. This function accepts two parameters: element and text.

The element parameter represents an HTML element to which the event listener will be attached. The text parameter represents a string that will be logged to the console when the event is triggered.

Within the setupHandler function, an event listener is added to the element with the addEventListener method. This method takes in two arguments: the type of event to listen for and a function to execute when the event occurs. Here, the event type is 'click', and the function to execute is an anonymous function that logs the text parameter to the console.

Next, a new button element is created with document.createElement('button'). This method creates an HTML element specified by the argument, in this case, a button.

The newly created button is then appended to the body of the document using document.body.appendChild(button). The appendChild method adds a node to the end of the list of children of a specified parent node. In this case, the button is added as the last child node of the body of the document.

Finally, the setupHandler function is invoked with the button and a string 'Button clicked!' as arguments. This attaches a click event listener to the button. Now, whenever the button is clicked, the text 'Button clicked!' will be logged to the console.

This code snippet is a simple demonstration of how to interact with HTML elements using JavaScript, specifically how to create elements, append them to the document, and attach event listeners to them.

5.4.3 Understanding Memory Implications

Closures are indeed powerful tools in the world of programming, however, they also come with significant memory implications. This is mainly because closures, by their very design, retain references to the variables of the outer function in which they are defined. Because of this inherent characteristic, it becomes extremely important to manage them carefully in order to avoid the pitfalls of memory leaks.

Best Practices for Closures: A Comprehensive Guide

- One of the key strategies to manage closures is to minimize their usage, especially in large-scale applications where numerous functions are being created. This is primarily due to the fact that each closure you create retains a unique link to its outer scope. This can eventually lead to memory bloat if not properly managed, hence the need for restraint in their usage.
- Another crucial point to consider when working with closures is related to event listeners. Often, closures are used when setting up these event listeners. It's vital, therefore, to ensure that you also have a mechanism in place to remove these listeners when they have served their purpose. This is because if these listeners are not removed, they can continue to occupy memory space even when they are no longer needed, leading to unnecessary memory usage. Hence, it's important to free up that memory to ensure efficient performance of your application.

Closures are a versatile and essential feature of JavaScript, providing powerful ways to manipulate data and functions with increased flexibility and privacy. By understanding and utilizing closures effectively, you can build more robust, secure, and maintainable JavaScript applications. Whether it's through creating private data, function factories, or managing event handlers, closures offer a range of practical benefits that can enhance any developer's toolkit.

5.4.4 Memoization with Closures

Memoization is a highly efficient optimization technique utilized in computer programming. It revolves around the concept of storing the results of complex and time-consuming function calls. This way, when these function calls are made again with the same inputs, the program does not have to perform the same calculations all over again.

Instead, the pre-stored, or cached, result is returned, thereby saving significant computational time and resources. An intriguing aspect of this technique is that it can be effectively implemented using closures.

Closures, a fundamental concept in many programming languages, allow functions to have access to variables from an outer function that has already completed its execution. This ability makes closures particularly suitable for implementing memoization, as they can store and access previously computed results efficiently.

Example: Memoization with Closures

```
function memoize(fn) {
    const cache = {};
    return function (...args) {
        const key = JSON.stringify(args);
        if (!cache[key]) {
            cache[key] = fn.apply(this, args);
        }
        return cache[key];
    };
}

const fib = memoize(n => n <= 1 ? n : fib(n - 1) + fib(n - 2));
console.log(fib(10));   // Outputs: 55
```

In this example, a memoize function is created which uses a closure to store the results of function calls. This is particularly useful for recursive functions like calculating Fibonacci numbers.

This example demonstrates the concept of memoization. The function "memoize" takes in a function "fn" as an argument and uses an object "cache" to store the results of the function calls. It returns a new function that checks if the result for a certain argument is already in the cache. If it is, it returns the cached result, otherwise, it calls "fn" with the arguments and stores the result in the cache before returning it.

The code then defines a memoized version of a function to compute Fibonacci numbers, called "fib". The Fibonacci function is defined recursively: if the input "n" is 0 or 1, it returns "n"; otherwise, it returns the sum of the two previous Fibonacci numbers.

The function call "fib(10)" computes the 10th Fibonacci number and logs it to the console, which is 55.

5.4.5 Closures in Event Delegation

Closures, a powerful concept in programming, can be particularly useful in the context of event delegation. Event delegation is a process where instead of assigning separate event listeners to every single child element, you assign a single, unified event listener to the parent element.

This parent element then manages events from its children, making the code more efficient. The advantage of using closures in this scenario is that they provide an excellent way to associate specific data or actions with a particular event or element.

This is often achieved by enclosing the data or actions within a closure, hence the name. Therefore, through the use of closures in such a context, one can manage multiple events efficiently and effectively.

Example: Using Closures for Event Delegation

```javascript
document.getElementById('menu').addEventListener('click', function(event) {
    if (event.target.tagName === 'LI') {
        handleMenuClick(event.target.id);  // Using closure to access specific element id
    }
});

function handleMenuClick(itemId) {
    console.log('Menu item clicked:', itemId);
}
```

This setup reduces the number of event listeners in the document and leverages closures to handle specific actions based on the event target, enhancing performance and maintainability.

The first line of the script selects an HTML element with the id 'menu' using the document.getElementById method. This method returns the first element in the document with the specified id. In this case, it's assumed that 'menu' is a container element that holds a list of 'LI' elements (usually used to represent menu items in a navigation bar or a dropdown menu).

An event listener is then attached to this 'menu' element using the addEventListener method. This method takes two arguments: the type of the event to listen for ('click' in this case), and a function to be executed whenever the event occurs.

The function that's set to execute on click is an anonymous function that receives an event parameter. This event object contains a lot of information about the event, including the specific element that triggered the event which can be accessed via event.target.

Inside this function, there is a condition that checks if the clicked element is a 'LI' element using event.target.tagName. If the clicked element is a 'LI', it calls another function named 'handleMenuClick' and passes the id of the clicked 'LI' element as an argument (event.target.id).

Here, the power of closures comes into play. The anonymous function creates a closure that encapsulates the specific 'LI' element id (event.target.id) and passes it to the 'handleMenuClick' function. This allows the 'handleMenuClick' function to handle the click event for a specific 'LI' element, even though the event listener was attached to the parent 'menu' element. This is an example of event delegation, which is a more efficient approach to event handling especially when dealing with a large number of similar elements.

The 'handleMenuClick' function takes an 'itemId' parameter (which is the id of the clicked 'LI' element) and logs a message along with this id to the console. This function essentially acts as an event handler for click events on 'LI' elements within the 'menu' element.

To summarize, this code attaches a click event listener to a parent 'menu' element, uses a closure to capture the id of a specific clicked 'LI' element, and passes it to another function that handles the click event. This approach reduces the number of event listeners in the document and leverages the power of closures to handle specific actions based on the event target, enhancing both performance and maintainability of the code.

5.4.6 Using Closures for State Encapsulation in Modules

Closures are a remarkable and powerful feature in JavaScript. They are particularly excellent for creating and maintaining a private state within modules or similar constructs. This ability to keep state private is a fundamental aspect of the module pattern in JavaScript.

The module pattern allows for public and private access levels. Closures provide a way to create functions with private variables. They help to encapsulate and protect variables from going global, reducing the chances of naming clashes.

This closure mechanism, in essence, provides an excellent way to achieve data privacy and encapsulation, which are key principles in object-oriented programming.

Example: Module Pattern Using Closures

```
const counterModule = (function() {
    let count = 0;  // Private state
    return {
        increment() {
            count++;
            console.log(count);
        },
        decrement() {
            count--;
            console.log(count);
        }
    };
})();

counterModule.increment();  // Outputs: 1
counterModule.decrement();  // Outputs: 0
```

This pattern uses an immediately invoked function expression (IIFE) to create private state (count) that cannot be accessed directly from outside the module, only through the exposed methods.

This is an example code snippet that utilizes a well-known design pattern called the Module Pattern. In this pattern, an Immediately Invoked Function Expression (IIFE) is used to create a private scope, effectively creating a private state that can only be accessed and manipulated through the module's public API.

In the code, the module is named 'counterModule'. The IIFE creates a private variable called 'count', initialized to 0. This variable is not accessible directly from outside the function due to JavaScript's scoping rules.

However, the IIFE returns an object that exposes two methods to the outer scope: 'increment' and 'decrement'. These methods provide the only way to interact with the 'count' variable from outside the function.

The 'increment' method, when invoked, increases the value of 'count' by one and then logs the updated count to the console. On the other hand, the 'decrement' method decreases the value of 'count' by one and then logs the updated count to the console.

The 'counterModule' is immediately invoked due to the parentheses at the end of the function declaration. This results in the creation of the 'count' variable and the return of the object with the 'increment' and 'decrement' methods. The returned object is assigned to the 'counterModule' variable.

The 'counterModule.increment()' and 'counterModule.decrement()' lines demonstrate how to use the public API of the 'counterModule'. When 'increment' is called, the count is increased by 1 and the updated count (1) is logged to the console. When 'decrement' is subsequently called, the count is decreased by 1, bringing it back to 0, and the updated count (0) is logged to the console.

This pattern is powerful as it enables encapsulation, one of the key principles of object-oriented programming. It allows the creation of public methods that can access private variables, thereby controlling the way these variables are accessed and modified. It also prevents these variables from cluttering the global scope, thus reducing the chance of variable naming collisions.

5.4.7 Best Practices for Using Closures

- **Avoid Unnecessary Closures**: Closures are indeed powerful tools in the realm of programming, but their misuse can lead to an undesirable increase in memory usage. They should be used with caution, especially in contexts where they are created within loops or inside functions that are called frequently. It is crucial to evaluate the necessity of creating a closure in each instance.
- **Debugging Closures**: One of the challenges of working with closures is that they can be difficult to debug due to their inherent capability to encapsulate external scope. To overcome this hurdle, it is beneficial to use advanced debug tools that allow for the inspection of closures. These tools can provide a comprehensive understanding of the scope and closures present in your application's stack traces.
- **Memory Leaks**: When using closures, it is essential to be vigilant of potential memory leaks. These are particularly problematic in large applications or when closures capture extensive contexts. To prevent this, it is important to manage closures effectively and release them when they are no longer needed. Doing so can free up valuable resources and ensure the smooth operation of your application.

Closures are a fundamental concept in JavaScript that provide powerful capabilities for managing privacy, state, and functional behavior in your applications. By understanding how to use closures effectively, you can write cleaner, more efficient, and more secure JavaScript code. Whether you are implementing memoization, managing event handlers, or creating module patterns, closures offer a versatile set of tools for enhancing your programming projects.

Practical Exercises

To reinforce the concepts discussed in this chapter on advanced functions, here are several practical exercises. These exercises are designed to test your understanding of arrow functions, callbacks and promises, async/await, and closures. Each exercise includes a solution to help you verify your implementation.

Exercise 1: Convert to Arrow Functions

Convert the following traditional function expressions into arrow functions.

Traditional Function Expressions:

```javascript
function add(x, y) {
    return x + y;
}

function filterNumbers(arr) {
    return arr.filter(function(item) {
        return item > 5;
    });
}
```

Solution:

```javascript
const add = (x, y) => x + y;

const filterNumbers = arr => arr.filter(item => item > 5);
```

Exercise 2: Implement a Simple Promise

Create a function multiply that returns a promise which resolves with the product of two numbers passed as arguments.

Solution:

```javascript
function multiply(x, y) {
    return new Promise((resolve, reject) => {
        if (typeof x !== 'number' || typeof y !== 'number') {
            reject(new Error("Invalid input"));
        } else {
```

```
            resolve(x * y);
        }
    });
}

multiply(5,      2).then(result     =>     console.log(result)).catch(error     =>
console.error(error));
```

Exercise 3: Using Async/Await

Write an async function that uses the multiply function from Exercise 2 to find the product of two numbers, then logs the result. Include error handling.

Solution:

```
async function calculateProduct(x, y) {
    try {
        const result = await multiply(x, y);
        console.log('Product:', result);
    } catch (error) {
        console.error('Error:', error.message);
    }
}

calculateProduct(10, 5); // Outputs: Product: 50
```

Exercise 4: Create a Closure

Create a closure that maintains a private counter variable and exposes methods to increase and decrease the counter.

Solution:

```
function createCounter() {
    let counter = 0;

    return {
        increment() {
            counter++;
            console.log('Counter:', counter);
        },
        decrement() {
            counter--;
            console.log('Counter:', counter);
        }
    };
}

const myCounter = createCounter();
myCounter.increment(); // Counter: 1
```

```
myCounter.increment(); // Counter: 2
myCounter.decrement(); // Counter: 1
```

Exercise 5: Memoization with Closures

Implement a memoization function that caches the results of a function based on its parameters to optimize performance.

Solution:

```javascript
function memoize(fn) {
    const cache = {};
    return function(...args) {
        const key = JSON.stringify(args);
        if (!cache[key]) {
            cache[key] = fn.apply(this, args);
        }
        return cache[key];
    };
}

const factorial = memoize(function(x) {
    if (x === 0) {
        return 1;
    } else {
        return x * factorial(x - 1);
    }
});

console.log(factorial(5));  // Outputs: 120
console.log(factorial(5));  // Outputs: 120 (from cache)
```

These exercises provide hands-on practice with the key concepts from this chapter, helping you solidify your understanding of advanced JavaScript functions and their applications in real-world scenarios.

Chapter Summary

In Chapter 5 of "JavaScript from Scratch: Unlock your Web Development Superpowers," we delved deeply into advanced function concepts that are pivotal for mastering JavaScript and building sophisticated applications. This chapter covered a range of topics including arrow functions, callbacks, promises, async/await, and closures, each essential for effective asynchronous programming and functional JavaScript development.

Arrow Functions

We began by exploring arrow functions, a concise syntax introduced in ES6, which simplifies writing smaller function expressions. Arrow functions not only reduce syntactic clutter but also handle this differently than traditional functions.

They inherit this from the surrounding context, making them ideal for scenarios where function scope can become an issue, such as in callbacks for timers, event handlers, or array methods. The adoption of arrow functions can lead to cleaner, more readable code, especially in functional programming patterns or when used in array transformations.

Callbacks and Promises

Next, we discussed callbacks, which are fundamental to JavaScript's asynchronous nature. Despite their wide use, callbacks can lead to complex nested structures, often referred to as "callback hell."

To address these challenges, we examined promises, which provide a more robust way to handle asynchronous operations. Promises represent a value that may not be known when the promise is created, but promises streamline asynchronous logic by providing a clearer and more flexible way to handle future outcomes. They allow developers to chain operations and handle asynchronous results or errors more gracefully with .then(), .catch(), and .finally() methods.

Async/Await

Building on promises, async/await syntax was introduced as a revolutionary feature that simplifies working with promises even further, allowing asynchronous code to be written with a synchronous style. This syntactic sugar makes it easier to read and debug complex chains of promises and is particularly powerful in handling sequential asynchronous operations. The use of async/await enhances code clarity and error handling, making asynchronous code less cumbersome and more intuitive.

Closures

We also covered closures, a powerful feature of JavaScript where a function has access to its own scope, the scope of the outer function, and global variables. Closures are crucial for data privacy and encapsulation, allowing developers to create private variables and methods. We explored practical applications of closures in creating function factories, memoizing expensive operations, and managing state in event handlers or with modular code.

Reflection

This chapter not only enhanced your understanding of JavaScript functions but also equipped you with essential tools to tackle complex programming challenges. These concepts are not just theoretical; they have practical implications in everyday coding tasks, from handling user interactions and managing state to performing network requests and processing data asynchronously.

Looking Ahead

As we move forward, the skills acquired in this chapter will serve as a foundation for more advanced topics in JavaScript and web development. Understanding these advanced function techniques is pivotal as they form the backbone of modern JavaScript frameworks and libraries. The ability to effectively use these patterns will open up numerous possibilities for creating more efficient, effective, and robust web applications.

By mastering these advanced functions, you are now better prepared to write clean, efficient, and maintainable code, tackle more complex problems, and ultimately become a more proficient JavaScript developer.

Chapter 6: Object-Oriented JavaScript

Welcome to the comprehensive exploration of Chapter 6, titled "Object-Oriented JavaScript". In this enlightening chapter, we are going to delve into the fascinating world of object-oriented programming (OOP) in relation to JavaScript. This chapter has been meticulously crafted to deepen and broaden your understanding of how JavaScript, a language that stands out from traditional class-based languages, handles and deals with object-oriented concepts.

In order to thoroughly understand these concepts, we will dive into the intriguing aspects of object constructors, an integral part of JavaScript. Moreover, we will explore the concept of prototypes, and the class syntax that was introduced in the significant update of ES6. We won't stop there; we will also learn about inheritance, a powerful tool in object-oriented programming, and various design patterns that elegantly leverage these features to create efficient and effective code.

Object-oriented programming in JavaScript is not merely a programming style; it is a powerful tool that can significantly enhance the modularity, reusability, and maintainability of your code. It provides a structured approach that is immensely beneficial when dealing with complex systems that require careful management of numerous moving parts.

The understanding and application of these concepts are not just important, but crucial for building scalable, efficient, and powerful web applications. The knowledge gained in this chapter will be your stepping stone towards mastering complex systems and creating robust web applications.

6.1 Object Constructors and Prototypes

JavaScript is a distinctive programming language that is characterized by its prototype-based structure. This unique approach sets it apart from other languages such as Java or C#, which predominantly use classical classes.

The prototype-based nature of JavaScript means that it relies on constructors and prototypes to offer object-oriented functionality, as opposed to the more traditional class-based object orientation. This section of our discussion will delve into a comprehensive explanation of how to create object constructors within the JavaScript language.

Furthermore, we will explore the way in which prototypes are used to extend the properties and methods of objects, thereby enhancing functionality and flexibility. This understanding will provide a solid foundation for effectively using and navigating the dynamic world of JavaScript programming.

6.1.1 Object Constructors

In JavaScript, the role of constructors is unparalleled and exceedingly significant. Though they might appear to be merely functions in their raw form, the vital purpose they serve sets them distinctively apart from the rest of the elements. Constructors are specifically utilized for the creation and initialization of instances of objects, thereby playing an absolutely crucial role in the realm of object-oriented programming.

One of the key conventions in JavaScript is to commence the name of these constructor functions with a capital letter. This particular naming convention is not just a tradition followed in the programming world, but it serves a practical purpose. It is a highly useful way to clearly distinguish these special functions from other types of common functions present in the code.

As a result, this practice greatly enhances the readability of the code, making it significantly easier for programmers to read, understand, and debug if necessary. This ultimately leads to more efficient and effective programming, saving valuable time and effort.

Example: Creating a Constructor Function

```
function Car(make, model, year) {
    this.make = make;
    this.model = model;
    this.year = year;
}

const myCar = new Car('Toyota', 'Corolla', 1997);
console.log(myCar.model);  // Outputs: 'Corolla'
```

In this example, Car is a constructor function that initializes a new object with properties make, model, and year. The new keyword is used to create an instance of Car, resulting in a new object that myCar references.

This code defines a constructor function called "Car". This function is used to create new objects with the properties 'make', 'model', and 'year'. Then, a new object 'myCar' is created using the "Car" function with 'Toyota', 'Corolla', and 1997 as arguments. Finally, the model of 'myCar' is logged out, which results in 'Corolla'.

6.1.2 Prototypes

In the realm of JavaScript, every object therein has a prototype, which is, in itself, an object as well. The crucial concept to understand here is that every single JavaScript object inherits its properties and methods from this prototype. This inheritance from the prototype is a fundamental characteristic of JavaScript objects.

The prototype of the constructor function plays a vital role in this inheritance process. By modifying or altering the prototype of this constructor function, an impactful change occurs: all the instances that have been, or will be, created from this constructor function will gain access to these modified properties and methods.

This means that the changes to the prototype have a cascading effect, impacting all instances derived from the constructor function. This highlights the powerful influence of the prototype in JavaScript object creation and function.

Example: Extending Constructors with Prototypes

```
Car.prototype.getAge = function() {
    return new Date().getFullYear() - this.year;
};

console.log(myCar.getAge());  // Calculates the age of 'myCar' based on the current
year
```

By adding the getAge method to Car's prototype, every instance of Car now has access to this method. This is a powerful feature of JavaScript's prototype-based inheritance, allowing for efficient memory management and sharing of methods across all instances.

The Car.prototype.getAge declaration is an addition of a method to the 'Car' constructor's prototype. Prototypes in JavaScript are a mechanism that allows objects to inherit features from other objects. Adding methods and properties to an object's prototype is an efficient way to conserve memory resources and keep the code DRY (Don't Repeat Yourself).

In this case, the getAge method is added to the Car prototype, which means this method will now be accessible by all instances of Car. The getAge method calculates the age of a car by subtracting the car's manufacturing year (stored in this.year) from the current year. new Date().getFullYear() gets the current year.

Finally, console.log(myCar.getAge()) prints the result of this method when called on the myCar object to the console. This line is calculating the age of myCar by calling the getAge method we added to the Car prototype and then logging that result to the console.

This is a demonstration of a powerful feature of JavaScript's prototype-based inheritance, which allows for efficient memory management and the sharing of methods across all instances of an object.

Why Use Prototypes?

The utilization of prototypes comes with a myriad of advantages:

Memory Efficiency

In traditional object-oriented programming, each instance of an object would store its own unique copy of functions, which could lead to considerable memory usage. However, when using prototypes, all instances of an object share the same set of functions through a common prototype.

This means that the functions only need to be stored once, instead of once per instance. As a result, memory usage can be significantly reduced, thereby enhancing the performance and speed of your code.

Dynamic Updates

Another profound benefit of using prototypes is their ability to facilitate dynamic updates. In a scenario where a method is added to a prototype after instances have already been created, all instances will still be able to access that newly added method. This is due to the fact that they all share the same prototype.

This feature provides unprecedented flexibility in how objects are extended and modified. It allows for dynamic changes to the functionality of all instances of an object, without the need to manually update each instance individually. This can be particularly beneficial in large-scale software projects where changes may need to be made frequently or on-the-fly.

Understanding object constructors and prototypes is fundamental to leveraging JavaScript's capabilities in an object-oriented manner. These features provide powerful tools for developers to build more structured and efficient applications.

6.1.3 Customizing Constructors

While the basic constructor pattern proves to be quite powerful for defining objects in JavaScript, the language provides flexibility for defining more sophisticated behaviors within these constructors.

This is achieved through the use of closures, which allow for the encapsulation of functionality, thereby enabling the creation of private variables and methods. This adds an extra layer of security and control to our objects, as private variables and methods cannot be accessed directly from outside the object.

Instead, they can only be accessed through public methods, providing a more robust and secure approach to object-oriented programming in JavaScript.

Example: Encapsulating Private Data in Constructors

```javascript
function Bicycle(model, color) {
    let speed = 0;  // Private variable

    this.model = model;
    this.color = color;

    this.accelerate = function(amount) {
        speed += amount;
        console.log(`Accelerated to ${speed} mph`);
    };

    this.getSpeed = function() {
        return speed;
    };
}

const myBike = new Bicycle('Trek', 'blue');
myBike.accelerate(15);
console.log(myBike.getSpeed());  // Outputs: 15
console.log(myBike.speed);       // Outputs: undefined (private)
```

In this example, the speed variable is private to the Bicycle instance. This pattern leverages closures to keep speed accessible only through the methods defined in the constructor, ensuring encapsulation and protection of the internal state.

This example code is a demonstration of how constructors can be used in object-oriented programming (OOP) in JavaScript. It defines a constructor function named 'Bicycle'.

A constructor function is a special type of function that is used to initialize new objects. In this case, the 'Bicycle' constructor function is used to create new 'Bicycle' objects. The constructor takes two parameters: 'model' and 'color', which represent the model and color of the bicycle respectively.

Inside the constructor, a variable 'speed' is declared with an initial value of 0. This variable is local to the constructor and hence, acts as a private variable to each 'Bicycle' instance. This means 'speed' is not directly accessible from the outside of the object and can only be manipulated through the object's methods.

The constructor also defines two methods: 'accelerate' and 'getSpeed'. The 'accelerate' method takes an amount as a parameter and adds it to the 'speed' variable, effectively increasing the speed of the bicycle. It also logs a message to the console indicating the new speed. The 'getSpeed' method, on the other hand, is a simple getter function that returns the current speed of the bicycle.

The code then creates a new 'Bicycle' object named 'myBike' with the model 'Trek' and color 'blue'. The 'accelerate' method is called on 'myBike' with an argument of 15, increasing the speed of 'myBike' to 15. The current speed of 'myBike' is then logged to the console by calling the 'getSpeed' method, which outputs 15.

Interestingly, when the code tries to log 'myBike.speed' directly, it outputs 'undefined'. This is because 'speed' is a private variable and cannot be accessed directly from outside the object. This encapsulation of 'speed' is a fundamental aspect of object-oriented programming, providing a way to safeguard data from being manipulated directly.

6.1.4 Prototypical Inheritance

Prototypes in programming have a fascinating feature that makes them particularly powerful: the ability to create inheritance chains. This essentially means that an object can inherit properties and methods from another object.

In a broader context, this feature is what enables the principle of object-oriented programming, where objects that share common characteristics can inherit from each other, making the code more efficient and reusable.

This can drastically reduce the amount of code required and make the codebase easier to maintain, enhancing the overall process of software development.

Example: Inheriting from a Prototype

```
function Vehicle(type) {
    this.type = type;
}

Vehicle.prototype.drive = function() {
    console.log(`Driving a ${this.type}`);
};

function Car(make, model) {
    Vehicle.call(this, 'car');  // Call the parent constructor with 'car' as type
    this.make = make;
    this.model = model;
}

Car.prototype = Object.create(Vehicle.prototype);  // Inherit from Vehicle
Car.prototype.constructor = Car;  // Set the constructor property to Car

Car.prototype.display = function() {
    console.log(`${this.make} ${this.model}`);
};

const myCar = new Car('Toyota', 'Corolla');
myCar.drive();  // Outputs: Driving a car
myCar.display();  // Outputs: Toyota Corolla
```

This example demonstrates how Car can inherit the drive method from Vehicle through prototype chaining, while also defining its specific properties and methods.

The example code is a demonstration of object-oriented programming (OOP) principles in JavaScript, more specifically constructor functions and prototype-based inheritance. Let's dissect it in detail:

1. **Defining the Vehicle constructor**: The code starts with the declaration of a function named Vehicle. In this context, Vehicle is not just a regular function, but it's a constructor function. A constructor function is a special kind of function used to

initialize new objects. This Vehicle constructor takes one parameter, type, and assigns it to the this.type property. The this keyword is a special identifier in JavaScript that inside a constructor function refers to the new object that's being created.

2. **Adding a method to the Vehicle prototype**: The next part is Vehicle.prototype.drive = function() {...}. Here, a method named drive is being added to the Vehicle's prototype. A prototype is an object from which other objects inherit properties. In JavaScript, each object has a prototype and the properties of the prototype can be accessed by all objects that are linked to it. The drive method logs a string to the console that includes the type of the vehicle.

3. **Defining the Car constructor and inheriting from Vehicle**: The Car function is another constructor that creates a Car object. It takes two parameters, make and model. Inside the constructor, Vehicle.call(this, 'car') is used to call the parent constructor (Vehicle). This is a way of implementing inheritance in JavaScript. By calling the parent constructor, Car is effectively inheriting all properties and methods from Vehicle. It also adds two of its own properties, make and model.

4. **Setting the Car prototype and constructor**: Car.prototype = Object.create(Vehicle.prototype); sets the Car's prototype to be the Vehicle's prototype, meaning that Car inherits from Vehicle. The line Car.prototype.constructor = Car; then sets the constructor property of Car.prototype back to Car, as it was overwritten in the previous line.

5. **Adding a method to the Car prototype**: Car.prototype.display = function() {...} adds a display method to Car's prototype. This method logs the make and model of the car to the console.

6. **Creating an instance of Car and calling its methods**: Finally, the code creates an instance of Car named myCar with 'Toyota' as its make and 'Corolla' as its model. After this, it calls the drive and display methods on myCar. Since Car inherits from Vehicle, myCar can access both the drive method from Vehicle and the display method from Car. The result of these method calls is "Driving a car" and "Toyota Corolla" respectively.

6.1.5 Performance Considerations

Prototypes, while tremendously powerful, need to be handled with care in the context of their impact on performance, particularly in the case of expansive applications:

- **Costs of Prototype Lookup**: The process of accessing properties that are not directly located on the object, but instead exist on the prototype chain, incurs lookup costs. This can have a detrimental effect on performance if it is a practice that is excessively utilized. This is because each lookup operation requires time and computing power, and in a large-scale application where such lookups could potentially occur numerous times, this can add up to a significant performance cost.

- **Modifying Prototypes at Runtime**: The act of modifying an object's prototype while the program is running, especially after instances of that object have already been created, can result in substantial performance penalties. This happens due to the way JavaScript engines optimize the access to objects. When the structure of an object, such as its prototype, is altered after it has been instantiated, the JavaScript engines need to re-optimize for this new structure, which can be a heavy operation and negatively impact performance.

Understanding and effectively using constructors and prototypes are crucial for applying object-oriented principles in JavaScript. These concepts not only facilitate code organization and reuse but also allow for the creation of complex inheritance structures that can mimic the capabilities found in more traditional OOP languages.

6.2 ES6 Classes

Introduced as a key feature in ECMAScript 2015, also known as ES6, the class system in JavaScript brought a revolutionary change to the language. Classes in JavaScript provide an alternative, more traditional syntax for generating object instances and handling inheritance. This additional feature was a significant departure from the prototype-based approach that was utilized in earlier versions of the language.

The prototype-based approach, while effective, was often seen as convoluted and difficult to grasp, especially for developers coming from a more classical object-oriented programming background. The introduction of classes was a breath of fresh air, bringing a familiar syntax and structure to JavaScript.

It's important to note that, despite the introduction of classes, JavaScript's underlying mechanism for creating objects and dealing with inheritance did not change. In essence, JavaScript classes are syntactical sugar over JavaScript's existing prototype-based inheritance system. This means that classes do not introduce a new object-oriented inheritance model to JavaScript but rather provide a simpler syntax to create objects and deal with inheritance.

The introduction of this feature has been widely acknowledged as a positive step in the language's evolution, as it offers a clearer, more concise syntax for creating objects and dealing with inheritance. This has the effect of making your code more clean, streamlined, and readable. It allows developers to write intuitive and well-structured code, which is especially beneficial in larger codebases and team projects where readability and maintainability are paramount.

6.2.1 Understanding ES6 Classes

Classes in JavaScript serve as a fundamental blueprint for constructing objects with specific, pre-defined characteristics and functionalities. They encapsulate, or securely contain, the data pertaining to the object, thereby ensuring that it remains unaltered and intact.

In addition to containing data, classes provide a comprehensive blueprint for creating numerous instances of the object, each of which will adhere to the structure and behavior defined in the class.

This is a crucial aspect of object-oriented programming in JavaScript as it allows for the creation of multiple objects of the same type, each with its own set of properties and methods, thereby promoting reusability and efficiency in your code.

Through the encapsulation of data and provision of a blueprint for object creation, classes help in making your object-oriented JavaScript code simpler, more intuitive, and easier to manage.

Basic Class Syntax

Let's delve into the concept of defining a class in JavaScript, a fundamental object-oriented programming concept. In JavaScript, a class is a type of function, but instead of using the keyword 'function', you'd use the keyword 'class', and the properties are assigned inside a constructor() method. Here's an example of how you can define a simple class in JavaScript:

Example: Defining a Simple Class

```javascript
class Car {
    constructor(make, model, year) {
        this.make = make;
        this.model = model;
        this.year = year;
    }

    display() {
        console.log(`This is a ${this.make} ${this.model} from ${this.year}.`);
    }
}

const myCar = new Car('Honda', 'Accord', 2021);
myCar.display();  // Outputs: This is a Honda Accord from 2021.
```

In this example, the Car class has a constructor method that initializes the new object's properties. The display method is an instance method that all instances of the class can call.

This code illustrates Object-Oriented Programming (OOP) through the use of classes, introduced in ECMAScript 6 (ES6). The code presents a simple 'Car' class, which acts as a blueprint for creating 'Car' objects.

The class is defined using the class keyword, followed by the name of the class, which in this case is 'Car'. Following the class declaration is a pair of braces {} which contain the class body.

Within the class body, a constructor method is defined. This is a special method that gets called whenever a new object is created from this class. The constructor takes three parameters: 'make', 'model', and 'year'. Within the constructor, these parameters are assigned to instance variables, denoted by this.make, this.model, and this.year. The this keyword refers to the instance of the object being created.

Following the constructor, a method named display is defined. This is an instance method, meaning it can be called on any object created from this class. The display method uses the console.log function to print a string to the console that includes the make, model, and year of the car.

After the class is defined, an instance of 'Car' is created using the new keyword followed by the name of the class and a set of parentheses containing arguments that match the parameters defined in the class constructor. In this case, a new 'Car' object named 'myCar' is created with 'Honda' as the make, 'Accord' as the model, and 2021 as the year.

Finally, the display method is called on the myCar object, which outputs: "This is a Honda Accord from 2021." to the console.

This piece of code is a simple yet effective demonstration of how classes can be used in JavaScript to create objects and define methods that can perform actions related to those objects. The use of classes makes the code more structured, organized, and easier to understand, especially when dealing with a large number of objects that share common properties and behaviors.

6.2.2 Advantages of Using Classes

Simpler Syntax for Inheritance: One of the key advantages of using extends and super is that classes can inherit from one another with ease, significantly simplifying the code required to create an inheritance hierarchy. This means less time and effort spent on writing complex lines of code, thereby increasing efficiency.

Class Definitions are Block-Scoped: Unlike function declarations, which are hoisted and can therefore be used before they are declared, class declarations are not hoisted. This makes them block-scoped, aligning more closely with other block-scoped declarations like let and const. This provides a more predictable and easier-to-understand behavior.

Method Definitions are Non-Enumerable: Another notable feature of classes is that method definitions are non-enumerable. This is a significant improvement over the function prototype pattern, where methods are enumerable by default and must be manually defined as non-enumerable if needed. This makes the code more secure and less prone to unwanted side effects.

Classes Use Strict Mode: All code written in the context of a class is executed in strict mode implicitly. This means there's no way to opt-out of it. The benefit of this is twofold: it helps in catching common coding mistakes early, and it makes the code safer and more robust. This is especially useful for those new to JavaScript, as it prevents them from making some common mistakes.

Example: Inheritance in Classes

```
class ElectricCar extends Car {
    constructor(make, model, year, batteryCapacity) {
        super(make, model, year);  // Call the parent class's constructor
        this.batteryCapacity = batteryCapacity;
    }

    charge() {
        console.log(`Charging ${this.make} ${this.model}`);
    }
}

const myElectricCar = new ElectricCar('Tesla', 'Model S', 2020, '100kWh');
myElectricCar.display();  // Outputs: This is a Tesla Model S from 2020.
myElectricCar.charge();   // Outputs: Charging Tesla Model S
```

In this example, ElectricCar extends Car, inheriting its methods and adding new functionality. The super keyword is used to call the constructor of the parent class.

The code snippet utilizes the ES6 class syntax to define a class called ElectricCar. This class extends from a parent class, denoted as Car. This is an example of inheritance in object-oriented programming, where a 'child' class (in this case, ElectricCar) inherits the properties and methods of a 'parent' class (Car).

The ElectricCar class includes a constructor method that takes four parameters: make, model, year, and batteryCapacity. These parameters represent the make and model of the car, the year of manufacture, and the capacity of the battery, respectively.

Inside the constructor, super(make, model, year) is used to call the constructor of the parent Car class with the make, model, and year parameters. The super keyword is used in class methods to refer to parent class methods. In the constructor, it's mandatory to call the super method before using this, as super is responsible for initializing this.

Additionally, the ElectricCar class defines a new property batteryCapacity and assigns it to this.batteryCapacity. The this keyword refers to the instance of the object being created.

The ElectricCar class also includes a charge method, which does not take any parameters. This method uses the console.log function to output a string to the console indicating that the make and model of the car are charging.

After the ElectricCar class is defined, an instance of this class is created with the name myElectricCar. The new keyword is used to instantiate a new object, and the arguments 'Tesla', 'Model S', 2020, and '100kWh' are passed to match the parameters required by the ElectricCar constructor.

Finally, the display and charge methods are called on the myElectricCar object. The display method comes from the parent Car class and outputs a string indicating the make, model, and year of the car. The charge method, specific to the ElectricCar class, signals that the car is charging.

This code provides an example of how classes in JavaScript can be used to create objects with specific properties and behaviors, as well as how inheritance allows for properties and methods to be shared and extended across classes. It demonstrates the principles of object-oriented programming, including encapsulation, inheritance, and polymorphism.

6.2.3 Practical Considerations

Classes in JavaScript bring a plethora of syntactical and practical advantages, but it's crucial to comprehend that they are essentially a more user-friendly veneer over JavaScript's pre-existing prototype-based inheritance system. There are a couple of key points to keep in mind:

- **Comprehending the Prototype Chain**: While classes simplify the process of working with objects, they don't replace the need to understand prototypes in JavaScript. Gaining a solid understanding of how prototypes work is fundamental for those times when things don't go as expected, or when you are required to debug complex problems that involve the creation and inheritance of objects.
- **Efficient Memory Usage**: In terms of memory usage, classes behave much like constructor functions. Methods that are defined inside a class don't get duplicated for each instance of the class. Rather, they are shared on the prototype object. This means that no matter how many instances of a class you create, the methods will only exist once in memory, leading to a more efficient use of system resources.

ES6 classes offer a more elegant and accessible way to deal with object construction and inheritance in JavaScript. By providing a familiar syntax for those coming from class-based languages, JavaScript classes help streamline the transition to and adoption of JavaScript for large-scale application development.

They allow developers to structure their code more cleanly and focus more on developing functionality rather than managing the nuances of prototype-based inheritance. As you incorporate classes into your JavaScript repertoire, they can significantly tidy up your codebase and improve maintainability.

6.2.4 Static Methods and Properties

In JavaScript, classes have a feature where they can support static methods and properties. This means these methods and properties are not called on instances of the class, but rather, they are directly called on the class itself.

This is particularly beneficial for utility functions, which are associated with the class and are an integral part of its functionality, but don't necessarily interact or operate on individual instances of the class. These utility functions can perform operations that are relevant to the class as a whole, rather than specific instances, making static methods and properties a valuable tool within JavaScript programming.

Example: Static Methods and Properties

```
class MathUtility {
    static pi = 3.14159;

    static areaOfCircle(radius) {
        return MathUtility.pi * radius * radius;
    }
}
```

```
console.log(MathUtility.areaOfCircle(10));  // Outputs: 314.159
console.log(MathUtility.pi);  // Outputs: 3.14159
```

This example shows how static methods and properties can be used to group related functionality under a class without needing to create an instance of the class.

The example defines a class named 'MathUtility'. A class is a blueprint for creating objects of the same type in Object-Oriented Programming (OOP).

In this class, there are two static elements: a property called 'pi' and a method called 'areaOfCircle'. Static elements are those that are attached to the class itself, and not to instances of the class. They can be accessed directly on the class, without the need to create an instance of the class.

The property 'pi' is set to the value of 3.14159, representing the mathematical constant Pi, which is the ratio of the circumference of any circle to its diameter.

The 'areaOfCircle' method is a function that calculates the area of a circle given its radius. This is done using the formula 'pi * radius * radius'. Since 'pi' is a static property of the class, it is accessed within the method as 'MathUtility.pi'.

Finally, the code includes two 'console.log' statements. These are used to print the output of the 'areaOfCircle' method when the radius is 10, and the value of 'pi' respectively. These values are accessed directly on the MathUtility class, demonstrating that static properties and methods can be used without creating an instance of the class.

Overall, this code snippet provides a useful example of how static properties and methods can be used within a class in JavaScript. Static properties and methods can be particularly useful for grouping related utility functions or constants under a common namespace, making the code more organized and easier to read.

6.2.5 Getters and Setters

Getters and setters are uniquely designed methods in programming that provide you with a mechanism to access (get) and modify (set) the properties, or attributes, of an object. They serve as a bridge between the internal implementation of an object and the outside world.

The beauty of these methods is their ability to incorporate additional functionality or apply certain rules when a property is accessed or modified. For instance, they can be particularly useful when you want to execute some specific code each time a property is accessed or set, allowing for more control and flexibility.

This makes getters and setters a key component in maintaining the integrity and consistency of an object's state.

Example: Using Getters and Setters

```
class User {
    constructor(firstName, lastName) {
        this.firstName = firstName;
        this.lastName = lastName;
    }

    get fullName() {
        return `${this.firstName} ${this.lastName}`;
    }

    set fullName(name) {
        [this.firstName, this.lastName] = name.split(' ');
    }
}

const user = new User('John', 'Doe');
console.log(user.fullName);   // Outputs: John Doe

user.fullName = 'Jane Smith';
console.log(user.fullName);   // Outputs: Jane Smith
```

This example demonstrates how getters and setters can be used to manage data access in a controlled manner, providing an interface to interact with the properties of an object.

The code snippet demonstrates the concept of classes, along with getters and setters, in ES6 syntax.

It defines a class named 'User' using the class keyword, which is a fundamental aspect of object-oriented programming in JavaScript. A class is a blueprint for creating objects that share common properties and behaviors.

Inside the 'User' class, a constructor method is defined with two parameters: 'firstName' and 'lastName'. The constructor method is a special function that gets executed whenever a new instance of the class is created. The parameters represent the first and last name of a user, and are assigned to the instance of the object being created using the 'this' keyword.

The class also includes a getter and a setter for a property called 'fullName'. The getter, get fullName(), is a method that when called, returns the full name of the user, which is a concatenation of the 'firstName' and 'lastName' properties. The setter, set fullName(name), is a method that allows you to change the value of the 'firstName' and 'lastName' properties. It does this by taking a string 'name', splitting it into two parts around the space character, and assigning the resulting values to 'firstName' and 'lastName' respectively.

Once the class is defined, an instance of the 'User' class is created using the new keyword, followed by the 'User' class and the arguments for the constructor enclosed in parentheses. In this case, a new 'User' object named 'user' is created with 'John' as the first name and 'Doe' as the last name.

The getter is then used to log the full name of the user to the console, which results in 'John Doe'. After that, the setter is used to change the full name of the 'user' object to 'Jane Smith', and the getter is used again to log the new full name to the console, resulting in 'Jane Smith'.

This examplt is a concise yet effective illustration of how classes, constructors, getters, and setters work in JavaScript. It shows how you can encapsulate related data and behavior within a class, and control access to an object's properties, making your code more structured, maintainable, and secure.

6.2.6 Private Methods and Fields

In the realm of programming, one of the most noteworthy improvements found in the latest iterations of JavaScript is the introduction of support for private methods and fields. This noteworthy development stands as a substantial upgrade in terms of encapsulation.

The principle of encapsulation is a cornerstone concept in object-oriented programming, and it revolves around the idea of restricting direct access to certain components of an object. With the introduction of private methods and fields in JavaScript, this crucial concept has been significantly bolstered.

This enhancement ensures that specific details inherent to a class are securely hidden and shielded from external access, thus preserving the integrity of the data and enhancing the overall security and robustness of the code.

Example: Private Fields and Methods

```
class Account {
    #balance = 0;
```

```javascript
    constructor(initialDeposit) {
        this.#balance = initialDeposit;
    }

    #updateBalance(amount) {
        this.#balance += amount;
    }

    deposit(amount) {
        if (amount < 0) throw new Error("Invalid deposit amount");
        this.#updateBalance(amount);
    }

    get balance() {
        return this.#balance;
    }
}

const acc = new Account(100);
acc.deposit(50);
console.log(acc.balance);   // Outputs: 150
```

In this example, #balance and #updateBalance are private, meaning they cannot be accessed outside of the Account class, thereby safeguarding the integrity of the internal state of the class instances.

The code example defines a class named 'Account'. This class acts as a blueprint for creating account objects according to object-oriented programming (OOP) principles.

The 'Account' class has a private field named '#balance'. In JavaScript, private fields are denoted by a hash '#' symbol before their names. They are private because they can only be accessed or modified within the class they are defined in. By default, this '#balance' field is initialized to 0, signifying that a new account will have a balance of 0 if no initial deposit is provided.

The class also has a constructor method. In OOP, the constructor method is a special method that is automatically called whenever a new object is created from a class. In this case, the constructor method takes one parameter, 'initialDeposit'. Inside the constructor, the private field '#balance' is set to the value of 'initialDeposit', indicating that whenever a new 'Account' object is created, its balance will be set to the value of the initial deposit.

Next, a private method '#updateBalance' is defined. This method takes one parameter, 'amount', and adds this amount to the current balance. The purpose of this method is to update the balance of the account after a deposit operation.

Then, a public method 'deposit' is defined. This method also takes one parameter, 'amount'. Inside this method, there's an 'if' statement that checks if the deposit amount is less than 0. If it is, an error is thrown with the message "Invalid deposit amount". This ensures that only valid amounts are deposited into the account. If the deposit amount is valid, the '#updateBalance' method is called with the deposit amount to update the account balance.

The class also includes a getter method for the 'balance' field. In JavaScript, getter methods allow you to retrieve the value of an object's property. In this case, the 'balance' getter method returns the current balance of the account.

After the 'Account' class is defined, an instance of the class is created using the 'new' keyword. This instance, named 'acc', is created with an initial deposit of 100. Then, the 'deposit' method is called on 'acc' to deposit an additional 50 into the account.

Finally, the current balance of the account is logged to the console using 'console.log'. Because the 'balance' field is private and cannot be accessed directly, the 'balance' getter method is used to retrieve the balance. The output of this operation is 150, which is the sum of the initial deposit and the subsequent deposit.

In summary, this example demonstrates how classes, private fields, constructor methods, private methods, public methods, and getter methods can be used in JavaScript to create and manipulate objects, following the principles of object-oriented programming.

Comprehensive Guide on Best Practices for Using Classes

- When dealing with structured, complex data types that necessitate the use of methods and inheritance, classes become an indispensable tool. They provide a framework that allows you to organize and manipulate data in a structured and systematic manner.
- While inheritance can be useful, it's generally recommended to prefer composition over inheritance whenever feasible. This approach can significantly reduce the complexity of your code while increasing modularity. It promotes code reuse and can make your programs easier to read and maintain.
- The use of getters and setters is a common practice in object-oriented programming. These functions control access to the properties of a class. This is especially handy when validation or preprocessing is needed before getting or setting a value. It adds a layer of protection for the data, ensuring that it remains consistent and valid throughout its lifecycle.
- Finally, take advantage of static properties and methods when dealing with functionality that does not depend on class instance data. Static methods and properties belong to the class itself, rather than an instance of the class. This means

they are shared across all instances and can be called without creating an instance of the class.

ES6 classes provide a clear, syntactical and functional benefit for structuring programs, particularly when coming from languages with classical OOP models. By understanding and utilizing advanced features like static properties, getters and setters, and private fields, you can craft more secure, maintainable, and robust applications. As JavaScript continues to evolve, these features are likely to become fundamental in the development of complex client-side and server-side applications.

6.3 Inheritance and Polymorphism

Inheritance and polymorphism stand as foundational concepts in the realm of object-oriented programming. They contribute significantly to the creation of code structures that are more organized, logical, and maintainable. By embracing these concepts, programmers can create code that is easier to understand, correct, and modify. In essence, inheritance and polymorphism are principles that enable the extension of functionality and the reuse of existing code.

This capability of extending and reusing code can drastically reduce complexity in software development, leading to more efficient, robust, and scalable applications. Code that uses inheritance and polymorphism can be modified or extended without having a ripple effect on the rest of the program, thereby reducing the likelihood of introducing new bugs when changes are made.

In the following section, we will delve deep into how JavaScript, one of the most widely-used programming languages globally, handles inheritance and polymorphism. We will critically examine how ES6, the sixth edition of the ECMAScript standard that JavaScript is based on, has enabled these features in a more intuitive and powerful way. ES6 classes have been instrumental in bringing a more traditional object-oriented approach to JavaScript, and we will explore how they have transformed the landscape of JavaScript programming.

6.3.1 Inheritance in JavaScript

Inheritance, a key concept in object-oriented programming, allows one class to inherit or acquire the properties and methods of another class. This means that an object can have properties of another object, allowing for code reusability and making the code much cleaner and easier to work with.

In JavaScript, a dynamic object-oriented programming language, this is traditionally achieved through prototypes. Prototypes are essentially a blueprint of an object, allowing for the creation of object types which can inherit properties and methods from each other.

However, with the introduction of ES6, a new version of JavaScript, class syntax was introduced which simplifies the creation of inheritance chains even further. This new syntax provides a more straightforward and clearer syntax for creating objects and dealing with inheritance.

Understanding Basic Inheritance with ES6 Classes in JavaScript

As discussed, Inheritance is a fundamental concept in Object-Oriented Programming (OOP) that helps to build complex applications with reusable and maintainable code. One of the great features of JavaScript ES6 is the ability to use classes for more complex OOP tasks.

In this context, let's explore how you can define a class that inherits properties and methods from another class, a capability that can significantly improve your efficiency and productivity as a developer. This is accomplished through the use of the 'extends' keyword in JavaScript:

Example: Creating a Subclass

```javascript
class Animal {
    constructor(name) {
        this.name = name;
    }

    speak() {
        console.log(`${this.name} makes a noise.`);
    }
}

class Dog extends Animal {
    constructor(name, breed) {
        super(name); // Call the parent class's constructor with 'name'
        this.breed = breed;
    }

    speak() {
        console.log(`${this.name} barks.`);
    }
}

const dog = new Dog('Max', 'Golden Retriever');
dog.speak(); // Outputs: Max barks.
```

In this example, Dog extends Animal. By using the extends keyword, Dog inherits all methods from Animal, including the constructor. The super function calls the parent's constructor, ensuring that Dog is initialized properly. The speak method in Dog overrides the one in Animal, demonstrating a simple form of polymorphism known as method overriding.

This code showcases the concept of inheritance. It achieves this by defining two classes: Animal and Dog.

The Animal class acts as the base or parent class. It uses a constructor, which is a special function in a class that gets executed whenever a new instance of the class is created. This constructor accepts one parameter, name, and assigns it to the this.name property of an instance of the class. Therefore, whenever an instance of Animal is created, it will always have a name property that can be accessed and used in other methods within the class.

One such method is the speak method. This is a simple function that generates a console log output. It uses a template literal to insert the name of the animal into a sentence, resulting in a string like 'Max makes a noise.' when the method is called on an instance of Animal.

The Dog class, on the other hand, is a derived or child class that extends the Animal class. This means that Dog inherits all the properties and methods of Animal, but it can also define its own properties and methods or override the inherited ones.

The Dog class also has a constructor, but this one accepts two parameters: name and breed. The name parameter is passed to the super function, which calls the constructor of the parent class, Animal. This ensures that the name property is set correctly in the Dog class. The breed parameter is then assigned to the this.breed property of the Dog instance.

The Dog class also overrides the speak method from Animal. Instead of saying that the dog 'makes a noise', this new speak method outputs that the dog 'barks'. This is an example of polymorphism, another key concept in object-oriented programming, where a child class can change the behavior of a method inherited from a parent class.

Finally, an instance of Dog is created using the new keyword, with 'Max' as the name and 'Golden Retriever' as the breed. This instance is stored in the dog variable. When the speak method is called on dog, it uses the Dog class's version of the method, not the Animal version. Therefore, it outputs 'Max barks.' to the console.

This example illustrates the power of inheritance in object-oriented programming, showing how you can create complex, hierarchical relationships between classes to share functionality and behavior while keeping your code DRY (Don't Repeat Yourself).

6.3.2 Polymorphism

Polymorphism, a fundamental concept in object-oriented programming, provides the ability for a method to exhibit varying behavior based on the object it is acting upon. Essentially, this means that a single method could perform different functionalities depending on the class or context of the object it is called upon.

This is a key feature of object-oriented programming as it enhances flexibility and promotes the reusability of code. For instance, when a method is invoked, the exact behavior or output it produces can differ based on the specific class or object that calls it. This dynamic nature of polymorphism is what makes it a crucial tool in the realm of object-oriented programming.

Method Overriding Example

In the example provided earlier, we can observe a case where the speak method was overridden specifically to alter the behavior for instances of the Dog class, distinguishing it from instances of the Animal class. The speak method, which exists within the Animal class, was redefined in the context of the Dog class to provide a different output or action.

This is a classic, straightforward example of the concept of polymorphism in object-oriented programming. The term 'polymorphism' refers to the ability of a variable, function or object to take on multiple forms. In this case, the interface - which is represented by the speak method - remains consistent.

However, its implementation varies significantly between different classes. This is the essence of polymorphism, where a single interface can map a different implementation depending on the specific class it is dealing with.

6.3.3 Using Inheritance and Polymorphism Effectively

Inheritance and polymorphism are undoubtedly formidable tools in the arsenal of a developer. They offer the ability to create interconnected and dynamic code structures. However, the power they wield should be handled judiciously to avert the creation of overly intricate class hierarchies, which can quickly escalate into labyrinthine structures difficult to navigate, manage, and understand.

Here are a few guidelines, drawn from best practices and professional experience, to follow when working with inheritance and polymorphism:

1. **Prefer Composition Over Inheritance**: This principle suggests that if a class requires to leverage the functionality of another class, it might be more beneficial to use the approach of composition, whereby it includes the needed class, instead of extending or inheriting from it. This methodology not only offers more flexibility by allowing the assembly of more complex objects from simpler ones, but it also significantly reduces dependencies and minimizes the risk of creating unnavigable class hierarchies.
2. **Use Polymorphism to Simplify Code**: In the realm of object-oriented programming, polymorphism stands as a key feature that allows one function to engage with objects of different classes. This can dramatically streamline your code, making it more readable, maintainable, and scalable. When in doubt, remember that polymorphism can be a powerful ally in writing cleaner, more efficient code.
3. **Keep Inheritance Hierarchies Shallow**: Although it might be tempting to create deep inheritance trees for the sake of thoroughness, they can inadvertently lead to code that is arduous to follow and debug. Therefore, it's recommended to keep inheritance hierarchies as shallow as possible. This practice helps to maintain a high level of clarity and simplicity in your code, making it easier for both you and others to work with.
4. **Ensure that Derived Classes Extend Base Classes Naturally**: When creating derived classes, it's important to make sure they are proper extensions of their base classes, strictly adhering to the "is-a" relationship. This means that the derived class should fundamentally be a type of the base class. For instance, a Dog is inherently an Animal. Therefore, it's logical and appropriate for Dog to extend Animal. This practice ensures that your inheritance structures remain intuitive and semantically correct.

Understanding and applying inheritance and polymorphism in JavaScript can greatly enhance your ability to write clean, effective, and maintainable object-oriented code. With ES6 classes, these concepts are more accessible and intuitive, allowing developers to build sophisticated systems that are easier to develop, test, and maintain.

6.3.4 Interfaces and Duck Typing

In contrast to languages such as Java or C#, JavaScript does not incorporate interfaces in its architecture. This is a feature that is often found in statically typed languages, where the interface acts as a contract to ensure a class behaves in a certain way. However, JavaScript, being a dynamically typed language, employs a different concept known as "duck typing".

In this paradigm, the determination of an object's suitability is not based on the actual type of the object, but rather by the presence of certain methods and properties. This approach grants JavaScript its flexibility, allowing objects to be used in a variety of contexts as long as they have the required attributes.

It is named after the phrase "If it looks like a duck, swims like a duck, and quacks like a duck, then it probably is a duck," reflecting the idea that an object's behavior determines its suitability, rather than its lineage or class inheritance.

Example: Duck Typing

```javascript
function makeItSpeak(animal) {
    if (animal.speak) {
        animal.speak();
    } else {
        console.log("This object cannot speak.");
    }
}

const cat = {
    speak() { console.log("Meow"); }
};

const car = {
    horn() { console.log("Honk"); }
};

makeItSpeak(cat);  // Outputs: Meow
makeItSpeak(car);  // Outputs: This object cannot speak.
```

This example shows how you can design functions that interact with objects based on their capabilities rather than their specific class, embodying the principle of "if it walks like a duck and it quacks like a duck, then it must be a duck."

The code example illustrates the concept of "Duck Typing". In Duck Typing, an object's suitability is determined by the presence of certain methods and properties, rather than the actual type of the object.

The code defines a function named makeItSpeak which accepts an object as a parameter. This function checks if the passed object has a method named speak. If the method exists, it's executed. If it does not exist, a message "This object cannot speak." is logged to the console.

Next, two objects are defined: cat and car. The cat object has a speak method which logs the string "Meow" to the console when called. The car object, on the other hand, does not have a speak method. Instead, it has a horn method that logs "Honk" to the console when called.

In the final part of the code, the makeItSpeak function is invoked twice, first with the cat object, and then with the car object. When the cat object is passed to makeItSpeak, the cat's speak method is found and called, resulting in "Meow" being logged to the console. However, when the car object is passed, since it doesn't have a speak method, the default message "This object cannot speak." is logged to the console.

This code example is a demonstration of Duck Typing in action. It shows that it's not the type of the object that determines if it can 'speak', but rather whether or not the object has a speak method. This reflects the saying "If it looks like a duck, swims like a duck, and quacks like a duck, then it probably is a duck", which is the principle behind Duck Typing. The makeItSpeak function doesn't care about the type of the object it receives, it only cares if the object can 'speak'.

6.3.5 Mixins for Multiple Inheritance

In JavaScript, a language that doesn't natively support multiple inheritance—where a class can inherit properties and methods from more than one class—a workaround exists that provides similar flexibility and functionality.

This solution is known as 'mixins'. Mixins essentially enable the combination and incorporation of behaviors from numerous sources. This equips developers with the ability to create more dynamic, multifaceted objects, thereby enhancing the robustness of their code without needing to rely on the traditional inheritance model.

Example: Creating Mixins

```
let SayMixin = {
    say(phrase) {
        console.log(phrase);
    }
};

let SingMixin = {
    sing(lyric) {
        console.log(lyric);
    }
};

class Person {
    constructor(name) {
        this.name = name;
    }
}

// Copy the methods
Object.assign(Person.prototype, SayMixin, SingMixin);

const john = new Person("John");
```

```
john.say("Hello");   // Outputs: Hello
john.sing("La la la");   // Outputs: La la la
```

This approach allows you to "mix" additional functionality into a class's prototype, enabling a form of multiple inheritance where a class can inherit methods from multiple mixin objects.

A mixin is essentially a class or object that contains methods that can be borrowed or "mixed in" with other classes. Mixins are a way to distribute reusable functionalities for classes. They are not intended to be used independently, but to be added to and used by other classes.

In this code, two mixins are created: SayMixin and SingMixin. Each mixin is an object that contains a single method—SayMixin contains the say() method, and SingMixin contains the sing() method. These methods simply log to the console the phrase or lyric that is passed to them as a parameter.

Next, a Person class is defined with a constructor that sets a name property. This class doesn't have any methods of its own at this point.

The mixins are then applied to the prototype of the Person class using the Object.assign() method. This essentially copies the properties from SayMixin and SingMixin onto Person.prototype, allowing instances of the Person class to use the say() and sing() methods.

An instance of the Person class, john, is then created using the new keyword. Because the mixins were applied to Person.prototype, john can use both the say() and sing() methods. The code demonstrates this by having john say "Hello" and sing "La la la", which are logged to the console.

In conclusion, this code provides a simple demonstration of how mixins can be used in JavaScript. Mixins are a powerful tool for sharing behavior between different classes, helping to keep code DRY (Don't Repeat Yourself) and organized.

6.3.6 Factory Functions

Factory functions represent an alternative pattern that can be employed in lieu of traditional classes for the purpose of creating objects. They are particularly beneficial as they can effectively encapsulate the logic behind the creation of objects.

This encapsulation results in a clear separation between the process of creation and the actual use of the objects, providing a level of abstraction that can aid in the understanding and maintenance of the code.

Additionally, factory functions leverage the power of closures to provide privacy, which is a feature that's not natively supported in JavaScript. This brings a new level of security and control over how data is accessed and manipulated, making it a viable alternative approach to using constructors and the class-based inheritance model that is typically found in object-oriented programming.

Example: Factory Function

```javascript
function createRobot(name, capabilities) {
    return {
        name,
        capabilities,
        describe() {
            console.log(`This robot can perform: ${capabilities.join(', ')}`);
        }
    };
}

const robo = createRobot("Robo", ["lift things", "play chess"]);
robo.describe();  // Outputs: This robot can perform: lift things, play chess
```

Factory functions provide flexibility and encapsulation, making them a powerful alternative to classes, especially when object creation does not fit neatly into a single inheritance hierarchy.

The example code showcases how to define a function that creates and returns an object. This is a common pattern in JavaScript and is often used when you need to create multiple objects with the same properties and methods.

The function in the code is named createRobot. It's designed to build "robot" objects, and it takes two arguments: name and capabilities.

The name argument represents the name of the robot. It's expected to be a string. For example, it could be "Robo", "CyberBot", "AlphaBot", etc.

The capabilities argument represents the abilities of the robot. It's expected to be an array of strings, with each string describing a capability. For instance, this could include tasks the robot can perform, such as "lift things", "play chess", "calculate probabilities", etc.

The createRobot function works by returning a new object. This object includes the name and capabilities provided as arguments, as well as a method called describe.

The describe method is a function that, when called, uses JavaScript's console.log function to output a string to the console. This string provides a description of what the robot can do, by joining all the capabilities with ", " and including them in a sentence.

After defining the createRobot function, the code then demonstrates how to use it. It creates a new robot named "Robo" that can "lift things" and "play chess". This is done by calling createRobot with the appropriate arguments and storing the returned object in a constant variable called robo.

Finally, the describe method is called on robo. This outputs a sentence to the console that describes the robot's capabilities, specifically: "This robot can perform: lift things, play chess".

In summary, this code provides a clear example of how to define a function that creates and returns objects in JavaScript. It also demonstrates how to use such a function to create an object, and how to call a method on that object. This is a common pattern in JavaScript and many other object-oriented programming languages, and understanding it is crucial to writing effective, object-oriented code.

By diving deeper into these advanced aspects of inheritance and polymorphism, you can develop a more nuanced understanding of object-oriented programming in JavaScript. Whether it's implementing duck typing, using mixins for multiple inheritance, or employing factory functions for object creation, these techniques can provide powerful tools for building flexible, scalable, and maintainable software.

6.4 Encapsulation and Abstraction

In the realm of object-oriented programming (OOP), there are two fundamental concepts that substantially contribute to the reduction of complexity and augmentation of code reusability. These essential principles are known as encapsulation and abstraction.

Encapsulation is the technique of enclosing or wrapping up data, represented by variables, and the associated methods, which are essentially functions that manipulate the encapsulated data. This packaging of data and corresponding methods is achieved within a singular unit or class. This mechanism ensures that the internal state of an object is protected from external interference, leading to a robust and controlled design.

On the contrary, abstraction aims to obscure the intricate details of reality while only exposing those parts of an object that are deemed necessary. It simplifies the representation of reality, making it easier for the programmer to handle complexity.

In the context of JavaScript, a scripting language widely used for client-side web development, these concepts can be implemented using a variety of techniques. These include classes, which provide a template for creating objects and encapsulating data and methods, closures which allow functions to have private variables, and module patterns which help in organizing code in a maintainable way. By employing these techniques, programmers can enhance the safety, robustness, and maintainability of their code, thereby improving the overall quality and reliability of the software.

6.4.1 Understanding Encapsulation

The principle of encapsulation is a fundamental aspect of object-oriented programming that enables an object to conceal its internal state, meaning that all interactions must be carried out through the object's methods.

This is more than just a way of structuring data; it's a robust approach to managing complexity in large-scale software systems. By providing a controlled interface to the object's data, encapsulation ensures that the internal workings of the object are shielded from the outside world.

This prevents the state of the object from being altered in unexpected ways, which can lead to bugs and unpredictable behavior. Additionally, encapsulation promotes modularity and separation of concerns, making the code easier to maintain and understand.

Example: Using Classes to Achieve Encapsulation

```javascript
class BankAccount {
    #balance;   // Private field

    constructor(initialBalance) {
        this.#balance = initialBalance;
    }

    deposit(amount) {
        if (amount < 0) {
            throw new Error("Amount must be positive");
        }
        this.#balance += amount;
        console.log(`Deposited $${amount}. Balance is now $${this.#balance}.`);
    }

    withdraw(amount) {
        if (amount > this.#balance) {
            throw new Error("Insufficient funds");
        }
```

```
        this.#balance -= amount;
        console.log(`Withdrew $${amount}. Balance is now $${this.#balance}.`);
    }

    getBalance() {
        return this.#balance;
    }
}

const account = new BankAccount(1000);
account.deposit(500);
account.withdraw(200);
console.log(`The balance is $${account.getBalance()}.`);
// Outputs: Deposited $500. Balance is now $1500.
//          Withdrew $200. Balance is now $1300.
//          The balance is $1300.
```

In this example, the #balance field is private, which means it cannot be accessed directly from outside the class. This encapsulation ensures that the balance can only be modified through the deposit and withdraw methods, which include validations.

The code snippet defines a class named 'BankAccount'. This class is a blueprint for creating 'BankAccount' objects, each representing a unique bank account.

The 'BankAccount' class contains one private field, '#balance'. This field is intended to store the balance of the bank account. It is marked as private, denoted by the '#' symbol, which means it can only be accessed directly within the class itself. This is a key aspect of encapsulation, a fundamental principle in object-oriented programming that restricts direct access to an object's properties for the purpose of maintaining the integrity of the data.

The class also defines a 'constructor' method. This special method is automatically called when a new 'BankAccount' object is created. It takes one parameter, 'initialBalance', which is used to set the initial balance of the bank account by assigning it to the '#balance' private field.

Three methods, 'deposit', 'withdraw', and 'getBalance', are defined in the 'BankAccount' class:

- The 'deposit' method takes an 'amount' as a parameter. It checks if the amount is less than zero, and if so, throws an error. Otherwise, it adds the amount to the '#balance' and prints a message showing the deposited amount and the new balance.
- The 'withdraw' method also takes an 'amount' as a parameter. It checks if the amount is more than the current '#balance', and if so, throws an error. Otherwise, it subtracts the amount from the '#balance' and prints a message showing the withdrawn amount and the new balance.

- The 'getBalance' method does not take any parameters. It simply returns the current '#balance'.

The last few lines of the code snippet demonstrate how to use the 'BankAccount' class. It creates a new 'BankAccount' object with an initial balance of 1000, deposits 500 into the account, withdraws 200 from the account, and finally prints the current balance of the account.

Thus, the 'BankAccount' class encapsulates the properties and methods related to a bank account, providing a way to manage the account's balance in a controlled manner. The balance can only be modified through the 'deposit' and 'withdraw' methods, and retrieved using the 'getBalance' method, ensuring the integrity of the balance.

6.4.2 Implementing Abstraction

Abstraction is a crucial programming concept that is designed with the explicit goal of concealing the intricate and often complex implementation details of a particular class, exposing only the essential components to the user.

This concept is an integral part of programming that provides a layer of simplicity and ease for the user while the complex processes are carried out behind the scenes. This fundamental principle can indeed be implemented in JavaScript, a robust and popular programming language.

The implementation of abstraction in JavaScript can be achieved by carefully controlling and limiting the exposure of properties and methods. By doing this, we ensure that a user only interacts with the necessary elements, thus providing a simpler, more streamlined programming experience.

Example: Using Function Constructors for Abstraction

```javascript
function Car(model, year) {
    this.model = model;
    let mileage = 0;  // Private variable

    this.drive = function (miles) {
        if (miles < 0) {
            throw new Error("Miles cannot be negative");
        }
        mileage += miles;
        console.log(`Drove ${miles} miles. Total mileage is now ${mileage}.`);
    };

    this.getMileage = function () {
```

```
        return mileage;
    };
}

const myCar = new Car("Toyota Camry", 2019);
myCar.drive(150);
console.log(`Total mileage: ${myCar.getMileage()}.`);
// Outputs: Drove 150 miles. Total mileage is now 150.
//          Total mileage: 150.
```

In this Car example, the mileage variable is not exposed directly; instead, it is accessed and modified through the methods drive and getMileage. This abstraction hides the details of how mileage is tracked and modified, which can prevent misuse or errors from direct manipulation.

The example code snippet demonstrates the creation of a 'Car' object using a function constructor, which is one of the ways to create objects in JavaScript.

In this example, the function constructor named 'Car' accepts two parameters, 'model' and 'year'. The 'model' parameter represents the model of the car, while the 'year' parameter indicates the manufacturing year of the car.

Inside this function, the 'this' keyword is used to assign the values of the 'model' and 'year' parameters to the respective properties of the Car object being created.

Next, a private variable 'mileage' is defined and initialized with a value of 0. In JavaScript, private variables are variables that are accessible only within the function where they are defined. In this case, 'mileage' is only accessible within the 'Car' function.

The 'Car' function further defines two methods, 'drive' and 'getMileage'.

The 'drive' method accepts a parameter 'miles', which represents the number of miles the car has driven. It then checks if 'miles' is less than 0, and if so, throws an error, because driving a negative number of miles is not possible. If 'miles' is not less than 0, it adds 'miles' to 'mileage', effectively increasing the car's total mileage, and then logs a message stating how many miles were driven and what the total mileage now is.

The 'getMileage' method, on the other hand, simply returns the current value of the 'mileage' variable. This allows us to check the car's total mileage without directly accessing the private 'mileage' variable.

After defining the 'Car' function, the code creates a new instance of the Car object, named 'myCar', with the model "Toyota Camry" and the year 2019. This is done using the 'new' keyword, which invokes the 'Car' function with the given arguments and returns a new Car object.

The 'myCar' object then calls the 'drive' method with an argument of 150, indicating that 'myCar' has driven 150 miles. This increases 'myCar's total mileage by 150 and logs a message about it.

Finally, the code logs the total mileage of 'myCar' by calling the 'getMileage' method on 'myCar'. This gives us the total mileage of 'myCar' after driving 150 miles.

In summary, this code snippet demonstrates how to create an object with public properties and methods, as well as a private variable, in JavaScript using a function constructor. It also shows how to create an instance of an object and call its methods.

6.4.3 Best Practices

- One of the fundamental principles you should follow is using encapsulation to safeguard the object's state from any unforeseen or unauthorized modifications. This will ensure the integrity of the data and prevent any accidental changes that could disrupt the functionality of the object.
- Another key practice is to employ abstraction to minimize complexity. By providing only the essential components of an object to the outside world, you can simplify the interaction with the object and reduce the risk of errors or misunderstandings. This approach helps to ensure that each object is understood in terms of its true essence, without unnecessary details distracting from its core functionality.
- Lastly, when designing classes and methods, aim to expose a clear and simple interface for interacting with the data. This means creating intuitive methods and properties that allow other developers to easily understand and use your object, without needing to know the intricate details of its internal workings. By doing so, you can improve the overall readability and maintainability of your code, making it easier for others to work with and extend.

Encapsulation and abstraction are essential for creating robust and maintainable code. By effectively using these concepts, you can write JavaScript programs that are secure, reliable, and easy to understand. These principles guide the design of interfaces that are both easy to use and hard to misuse, fundamentally enhancing the quality of your software.

6.4.4 Module Pattern for Encapsulation

The module pattern is a renowned and widely used design pattern in the realm of JavaScript. Its primary function is to encapsulate or wrap a set of interconnected functions, variables, or a combination of both into a singular, cohesive conceptual entity commonly referred to as a "module".

This sophisticated pattern can prove to be extremely effective and beneficial, especially when there is a necessity to maintain a neat and well-organised global namespace. By using this pattern, you can successfully prevent any unwanted pollution or cluttering of the global scope.

This ensures that the global scope remains uncontaminated, thus promoting better coding practices and improving the overall performance and readability of your JavaScript code.

Example: Module Pattern

```javascript
const CalculatorModule = (function() {
    let data = { number: 0 };   // Private

    function add(num) {
        data.number += num;
    }

    function subtract(num) {
        data.number -= num;
    }

    function getNumber() {
        return data.number;
    }

    return {
        add,
        subtract,
        getNumber
    };
})();

CalculatorModule.add(5);
CalculatorModule.subtract(2);
console.log(CalculatorModule.getNumber());   // Outputs: 3
```

In this example, the CalculatorModule encapsulates the data object and the functions add, subtract, and getNumber within an immediately invoked function expression (IIFE). The module exposes only the methods it wants to make public, thus controlling the access to its internal state.

This code is an example of a "Module Pattern", which is a design pattern used in JavaScript to bundle a group of related variables and functions together, providing a level of encapsulation and organization in your code.

In this specific example, the module is encapsulating a simple calculator logic. The code is defining a module called CalculatorModule. This module is defined as an Immediately-Invoked Function Expression (IIFE), which is a function that is defined and then immediately invoked or run.

Inside this CalculatorModule, there are several pieces:

- A private data object that stores a number property. This number is what the calculator will perform operations on. It's private because it's not exposed outside the module and can only be accessed and manipulated by the functions within the module.
- An add function that takes a number as input and adds it to the number property in the data object.
- A subtract function that takes a number as input and subtracts it from the number property in the data object.
- A getNumber function that returns the current value of the number property in the data object.

After defining these functions, the return statement at the end of the module specifies what will be exposed to the outside world. In this case, the add, subtract, and getNumber functions are made public, which means they can be accessed outside the CalculatorModule.

Following the definition and immediate invocation of the CalculatorModule, the example demonstrates how to use the module. It calls the add method to add 5 to the number (which starts at 0), then calls the subtract method to subtract 2, resulting in a final number of 3. It then calls getNumber to retrieve the current number, and logs it to the console, outputting 3.

This module pattern allows developers to organize related pieces of JavaScript code into a single, self-contained unit that provides a controlled and consistent interface for interacting with the module's functionality. This aids in the understanding and maintenance of the code, ensuring data integrity and security by hiding the internal data and exposing only the necessary functions.

6.4.5 Using ES6 Modules for Better Abstraction

With the introduction of ES6, also known as ECMAScript 2015, JavaScript now has built-in support for modules. This significant development enables developers to write modular code, which is a way of managing and organizing code in a more efficient and maintainable manner.

This modular code can be seamlessly imported and exported across different files, improving code reusability and reducing redundancy. Furthermore, this native module system supports crucial programming principles such as encapsulation and abstraction. These principles allow developers to hide the complexities of a module and expose only specific, necessary parts of it.

This leads to a cleaner, more readable, and more efficient codebase. In essence, with the built-in module support introduced in ES6, JavaScript programming has become more streamlined and programmer-friendly.

Example: ES6 Module

```
// file: mathUtils.js
let internalCount = 0;   // Private to this module

export function increment() {
    internalCount++;
    console.log(internalCount);
}

export function decrement() {
    internalCount--;
    console.log(internalCount);
}

// file: app.js
import { increment, decrement } from './mathUtils.js';

increment();   // Outputs: 1
decrement();   // Outputs: 0
```

This structure ensures that internalCount remains private to the mathUtils.js module, with only the increment and decrement functions exposed to other parts of the application.

In this example, we are demonstrating the use of ES6 modules. ES6 modules are a feature introduced in the ECMAScript 6 (ES6) version of JavaScript, which allows developers to write reusable pieces of code in one file and import them for use in another file. This helps in keeping the code organized and maintainable.

The first part of the code defines a module in a file named "mathUtils.js". This module contains a variable 'internalCount' and two functions: 'increment' and 'decrement'.

The variable 'internalCount' is declared with the 'let' keyword and initialized with a value of 0. This variable is private to the "mathUtils.js" module, which means it cannot be accessed directly from outside this module. Its value can only be manipulated by the functions within this module.

The 'increment' function is a simple function that increases the value of 'internalCount' by 1 each time it's called. After incrementing 'internalCount', it logs the new value to the console using the 'console.log()' function. This function is exported from the module, so it can be imported and used in other files.

Similarly, the 'decrement' function decreases the value of 'internalCount' by 1 each time it's called. It also logs the new value of 'internalCount' to the console after performing the decrement. Like 'increment', this function is also exported from the module.

In the second part of the code, the 'increment' and 'decrement' functions are imported into another file named "app.js". This is done using the 'import' keyword followed by the names of the functions to import, enclosed in curly braces, and the relative path to the "mathUtils.js" file.

Once imported, the 'increment' and 'decrement' functions are called in "app.js". The first call to 'increment' increases 'internalCount' to 1 and logs '1' to the console. The subsequent call to 'decrement' decreases 'internalCount' back to 0 and logs '0' to the console.

To summarize, this code example demonstrates the use of ES6 modules in JavaScript, showing how to define a module that exports functions, how to import those functions into another file, and how to call the imported functions. It also demonstrates the concept of private variables in modules, which are variables that can only be accessed and manipulated by the functions within the same module.

6.4.6 Proxy for Controlled Access

Proxies in JavaScript represent a robust tool that facilitates the creation of an abstraction layer over an object, thereby providing control over interactions with said object. This feature is particularly useful as it allows developers to manage and monitor how the object is accessed and manipulated. The applications of proxies are extensive and include but are not limited to logging, profiling, and validation.

For instance, they can be employed to log the history of operations performed on an object, perform profiling by measuring the time taken for operations, or enforce validation rules before any changes are made to the object. Therefore, understanding and utilizing JavaScript proxies can significantly enhance the functionality and security of your code.

Example: Using Proxy for Validation

```
let settings = {
    temperature: 0
};

let settingsProxy = new Proxy(settings, {
    get(target, prop) {
        console.log(`Accessing ${prop}: ${target[prop]}`);
        return target[prop];
    },
    set(target, prop, value) {
        if (prop === 'temperature' && (value < -273.15)) {
            throw new Error("Temperature cannot be below absolute zero!");
        }
        console.log(`Setting ${prop} to ${value}`);
        target[prop] = value;
        return true;
    }
});

settingsProxy.temperature = -300;  // Throws Error
settingsProxy.temperature = 25;   // Setting temperature to 25
console.log(settingsProxy.temperature);  // Accessing temperature: 25, Outputs: 25
```

In this example, the Proxy is used to control access to the settings object, adding checks and logs that enrich functionality and enforce constraints, showcasing a practical application of abstraction.

The code is a demonstration of using a JavaScript Proxy object to add custom behavior to basic operations performed on an object. In this case, the object being proxied is settings, which is a simple JavaScript object containing a single property called temperature that is initialized to 0.

A Proxy object is created with two arguments: the target object and a handler. The target is the object which the proxy virtualizes and the handler is an object whose methods define the custom behavior of the Proxy.

In this example, the target object is settings and the handler is an object with two methods, get and set. These methods are called "traps" because they "intercept" operations, providing an opportunity to customize the behavior.

The get trap is a method that is called when a property of the target object is accessed. This trap receives the target object and the property being accessed as parameters. In the handler object, the get trap is defined to log a message to the console that specifies which property is being accessed and what the current value of that property is. After logging the message, it returns the value of the property.

The set trap, on the other hand, is a method that is called when a property of the target object is modified. This trap receives the target object, the property being modified, and the new value as parameters. In the handler object, the set trap is defined to first check if the property being modified is 'temperature' and if the new value is below -273.15 (which is absolute zero in Celsius). If both conditions are true, it throws an Error, because the temperature in Celsius cannot be below absolute zero. If either of the conditions is not true, it logs a message to the console specifying the property being modified and the new value. It then updates the property with the new value and returns true to indicate that the property was successfully modified.

The final three lines of the script demonstrate how to use the settingsProxy object. First, it attempts to set the temperature property to -300. This operation results in an Error because -300 is below absolute zero. Next, it sets the temperature property to 25. This operation is successful and results in a console message indicating that the temperature property was set to 25. Finally, it accesses the temperature property, which results in a console message indicating that the temperature property was accessed and displaying its current value, which is 25.

In conclusion, the Proxy object provides a powerful way to add custom behavior to basic operations performed on an object, such as accessing or modifying properties. This can be used for various purposes, such as logging, validation, or implementing business rules.

Encapsulation and abstraction are foundational concepts in building robust and maintainable software. By leveraging JavaScript's capabilities for implementing these principles—whether through design patterns, modern syntax, or advanced features—you can ensure your applications are well-structured and secure. These techniques not only enhance code quality but also foster development practices that scale effectively as applications grow in complexity.

Practical Exercises

This set of exercises is designed to reinforce the concepts discussed in Chapter 6, focusing on object-oriented programming principles such as encapsulation, abstraction, inheritance, and the use of ES6 classes in JavaScript. Each exercise includes a challenge and a solution to help you apply what you've learned in practical coding scenarios.

Exercise 1: Implementing a Class with Encapsulation

Create a Person class that encapsulates an individual's name and age and provides methods to get and set each attribute. Ensure that the age cannot be set to a negative number.

Solution:

```
class Person {
    constructor(name, age) {
        this.name = name;
        this.age = age;
    }

    getName() {
        return this.name;
    }

    setName(name) {
        this.name = name;
    }

    getAge() {
        return this.age;
    }

    setAge(age) {
        if (age < 0) {
            throw new Error("Age cannot be negative");
        }
        this.age = age;
    }
}

const person = new Person("John", 30);
console.log(person.getName()); // Outputs: John
console.log(person.getAge()); // Outputs: 30

person.setAge(25);
console.log(person.getAge()); // Outputs: 25

// Attempting to set a negative age
try {
    person.setAge(-5);
} catch (e) {
    console.log(e.message); // Outputs: Age cannot be negative
}
```

Exercise 2: Implementing Inheritance and Method Overriding

Extend the Person class to create a new class called Employee that includes everything from Person and adds employeeId and a method to display all details of the employee.

Solution:

```
class Employee extends Person {
    constructor(name, age, employeeId) {
        super(name, age); // Calls the constructor of the base class
        this.employeeId = employeeId;
    }

    display() {
        console.log(`Name:    ${this.name},    Age:    ${this.age},    Employee    ID:
${this.employeeId}`);
    }
}

const employee = new Employee("Alice", 28, "E12345");
employee.display(); // Outputs: Name: Alice, Age: 28, Employee ID: E12345
```

Exercise 3: Using Static Methods

Create a class Calculator with static methods for basic arithmetic operations (add, subtract, multiply, divide) that take two numbers and return the result.

Solution:

```
class Calculator {
    static add(a, b) {
        return a + b;
    }

    static subtract(a, b) {
        return a - b;
    }

    static multiply(a, b) {
        return a * b;
    }

    static divide(a, b) {
        if (b === 0) {
            throw new Error("Division by zero");
        }
```

```
        return a / b;
    }
}

console.log(Calculator.add(10, 5)); // Outputs: 15
console.log(Calculator.subtract(10, 5)); // Outputs: 5
console.log(Calculator.multiply(10, 5)); // Outputs: 50
console.log(Calculator.divide(10, 5)); // Outputs: 2
```

Exercise 4: Abstraction with Proxies

Create a proxy for a settings object that validates changes to properties. Specifically, ensure that volume is between 0 and 100.

Solution:

```
let settings = {
    volume: 30
};

let settingsProxy = new Proxy(settings, {
    set(target, prop, value) {
        if (prop === 'volume') {
            if (value < 0 || value > 100) {
                throw new Error("Volume must be between 0 and 100");
            }
        }
        target[prop] = value;
        return true;
    }
});

settingsProxy.volume = 90; // Works fine
console.log(settings.volume); // Outputs: 90

// Attempting to set volume out of range
try {
    settingsProxy.volume = 101;
} catch (e) {
    console.log(e.message); // Outputs: Volume must be between 0 and 100
}
```

These exercises provide practical applications of the object-oriented programming features discussed in Chapter 6, helping you solidify your understanding of these concepts through coding challenges that reflect real-world scenarios.

Chapter Summary

In Chapter 6, we embarked on an exploratory journey through the principles of object-oriented programming (OOP) as implemented in JavaScript. This chapter aimed to demystify the concepts of encapsulation, abstraction, inheritance, and polymorphism, which are pivotal for developing scalable and maintainable applications. By diving into these core principles, we've equipped you with the knowledge to leverage JavaScript's capabilities to create robust and efficient web applications.

Encapsulation and Abstraction

We began by addressing encapsulation, a fundamental OOP concept that involves bundling the data (variables) and the methods (functions) that act on the data into single units called classes. Encapsulation protects an object's internal state from unwanted external interference and misuse, which enhances data integrity and security. We saw how JavaScript uses function scopes and closures to maintain private states within objects, a practice crucial in preserving the integrity and security of applications.

Abstraction was another focus area, simplifying complex systems by hiding the irrelevant details from the users and exposing only the necessary parts of the objects. This helps in reducing programming complexity and increasing efficiency. JavaScript's support for abstraction comes through its ability to create objects that expose only selected attributes and methods to the outside world, allowing developers to change and refactor an object's internal workings without altering how other code interacts with it.

Inheritance and Polymorphism

Inheritance in JavaScript, traditionally achieved through prototypes and more recently through class syntax introduced in ES6, allows objects to inherit properties and methods from other objects. We explored how to create class hierarchies that reflect real-world relationships, enabling you to write less code while increasing functionality. By using the extends keyword for class inheritance and the super constructor for initializing the parent's constructor, JavaScript simplifies the prototype chain management, making it more accessible to developers familiar with traditional OOP languages.

Polymorphism was discussed as a mechanism allowing objects to be treated as instances of their parent class, with the ability to override methods to perform different functionalities. This aspect of OOP in JavaScript enables you to call the same method on different objects, each responding in a way appropriate to the object's type, thus enhancing code flexibility and reusability.

Practical Applications and Best Practices

Throughout the chapter, practical examples demonstrated how to implement these OOP principles in real-world scenarios. From creating classes and managing inheritance trees to applying encapsulation and abstraction for better data management and system design, the exercises provided a hands-on approach to solidifying your understanding of these concepts.

Conclusion

This chapter has laid a solid foundation for understanding and applying object-oriented principles in JavaScript. By mastering these concepts, you are now better equipped to tackle complex development challenges, create highly reusable code, and design your applications with better structure and maintainability. As you continue to delve deeper into JavaScript and its object-oriented features, remember that these principles are not just theoretical but are essential tools that can significantly enhance the functionality and quality of your software projects.

Chapter 7: Web APIs and Interfaces

Welcome to the comprehensive Chapter 7, "Web APIs and Interfaces." In this enlightening chapter, we delve deep into the versatile interfaces and APIs that today's web browsers so generously offer. With these APIs at our disposal, we, as web developers, have the power to craft rich, interactive web applications that fully utilize not only the capabilities of the browser but also the potential of the underlying operating system.

This chapter promises to cover a wide range of web APIs, from those primarily involved in making HTTP requests, to those that are adept at handling files, managing a variety of media, and even those that interact directly with device hardware.

In the rapidly evolving digital age, web applications frequently find themselves needing to establish communication with external servers, fetch crucial data, send timely updates, and interact dynamically and seamlessly with users. Having a solid understanding and effective usage of web APIs becomes a critical skill in building such responsive applications. To equip you with this essential skill, we kickstart this chapter with an in-depth exploration of the Fetch API - a contemporary and flexible approach to making HTTP requests, designed for the modern web.

7.1 Fetch API for HTTP Requests

The Fetch API represents a modern and sophisticated interface that offers the capability to make network requests akin to what is possible with XMLHttpRequest (XHR). However, compared to XMLHttpRequest, the Fetch API brings to the table a much more potent and flexible range of features. One of the key enhancements of the Fetch API is its use of promises.

Promises are a contemporary approach to managing asynchronous operations, which are operations that do not have to be completed before other code can run. By using promises, the Fetch API allows for code to be written and read in a much cleaner and more streamlined manner, thereby enhancing the efficiency of the coding process and leading to more maintainable code in the long run.

7.1.1 Basic Usage of Fetch API

The fetch() function serves as the backbone of the Fetch API, a significant tool in modern web development. It is an incredibly versatile function that permits a wide spectrum of network requests like GET, POST, PUT, DELETE, among other request types. These requests are essential in interacting with servers and manipulating data on the web.

The GET request, for instance, is frequently used to retrieve data from a server in the JSON format. The data retrieved can be any kind of information that is stored on the server. Here is a brief demonstration of how you can use the fetch() function to execute a straightforward HTTP GET request in order to obtain JSON data from a server.

Example: Fetching JSON Data

```
fetch('<https://api.example.com/data>')
    .then(response => {
        if (!response.ok) {
            throw new Error('Network response was not ok ' + response.statusText);
        }
        return response.json();
    })
    .then(data => console.log(data))
    .catch(error => console.error('There was a problem with your fetch operation:',
error));
```

This code snippet demonstrates the use of the Fetch API. The Fetch API utilizes promises, a contemporary approach to managing asynchronous operations, which are operations that don't have to finish before the rest of the code can run. This allows for more efficient, maintainable, and readable code.

In this example, we're using the Fetch API to make an HTTP GET request to a server. The GET request is often used to retrieve data from a server, and in this case, the data is expected to be in JSON format. The server we're requesting data from is specified by the URL 'https://api.example.com/data'.

The fetch operation starts with the fetch() function, which returns a promise. This promise resolves to the Response object representing the response to the request. The Response object contains information about the server's response, including the status of the request.

The first then() method in the promise chain handles the response from the fetch operation. Inside this method, we're checking if the response was successful by using the ok property of the Response object. This property returns a Boolean that indicates whether the response's status was within the successful range (200-299).

If the response was not OK, we throw an Error with a custom message that includes the status text of the response. The status text provides a human-readable explanation of the status of the response, such as 'Not Found' for a 404 status.

If the response was OK, we return the response body parsed as JSON using the json() method. This method reads the response stream to completion and parses the result as a JSON object.

The second then() method in the promise chain receives the parsed JSON data from the previous then(). Here, we're simply logging the data to the console.

The catch() method at the end of the promise chain is used to catch any errors that might occur during the fetch operation or during the parsing of the JSON data. If an error is caught, we're logging it to the console with a custom error message.

In summary, this code snippet demonstrates the basic usage of the Fetch API to make an HTTP GET request, check if the response was successful, parse the response data as JSON, and handle any errors that might occur during the operation.

7.1.2 Making POST Requests with Fetch

When you need to send data to a server, one efficient method you can employ is making a POST request by using the Fetch API. This process entails the clear specification of the request method as 'POST'. In addition to this, the data you desire to transmit is required to be included in the body of the request.

This data can be various types, such as JSON or form data, depending on what the server is set up to receive. The Fetch API makes this process straightforward and intuitive, simplifying the task of sending data to servers and making your web development tasks more efficient.

In the realm of web development, when there's a requirement to send data to a server, one effective method is making a POST request. The Fetch API, a modern and flexible tool for making network requests, simplifies this process.

To make a POST request using the Fetch API, you need to specify 'POST' as the request method. This clear specification ensures that the server understands what kind of request is being made. Along with this, the data you intend to send to the server needs to be included in the body of the request.

The data that you send can be of various types, such as in the form of JSON (JavaScript Object Notation) or form data. The type of data you choose to send depends on the server's configuration and what it's prepared to receive. Therefore, it's vital to have a clear understanding of the server's specifications before sending the data.

The Fetch API has made this process of sending data to servers more straightforward and intuitive, simplifying the task for developers. It abstracts the complex underlying details, allowing developers to focus on the data they want to send rather than the technical details of the request. As a result, it significantly enhances the efficiency of web development tasks by eliminating unnecessary complexities.

By using the Fetch API for making POST requests, web developers can build more dynamic, responsive web applications that interact seamlessly with servers and external APIs. It's a critical skill in today's rapidly evolving digital landscape, where web applications frequently need to establish communication with external servers, fetch crucial data, send timely updates, and interact dynamically with users.

Example: Making a POST Request

```javascript
fetch('<https://api.example.com/data>', {
    method: 'POST',
    headers: {
        'Content-Type': 'application/json',
    },
    body: JSON.stringify({
        name: 'John',
        email: 'john@example.com'
    })
})
.then(response => {
    if (!response.ok) {
        throw new Error('Network response was not ok ' + response.statusText);
    }
    return response.json();
})
.then(data => console.log('Success:', data))
.catch(error => console.error('Error:', error));
```

This example is demonstrating the use of the Fetch API to perform an HTTP POST request to a certain URL, in this case, 'https://api.example.com/data'. The Fetch API is a modern, promise-based API for making network requests from within JavaScript applications, and offers more flexibility and features than the older XMLHttpRequest (XHR) method.

The script begins with the fetch() function, to which we pass the desired URL we want to send our request to. Immediately following the URL, an object is provided that configures the specifics of our request. This configuration object includes the method property set to 'POST', indicating we're sending data to the server, not just requesting data from it.

In the headers property of the configuration object, 'Content-Type' is set to 'application/json'. This tells the server that we're sending JSON formatted data.

The body property contains the data we're sending to the server, which must be turned into a string using JSON.stringify() because HTTP is a text-based protocol and requires that any data sent to the server be in string format. In this case, an object containing 'name' and 'email' properties is stringified and included in the body of the request.

After the fetch() function, a promise chain is constructed to handle the response from the server and any errors that might occur during the fetch operation. This is done using the .then() and .catch() methods that are part of the promise-based Fetch API.

The first .then() block receives the server's response as its argument. Inside this block, a check is performed to see if the response was successful using the ok property of the response object. If the response was not okay, an error is thrown with a custom message and the status text of the response. If the response is okay, it's returned as JSON using the json() method of the response object.

The second .then() block receives the parsed JSON data from the previous block. This is where we can interact with the data returned from the server, in this case, it's simply logged to the console with a 'Success:' message.

Lastly, the .catch() block at the end of the promise chain catches any errors that occur during the fetch operation or during the parsing of the JSON data. These errors are then logged to the console with an 'Error:' message.

7.1.3 Handling Errors

Effective error handling is crucial when making network requests in any programming project, as it ensures the resilience and reliability of your application. In this regard, the Fetch API is a powerful tool for developers.

The Fetch API provides a way to catch network errors, such as connectivity issues or server errors, and handle errors that might occur during data parsing. This includes issues that may arise when converting response data into a usable format. Using the Fetch API, you can implement a robust error handling mechanism for your network requests, thereby improving the user experience.

In web development, particularly when dealing with network operations like fetching or posting data to servers, various errors can arise due to network connectivity, server response, or during data parsing.

With the Fetch API, developers can capture both network and parsing errors and define custom responses. These could include logging the error for debugging, displaying a message to the user, or triggering corrective actions.

By using the Fetch API for error handling, you can build robust web applications. It allows the application to handle network operation issues gracefully, enhancing both its reliability and user experience.

Example: Error Handling

```
fetch('<https://api.example.com/data>')
    .then(response => {
        if (!response.ok) {
            throw new Error('Network response was not ok ' + response.statusText);
        }
        return response.json();
    })
    .then(data => console.log(data))
    .catch(error => {
        console.error('There was a problem with your fetch operation:', error);
    });
```

The code starts with a call to the fetch() function, passing in a URL string 'https://api.example.com/data'. This sends a GET request to the specified URL, which is expected to return some data.

This fetch() function returns a Promise. Promises in JavaScript represent a completion or failure of an asynchronous operation and its resulting value. They are used to handle asynchronous operations like this network request.

The returned Promise is then chained with a then() method. The then() method takes in a callback function that will be executed when the Promise resolves successfully. The callback function receives the response from the fetch operation as its argument.

Inside the callback, we first check if the response was successful by checking the ok property of the response object. If the response was not successful, an Error is thrown with a message saying 'Network response was not ok', along with the status text of the response.

If the response was successful, the callback returns another Promise by calling response.json(). This method reads the response stream to completion and parses the result as JSON.

The then() method is chained again to handle the resolved value of the response.json() Promise. This callback receives the parsed JSON data as its argument and logs the data to the console.

Finally, a catch() method is chained to the end of the Promise chain. The catch() method is used to handle any rejections of the Promises in the chain, including any errors that might occur during the fetch operation or the parsing of the JSON. If an error is caught, the error is logged to the console with a custom error message.

In summary, this code example demonstrates how to use the Fetch API to perform a network request, handle the response, parse the response data as JSON, and handle any errors that might occur during these operations.

7.1.4 Using Async/Await with Fetch

The Fetch API, a powerful and flexible tool for making network requests, can be used in conjunction with async and await to create a more synchronous style of handling asynchronous operations. This approach allows us to write code that is easier to understand and debug because it appears to be synchronous, even though it is actually being executed asynchronously.

This is particularly useful in scenarios where we need to wait for the response from one request before making another, for example, when chaining API requests. By using async and await with the Fetch API, we can greatly simplify the structure and readability of our code.

The Fetch API facilitates the fetching of resources across the network and is an integral part of modern web development, enabling a more flexible and powerful way to make HTTP requests compared to the traditional XMLHttpRequest.

In JavaScript, async and await are keywords that provide a way to write promise-based code in a more synchronous manner. They allow developers to handle asynchronous operations without getting into callback hell, improving code readability and maintainability.

When using async and await with the Fetch API, asynchronous operations such as network requests or file operations can be written in a way that appears to be blocking, but in reality, it's non-blocking. This means that while the asynchronous operation is being processed, the JavaScript engine can execute other operations without being blocked by the pending asynchronous operation.

Example: Using Async/Await with Fetch

The usage of async and await with the Fetch API looks something like this:

```
async function fetchData() {
    try {
        const response = await fetch('<https://api.example.com/data>');
        if (!response.ok) {
            throw new Error('Network response was not ok ' + response.statusText);
        }
        const data = await response.json();
        console.log(data);
    } catch (error) {
        console.error('There was a problem with your fetch operation:', error);
    }
}

fetchData();
```

In this example, the fetchData function is declared as async, indicating that the function will return a promise. Inside the fetchData function, the await keyword is used before the fetch method and response.json(). This tells JavaScript to pause the execution of the fetchData function until the promise from fetch and response.json() is settled, and then resumes the execution and returns the resolved value.

If an error occurs during the fetch operation or while parsing the JSON, it's caught in the catch block, preventing the program from crashing and providing an opportunity to handle the error gracefully.

Combining the Fetch API with async and await not only improves code readability but also makes it easier to handle errors and edge cases, making it a powerful tool for developing complex web applications that rely heavily on asynchronous operations.

7.1.5 Handling Timeouts with Fetch

The Fetch API does not natively support request timeouts. However, you can implement timeouts using JavaScript's Promise.race() function to enhance the robustness of your network requests, especially in environments with unreliable network conditions. Although the Fetch API provides a flexible and modern way of making network requests, it does not come with built-in support for request timeouts.

Request timeouts are important in managing network requests, especially in environments where network conditions may be unreliable or unstable. These timeouts help ensure that your application remains responsive and does not hang while waiting for a network request to complete, providing a better user experience.

To implement timeouts with the Fetch API, the document suggests using JavaScript's Promise.race() function. This function takes an array of promises and returns a promise that resolves or rejects as soon as one of the promises in the array resolves or rejects, hence the name "race."

By using Promise.race(), you can set up a race between the fetch request and a timeout promise. If the fetch request completes before the timeout, the fetch request's promise will resolve first, and its result will be used. If the timeout occurs before the fetch request completes, the timeout promise will reject first, allowing you to handle the timeout situation as needed.

This approach enhances the robustness of your network requests, giving you more control over their behavior and ensuring that your application can handle a wide range of network conditions effectively. This is particularly crucial in modern web applications, where smooth and responsive interaction with external servers and APIs is a key part of providing a high-quality user experience.

Example: Implementing Timeouts in Fetch

```
function fetchWithTimeout(url, options, timeout = 5000) {
    const fetchPromise = fetch(url, options);
    const timeoutPromise = new Promise((resolve, reject) => {
        setTimeout(() => reject(new Error("Request timed out")), timeout);
    });
    return Promise.race([fetchPromise, timeoutPromise]);
}

fetchWithTimeout('<https://api.example.com/data>')
    .then(response => response.json())
    .then(data => console.log(data))
    .catch(error => console.error('Failed:', error));
```

This JavaScript example snippet introduces a function named fetchWithTimeout, which is designed to send a network request to a specific URL with a timeout. This function is especially useful in scenarios where you're making network requests in environments with potentially unreliable or high-latency network connections, and you want to avoid having your application hang indefinitely waiting for a response that might never arrive.

The function's parameters are url, options, and timeout. The url is the endpoint you're sending the request to, options are any additional parameters or headers you want to include in your request, and timeout is the maximum number of milliseconds you want to wait for the response before giving up. The default timeout is set to 5000 milliseconds (or 5 seconds), but you can customize this value to your needs.

The function works by using the fetch API to make the request, which is a modern, promise-based method of making network requests in JavaScript. fetch provides a more powerful and flexible approach to making HTTP requests compared to older methods like XMLHttpRequest.

However, one limitation of the fetch API is that it does not natively support request timeouts. To work around this limitation, the function uses Promise.race to set the timeout. Promise.race is a method that takes an array of promises and returns a new promise that settles as soon as one of the input promises settles. In other words, it "races" the promises against each other and gives you the result of the fastest one.

In this case, we're racing the fetch request (which is a promise that settles when the request completes) against a timeout promise (which is a promise that automatically rejects after the specified timeout period). If the fetch request completes before the timeout, the fetch promise will settle first and its result will be used. If the timeout occurs before the fetch request completes, the timeout promise will reject first, and an Error with the message "Request timed out" will be thrown.

The usage of the function is demonstrated in the latter part of the code snippet. Here, it sends a GET request to 'https://api.example.com/data', attempts to parse the response as JSON using the .json() method, and then either logs the resulting data to the console if successful or logs an error message if it fails.

In this example, fetchWithTimeout races the fetch promise against a timeout promise, which will reject after a specified timeout period. This ensures that your application can handle situations where a request might hang longer than expected.

7.1.6 Streaming Responses

The Fetch API supports streaming of responses, allowing you to start processing data as soon as it begins to arrive. This is particularly useful for handling large datasets or streaming media.

The Fetch API is a powerful feature that gives developers the ability to start processing data as soon as it begins to arrive, rather than having to wait for the entire data set to be fully downloaded. This is particularly beneficial when you're dealing with large data sets or streaming media.

In traditional data transfer scenarios, you would typically have to wait for the entire data set to be downloaded before you could start processing it. This could result in significant delays, especially when dealing with large amounts of data or in scenarios where network connectivity is poor.

However, with the Fetch API's Streaming Responses feature, data can be processed in chunks as it arrives. This means that you can start working with the data almost immediately, improving the perceived performance of your application and providing a better user experience.

This feature can be particularly beneficial when developing applications that need to handle tasks such as live video streaming or real-time data processing, where waiting for the entire data set to download isn't practical or efficient.

The Fetch API abstracts away many of the complexities associated with streaming data, allowing developers to focus on building their applications without having to worry about the underlying details of data transmission and processing. With its support for Streaming Responses, the Fetch API is an invaluable tool for modern web development.

Example: Streaming a Response with Fetch

```javascript
async function fetchAndProcessStream(url) {
    const response = await fetch(url);
    const reader = response.body.getReader();
    while (true) {
        const { done, value } = await reader.read();
        if (done) break;
        console.log('Received chunk', value);
        // Process each chunk
    }
    console.log('Response fully processed');
}

fetchAndProcessStream('<https://api.example.com/large-data>');
```

Here's a step-by-step breakdown of what the function does:

1. async function fetchAndProcessStream(url) {: This line declares an asynchronous function named 'fetchAndProcessStream'. The 'async' keyword indicates that this function returns a Promise. The function takes a single argument 'url', which is the URL of the data resource you want to fetch.

2. const response = await fetch(url);: This line sends a fetch request to the specified URL and waits for the response. The 'await' keyword is used to pause the execution of the function until the Promise returned by the fetch() method is settled.
3. const reader = response.body.getReader();: This line gets a readable stream from the response body. A readable stream is an object that allows you to read data from a source in an asynchronous, streaming manner.
4. while (true) {: This line starts an infinite loop. This loop will continue until it's explicitly broken out of.
5. const { done, value } = await reader.read();: This line reads a chunk of data from the stream. The read() method returns a Promise that resolves to an object. The object contains two properties: 'done' and 'value'. 'done' is a boolean indicating if the reader has finished reading the data, and 'value' is the data chunk.
6. if (done) break;: This line checks if the reader has finished reading the data. If 'done' is true, the loop is broken, and the function stops reading data from the stream.
7. console.log('Received chunk', value);: This line logs each received chunk to the console. This is where you could add your code to process each chunk of data as it arrives.
8. console.log('Response fully processed');: After all data chunks have been received and processed, this line logs 'Response fully processed' to the console, indicating that the entire response has been handled.
9. fetchAndProcessStream('<https://api.example.com/large-data>');: The last line of the code is a call to the fetchAndProcessStream function, with the URL of a large data resource as an argument.

This function is particularly useful when you're dealing with large data sets or streaming data, as it allows for efficient, real-time processing of data as it arrives. Instead of waiting for the entire data set to download before starting processing, this function enables the application to start working with the data almost immediately, improving the perceived performance of the application and providing a better user experience.

This code example demonstrates how to read from a streaming response incrementally, which can improve the perceived performance of your web application when dealing with large amounts of data.

7.1.7 Fetch with CORS

Cross-Origin Resource Sharing (CORS) is a common requirement for web applications that make requests to domains different from the origin domain. Understanding how to handle CORS with Fetch is essential for modern web development.

CORS is a mechanism that permits or denies web applications to make requests to a domain that is different from their own origin domain. This is a common requirement in web development today as many web applications need to access resources, such as fonts, JavaScript, and APIs, which are hosted on a different domain.

The Fetch API is a modern, promise-based API built into JavaScript that provides a flexible and powerful way to make network requests. It's an upgrade to the older XMLHttpRequest and it allows developers to make requests to both same-origin and cross-origin destinations, hence making it a valuable tool for handling CORS.

Combining Fetch with CORS allows developers to make cross-origin requests directly from their web applications, providing a way to interact with other sites and services via their APIs. This can greatly expand the capabilities of a web application, enabling it to pull in data from various sources, integrate with other services, and interact with the broader web.

However, as with all things related to security and the web, it's important to use these tools wisely. CORS is a security feature designed to protect users and their data, so it's essential to understand how it works and how to use it properly. Fetch, while powerful and flexible, is a low-level API that requires a good understanding of HTTP and the same-origin policy to use effectively and securely.

The Fetch API and CORS are essential tools in modern web development. Understanding how they work together is key for building sophisticated web applications that can interact with the broader web while still protecting the user's security.

Example: Fetch with CORS

```javascript
fetch('<https://api.another-domain.com/data>', {
    method: 'GET',
    mode: 'cors',  // Ensure CORS mode is set if needed
    headers: {
        'Content-Type': 'application/json'
    }
})
.then(response => response.json())
.then(data => console.log(data))
.catch(error => console.error('CORS or network error:', error));
```

In this example, we are using the Fetch API, a built-in browser interface for making HTTP requests. It's promise-based, meaning it returns a Promise that resolves to the Response object representing the response to the request.

This is how it works:

1. fetch('<https://api.another-domain.com/data>', {...}): The fetch() function is called with the URL of the API we want to access. It takes two arguments, the input and the init (optional). The input is the URL we are fetching, and the init is an options object containing any custom settings that you want to apply to the request.

2. method: 'GET': This option indicates the request method, in this case, GET. The GET method is used to request data from a specified resource.

3. mode: 'cors': This is the mode of the request. Here, it is set to 'cors' which stands for Cross-Origin Resource Sharing. This is a mechanism that allows or blocks requested resources based on the origin domain. It's needed when we want to allow requests coming from different domains.

4. headers: {...}: Headers of the request are set in this section. The 'Content-Type' header is set to 'application/json', which means the server will interpret the sent data as a JSON object.

5. .then(response => response.json()): Once the request is made, the fetch API returns a Promise that resolves to the Response object. The .then() method is a Promise method used for callback functions for the success and failure of Promises. Here, the response object is passed into a callback function where it's converted into JSON format using the json() method.

6. .then(data => console.log(data)): After the conversion to JSON, the data is passed into another .then() callback where it's logged to the console.

7. .catch(error => console.error('CORS or network error:', error)): The catch() method here is used to catch any errors that might occur during the fetch operation. If an error occurs during the operation, it's passed into a callback function and is logged into the console.

In summary, this code sends a GET request to the specified URL and logs the response (or any error that might occur) to the console. The use of promises with .then() and .catch() methods allows for handling of asynchronous operations, making it possible to wait for the server's response and handle it once it's available.

7.2 Working with Files and Blobs

In the dynamic world of modern web development, the ability to handle files and what we refer to as binary large objects (or Blobs, for short) is a fundamental requirement. This is a common necessity across a broad range of tasks, whether it's for the purpose of uploading images to a server, processing files downloaded from the internet, or even saving data that has been generated within the workings of the web application itself.

JavaScript, a programming language that has become an integral part of the web development landscape, fortunately, provides us with robust and comprehensive APIs for dealing with files and Blobs. These APIs are designed to make it possible to handle such data both efficiently and securely, a critical aspect in today's era where data breaches are a real concern.

In this section, we will delve deeper into this topic. We'll explore how to work with files and Blobs in the context of your applications, offering practical examples and tips on how to manage such types of data effectively. We aim to provide you with a clear understanding and equip you with the necessary skills to handle these common yet crucial tasks in your web development journey.

7.2.1 Understanding Blobs

A Blob (Binary Large Object) represents immutable raw binary data, and it can hold large amounts of data. Blobs are typically used to handle data types such as images, audio, and other binary formats alongside text files.

A Blob or Binary Large Object in programming, and particularly in JavaScript, represents raw binary data that is immutable, meaning it cannot be changed once it's created. The ability to handle Blobs is a fundamental requirement in modern web development, especially when dealing with large amounts of data.

Blobs are immensely useful and versatile as they can hold large amounts of data, and are typically used to handle different data types. This includes images, audio, and other binary formats. They can also be used to handle text files. This makes them an essential tool in the world of web development, where handling and processing different types of data is a daily requirement.

You can create a Blob directly in JavaScript. This is done using the Blob() constructor. The Blob constructor takes an array of data and options as parameters. The array of data is the content that you want to store in the Blob, and the options object can be used to specify properties such as the type of data being stored.

Once a Blob is created, it can be manipulated in numerous ways depending on the requirements of your application. For example, you can read its contents, create a URL for it, or send it to a server. The Blob interface in JavaScript provides a number of methods and properties that you can use to carry out these operations.

Blobs represent a powerful way to handle large amounts of different types of data in JavaScript, and understanding how to create and manipulate them is an essential skill for any modern web developer.

Creating and Manipulating Blobs

You can create a Blob directly in JavaScript using the Blob() constructor, which takes an array of data and options as parameters.

Example:

Here's an example of how you might create a Blob:

```
const text = 'Hello, World!';
const blob = new Blob([text], { type: 'text/plain' });
```

In this example, a new Blob is created containing the text 'Hello, World!'. The type of data being stored is specified as 'text/plain'.

Blobs provide a powerful mechanism for handling large amounts of different types of data in JavaScript, and understanding how to create and manipulate them is an essential skill for any modern web developer.

Another Example: Creating a Blob

```
const data = new Uint8Array([0x48, 0x65, 0x6c, 0x6c, 0x6f]); // Binary data for 'Hello'
const blob = new Blob([data], { type: 'text/plain' });

console.log(blob.size); // Outputs: 5
console.log(blob.type); // Outputs: 'text/plain'
```

The ecode starts by creating a Uint8Array and assigns it to the variable data. A Uint8Array is a typed array that represents an array of 8-bit unsigned integers. The array holds the hexadecimal representation of the ASCII values for each character in the string 'Hello'. The ASCII value for 'H' is 0x48, 'e' is 0x65, and 'l' is 0x6c, and so on.

Next, a new Blob object is created with the new Blob() constructor. The constructor takes two arguments. The first argument is an array that contains the data you want to put in the Blob. In this case, it's the Uint8Array data. The second argument is an optional options object where you can set the type property to a MIME type that represents the kind of data you're storing. Here, the type is set to 'text/plain', which represents plain text data.

The size and type of the Blob object are then logged to the console using console.log(). The blob.size property returns the size of the Blob in bytes. In this case, it's 5, which corresponds to the number of characters in 'Hello'. The blob.type property returns the MIME type of the Blob. Here, it's 'text/plain', as set when the Blob was created.

This example showcases the creation and basic manipulation of Blobs in JavaScript, which is a very useful feature when working with binary data in web development.

7.2.2 Working with the File API

The File API extends the Blob interface, providing additional properties and methods to support user-generated file content. When a user selects files using an input element, you can access those files as a FileList object.

The File API extends the functionalities of the Blob interface, a component of JavaScript that represents raw binary data. The Blob interface is useful in handling different types of data, such as images, audio files, and other binary formats, as well as text files.

The File API takes this a step further by providing additional properties and methods to support user-generated file content. This is particularly useful in scenarios where a user is required to upload a file, such as an image or a document, through an input element in a web application.

When a user selects files using an input element in a web application, those files can be accessed as a 'FileList' object. The 'FileList' object is an array-like sequence of 'File' or 'Blob' objects, and it allows developers to access the details of each file in the list, such as its name, size, and type.

The 'FileList' object also enables developers to read the content of the files, manipulate them, or send them to a server. This is critical in scenarios such as when a user is required to upload a profile picture or a document on a web application.

For instance, a user may need to upload a profile picture on a social media application or a resume on a job portal. In these scenarios, the File API becomes an invaluable tool, allowing developers to handle the uploaded files, process them, and store them on a server.

The File API extends the Blob interface to provide powerful tools for developers to handle user-generated file content in web applications. It ensures that developers can effectively deal with files uploaded by users, from accessing and reading the files to processing and storing them on a server.

Example: Reading Files from an Input Element

```javascript
document.getElementById('fileInput').addEventListener('change', event => {
    const file = event.target.files[0]; // Get the first file
    if (!file) {
        return;
    }

    console.log(`File name: ${file.name}`);
    console.log(`File size: ${file.size} bytes`);
    console.log(`File type: ${file.type}`);
```

```
    console.log(`Last modified: ${new Date(file.lastModified)}`);
});
<!-- HTML to include a file input element -->
<input type="file" id="fileInput">
```

In this example, we add an event listener to a file input element. When files are selected, it logs details about the first file.

In the example code, we have two parts: a JavaScript code snippet and an HTML element. The HTML element is a file input form where a user can select a file from their computer. It has an ID of 'fileInput', which allows it to be selected and manipulated in JavaScript.

The JavaScript code snippet does the following:

1. It selects the file input element by using document.getElementById('fileInput'). This JavaScript function selects the HTML element with the ID 'fileInput', which in this case is our file input element.
2. It then adds an event listener to this file input element with .addEventListener('change', event => {...}). An event listener waits for a specific event to happen and then executes a function when that event occurs. In this case, the event is 'change', which is triggered when the user selects a file from the file input form.
3. Inside the event listener, it defines what should happen when the 'change' event occurs. The function takes an event object as a parameter.
4. It then gets the first file selected by the user with const file = event.target.files[0]. The files property is a FileList object that represents all the files selected by the user. It's an array-like object, so we can get the first file with index 0, files[0].
5. It checks if a file has indeed been selected with an if statement. If no file was selected (!file), it exits the function early with return.
6. If a file was selected, it logs the name, size in bytes, type, and the last modification date of the file using console.log().

This example is a simple implementation of a file input form where users can select a file from their computers. It then logs the details of the selected file, such as the file name, size, type, and the last modification date. This could be useful in scenarios where you need to handle file uploads and want to ensure that users upload the correct type and size of files, among other things.

7.2.3 Reading Files as Text, Data URLs, or Arrays

Once you have a reference to a File or Blob, you can read its content in various formats using the FileReader API. This is particularly useful for displaying file contents or further processing them in your web application.

This section delves into the process of reading files as text, Data URLs, or arrays once you have a reference to a File or Blob object in your web application.

Files and Blobs represent data in various formats such as images, audio, video, text, and other binary formats. Once you have obtained a reference to a File or Blob, the FileReader API can be used to read its content in different formats, depending on your needs.

The FileReader API is an interface provided by JavaScript that allows web applications to asynchronously read the contents of files or raw data buffers stored in the user's computer, using File or Blob objects to specify the file or data to read. It provides several methods for reading file data, including reading data as a DOMString (text), as a Data URL, or as an array buffer.

Reading a file as text is straightforward and useful when dealing with text files. The readAsText method of the FileReader object is used to start reading the contents of the specified Blob or File. When the read operation is finished, the onload event is triggered, and the result attribute contains the contents of the file as a text string.

Reading a file as a Data URL is useful for binary files such as images or audio. Data URLs are strings that contain a file's data as a base64-encoded string, preceded by the file's MIME type. The readAsDataURL method is used to start reading the specified Blob or File, and upon load, the file's data is represented as a Data URL string which can be used as a source for elements like <img>.

Reading a file as an array buffer is useful when dealing with binary data, as it allows you to manipulate the file's data at a byte level. The readAsArrayBuffer method is used here, and it starts reading the specified Blob or File. When the read operation is finished, the onload event is fired and the result attribute contains an ArrayBuffer representing the file's data.

The FileReader API is a powerful tool that allows web applications to read the contents of files or Blobs in a variety of formats, depending on the requirements of your application. It's particularly useful for displaying file contents on the web or for further processing them within your application.

Example: Reading a File as Text

```
function readFile(file) {
    const reader = new FileReader();
    reader.onload = function(event) {
        console.log('File content:', event.target.result);
    };
    reader.onerror = function(error) {
        console.error('Error reading file:', error);
    };
    reader.readAsText(file);
}

// This function would be called within the 'change' event listener above
```

This example features a JavaScript function named 'readFile'. It uses the FileReader API to read the contents of a file. The function is designed to take a file as its parameter.

Inside the function, a new instance of a FileReader is created and assigned to the variable 'reader'. The FileReader is an object that allows web applications to read the contents of files (or raw data buffers), which are stored on the user's computer.

Once the FileReader object is created, an 'onload' event handler is attached to it. This event handler is set to a function that gets triggered once the file is successfully read. Within this function, the contents of the file are logged to the console using console.log(). The contents of the file can be accessed through 'event.target.result'.

An 'onerror' event handler is also attached to the FileReader instance. If any error occurs while reading the file, this event handler gets triggered. Inside this function, the error is logged to the console using console.error(). The error can be accessed through 'error'.

Then, the function proceeds to initiate the reading of the file with 'reader.readAsText(file)'. This method is used to start reading the contents of the specified Blob or File. When the read operation is finished, the 'onload' event is fired and the 'result' attribute contains the contents of the file as a text string.

This function, 'readFile', is expected to be called within a 'change' event listener for a file input element in a HTML form, which means this function would be triggered every time the user selects a new file.

In summary, this JavaScript function 'readFile' is a simple yet effective demonstration of how to use the FileReader API to read the contents of a file selected by a user in a web application. It shows how to handle both the successful reading of a file and any potential errors that might occur during the process.

Example: Reading a File as a Data URL

```javascript
function readAsDataURL(file) {
    const reader = new FileReader();
    reader.onload = function(event) {
        // This URL can be used as a source for <img> or other elements
        console.log('Data URL of the file:', event.target.result);
    };
    reader.readAsDataURL(file);
}
```

The sample code provided is a JavaScript function named 'readAsDataURL'. Its purpose is to read the contents of a file using the FileReader API. This powerful JavaScript tool enables web applications to read the contents of files stored on a user's device.

In this function, a 'file' is passed as an argument. This 'file' here could be an image, text file, audio file or any other type of file.

The function begins by creating a new instance of a FileReader, which is stored in the 'reader' variable. The FileReader object provides a number of methods and event handlers that can be used to read the contents of the file.

Next, an event handler 'onload' is attached to the reader. This event gets triggered when the FileReader has successfully completed reading the file. Inside this event handler, a function is defined that logs the Data URL of the file to the console. The Data URL represents the file's data as a base64 encoded string, and can be used as a source for HTML elements like '<img>' or '<audio>'. This means the file can be rendered directly in the browser without needing to be separately downloaded or stored.

Finally, the function initiates reading the file as a Data URL by calling the method 'readAsDataURL' on the reader, passing in the file. This method starts the reading process, and when completed, the 'onload' event is triggered.

This function is a useful tool for handling files in JavaScript, especially in scenarios where you want to display a file's contents directly in the browser or manipulate the file's data further within your web application. It illustrates the power and flexibility of the FileReader API for handling files in web development.

7.2.4 Efficiently Handling Large Files

When dealing with large files, it's crucial to consider the performance and memory implications of the approach you choose. Large files can consume significant memory and processing power, which can degrade the performance of your application and lead to a poor user experience.

JavaScript provides the capability to handle large files in chunks. This approach is particularly beneficial when uploading or processing large files on the client side. By breaking down a large file into smaller chunks, the file can be managed more efficiently, preventing the user interface from becoming unresponsive or "freezing".

This chunking technique allows each piece of the file to be processed individually, instead of attempting to process the entire file at once, which could overwhelm the system's resources. Therefore, chunking is especially useful when working with files that are large enough to potentially exceed the available memory or when the processing of a file might take a significant amount of time.

By understanding and implementing these strategies, developers can ensure that their applications remain responsive and performant, even when dealing with large files.

Example: Chunking a File for Upload

```javascript
function uploadFileInChunks(file) {
    const CHUNK_SIZE = 1024 * 1024; // 1MB
    let start = 0;

    while (start < file.size) {
        let chunk = file.slice(start, Math.min(file.size, start + CHUNK_SIZE));
        uploadChunk(chunk); // Assume uploadChunk is a function to upload a chunk
        start += CHUNK_SIZE;
    }
}

function uploadChunk(chunk) {
    // Upload logic here
    console.log(`Uploaded chunk of size: ${chunk.size}`);
}
```

In this example code snippet, there are two functions: uploadFileInChunks and uploadChunk.

The uploadFileInChunks function is designed to handle the chunking of the file. It takes one parameter: file, which is the file to be uploaded. Inside the function, a constant CHUNK_SIZE is defined, which determines the size of each chunk. In this case, the chunk size is set to 1MB (1024 * 1024 bytes).

A start variable is also defined and initially set to 0. This variable keeps track of the starting index for each chunk.

A while loop is then implemented, which continues to run as long as the start index is less than the size of the file. Inside the loop, the slice method is used to create a chunk from the file, starting from the start index and ending at the smaller of either the file size or start + CHUNK_SIZE. This chunk is then passed to the uploadChunk function, which is assumed to handle the actual upload of each chunk.

After each iteration, the start index is incremented by the CHUNK_SIZE, effectively moving the starting point for the next chunk.

The uploadChunk function is the second function in the snippet. It takes one parameter: chunk, which is a piece of the file to be uploaded. Inside this function, the upload logic would typically be implemented. However, the provided code does not include the actual upload logic, and instead logs a message to the console indicating the size of the uploaded chunk.

By using this chunking approach, large files can be handled more efficiently, helping to improve the performance of the file upload process and prevent the user interface from becoming unresponsive.

7.2.5 Creating and Downloading Blobs

In addition to the basic tasks of reading and uploading files, there may be instances where you need to create new files directly on the client side. This could be necessary for a variety of reasons, perhaps to generate a report based on user activities, or to allow users to export their data in a usable format.

Whatever the reason, the creation of new files on the client side is an important aspect of many applications. Once these files have been created, it's often necessary to provide a way for users to download them. This could be accomplished by creating a Blob - a type of object that represents a chunk of bytes - which can hold large amounts of data. Once you've created your Blob, you can then use a data URL to represent the Blob's data.

Alternatively, you can use the URL.createObjectURL() method, which creates a DOMString containing a URL representing the object given in the parameter. This URL can then be used to facilitate the download of the Blob's data. Both of these methods are effective in enabling users to download files directly from the client side.

Example: Creating and Downloading a Text File

```javascript
function downloadTextFile(text, filename) {
    const blob = new Blob([text], {type: 'text/plain'});
    const url = URL.createObjectURL(blob);

    const a = document.createElement('a');
    a.href = url;
    a.download = filename;
    document.body.appendChild(a);
    a.click();

    document.body.removeChild(a);
    URL.revokeObjectURL(url); // Clean up the URL object
}

downloadTextFile('Hello, world!', 'example.txt');
```

In this example, a new text file is created from a string, turned into a Blob, and then a temporary link is created for downloading the file. This method efficiently handles the creation and cleanup of resources needed for the download.

The function named 'downloadTextFile' is designed to create a text file from a specified string of text and then trigger a download of that file to the user's computer.

The function takes two parameters: 'text' and 'filename'. 'text' is the content that will be written into the file, and 'filename' is the name that will be given to the downloaded file.

The function begins by creating a new Blob object, which is a way to handle raw data in JavaScript. This Blob object is created from the 'text' parameter and is of the type 'text/plain', indicating that it's plain text data.

Once the Blob object is created, a URL representing this object is created using the URL.createObjectURL() method. This method generates a URL which can be used to represent the Blob's data. This URL is then stored in a variable named 'url'.

Next, a new anchor (<a>) HTML element is created using document.createElement('a'). This anchor element is used to facilitate the download of the Blob's data. The 'href' attribute of the anchor element is set to the Blob URL, and the 'download' attribute is set to the 'filename' parameter. This ensures that when the anchor element is clicked, the Blob's data will be downloaded with the specified filename.

The anchor element is then appended to the body of the document using document.body.appendChild(a), and a click event is simulated on the anchor element using a.click(). This triggers the download of the file.

After the file is downloaded, the anchor element is removed from the document using document.body.removeChild(a), and the Blob URL is revoked using URL.revokeObjectURL(url). Revoking the Blob URL is important as it frees up system resources.

Finally, the function is called with the parameters 'Hello, world!' and 'example.txt', which creates a text file named 'example.txt' containing the text 'Hello, world!', and initiates a download of this file.

In summary, this function demonstrates a method of creating and downloading a text file purely on the client side, without needing to interact with a server. It showcases the use of Blob objects, object URLs, and the manipulation of HTML elements to achieve this functionality.

7.2.6 Security Considerations

When dealing with the process of allowing file uploads or performing any form of file manipulation, it is of utmost importance to take into consideration the several security implications that could potentially arise.

- **Validate Input**: Always make it a point to validate the input both from the client side as well as the server side. In the case of file uploads, it is crucial to verify the type of file, its size, and the content it carries. This is done in order to prevent the uploading of files that could potentially be harmful or pose a security threat.
- **Sanitize Data**: When the content of a file is being displayed, it is imperative to sanitize it thoroughly. This step is particularly important if the content comprises user-generated input, as it helps in preventing cross-site scripting (XSS) attacks, which can have serious security repercussions.
- **Use HTTPS**: When files are being uploaded or downloaded, it is essential to ensure that your connections are secured with HTTPS. This is to prevent any potential interception of the data during the process, thereby adding an extra layer of security to the file handling process.

7.2.7 Best Practices for File Handling

- **User Feedback**: In order to enhance user experience, it is essential to provide clear and concise feedback to the user regarding the progress of file uploads or any processing that is being done on the files. This feedback becomes even more important for operations that might take a significant amount of time. By doing so, users are kept informed and can manage their expectations about the task's completion time.
- **Error Handling**: The implementation of robust error handling is absolutely crucial. It is important to always anticipate potential failures that may occur during file operations. These anticipated failures should be handled gracefully within your application to prevent any negative impact on the user's experience or the application's performance.
- **Performance Optimization**: Consideration of the performance impact of your file handling operations is a must, particularly in web applications that are expected to handle large files or a high volume of file transactions. The speed and efficiency of these operations could considerably affect the overall performance of the application, and as such, strategies for optimization need to be carefully thought out and implemented.

7.3 The History API

The History API, a vital tool within the realm of modern web development, offers developers a unique opportunity to engage with the browser's session history. This sophisticated interface gives developers the power to manipulate the web browser's history stack—a crucial feature that has revolutionized how we interact with the web today.

This functionality has a particularly transformative effect on single-page applications (SPAs). In a traditional web browsing scenario, navigating to a different section of a website would typically require a complete page reload. However, with the advent of SPAs and the capabilities provided by the History API, browser navigation can now be handled more efficiently, without the need for a full page refresh.

In this comprehensive section, we'll delve deeper into the capabilities that the History API brings to the table. We will explore its functionalities, demonstrating how it can be harnessed to significantly enhance the navigation experience for users on your web applications.

By understanding and adopting the History API, developers can create more dynamic and user-friendly web applications. This not only improves the user experience but also results in a more effective and performant application overall.

7.3.1 Overview of the History API

The History API provides methods that enable the addition, removal, and modification of history entries. These features are beneficial for applications that need to dynamically change the URL without reloading the page, manage state based on user navigation, or restore the previous state when a user navigates within their browser.

Interacting with the browser's session history is a key feature of the History API, allowing manipulation of the web browser's history stack. This capability has significantly influenced how users interact with web applications today.

The History API is especially transformative for single-page applications (SPAs). Unlike traditional web browsing where navigating to different sections of a website necessitates a full page reload, SPAs, paired with the History API, enable a more efficient form of browser navigation that doesn't require a full page refresh.

Key Methods of the History API:

The History API's key methods are as follows:

- history.pushState(): This method adds an entry to the browser's history stack. It's useful when you want to track user navigation within your application.
- history.replaceState(): This method modifies the current history entry without adding a new one. This is handy when you want to update the state or URL of the current history entry.
- history.back(): This method navigates one step back in the history stack. It simulates the user clicking the back button in their browser.
- history.forward(): This method navigates one step forward in the history stack. It simulates the user clicking the forward button in their browser.
- history.go(): This method navigates to a specific point in the history stack. It can go forward or backward in the history stack relative to the current page.

Through these methods, the History API allows developers to add, remove, and modify history entries. This functionality is especially useful in applications where you need to dynamically change the URL without reloading the page, manage application state based on user navigation, or restore the previous state when a user navigates back and forth in their browser.

In essence, the History API allows developers to manage the history stack directly, providing them with the ability to control the user's navigation experience more finely. This not only improves the user experience by making web navigation more intuitive and efficient but also results in a more effective and performant application overall.

7.3.2 Using pushState and replaceState

These methods are essential for managing history entries. They both take similar arguments: a state object, a title (which is currently ignored by most browsers but should be included for future compatibility), and a URL. This is especially useful in single-page applications (SPAs), where the navigation experience can be significantly enhanced without the need for full page reloads.

Both pushState and replaceState take similar arguments. The first argument is a state object, which can contain any sort of data that you want to associate with the new history entry. This could be anything from the ID of a specific piece of content, the coordinates of a map view, or any other type of data that you need to restore the previous state of your application when the user navigates.

The second argument is a title. It's worth noting that this argument is currently ignored by most browsers due to legacy issues. However, it's recommended to include it for the sake of future compatibility, as some browsers may opt to use it in the future.

The third and final argument is a URL. This is the new URL that will be shown in the address bar of the browser. This URL should correspond to something that the user will expect to see when they navigate to the page, providing a consistent and predictable user experience.

In essence, the pushState method is used to add an entry to the browser's history stack and modify the URL shown in the address bar, without causing a page reload. On the other hand, replaceState is used to modify the current history entry, replacing it with the new state, title, and URL provided.

By effectively using these methods, developers can create a more dynamic, efficient, and user-friendly navigation experience, improving the overall performance and effectiveness of their web applications.

Example: Using pushState

```
document.getElementById('newPage').addEventListener('click', function() {
    const state = { page_id: 1, user_id: 'abc123' };
    const title = 'New Page';
    const url = '/new-page';
```

```
    history.pushState(state, title, url);
    document.title = title; // Update the document title
    // Load and display the new page content here
    console.log('Page changed to:', url);
});
```

This an example of how the history.pushState() method can be used to manipulate the browser's history stack. This is particularly useful in single-page applications (SPAs) to mimic the process of navigating to a new page without actually requiring a full page reload.

Here's a step-by-step breakdown of the code:

1. document.getElementById('newPage').addEventListener('click', function() {...});: This line of code adds a 'click' event listener to the HTML element with the id 'newPage'. When this element is clicked, the function enclosed within the event listener is executed.
2. Inside the function, a new state object is created with const state = { page_id: 1, user_id: 'abc123' };. This object can hold any data that is relevant to the new history entry. In this example, the state object contains a page_id and user_id.
3. The title for the new page is defined with const title = 'New Page';.
4. The URL for the new page is defined with const url = '/new-page';.
5. Then, the history.pushState(state, title, url); line uses the history.pushState() method to update the browser's history stack with the newly defined state object, title, and URL. This effectively adds a new entry to the history stack without reloading the page.
6. The document's title is then updated to match the new page title with document.title = title;.
7. It's assumed that the new page's content would be loaded and displayed at this point, although this isn't shown in the code snippet.
8. Finally, a message is logged to the console indicating that the page has changed to the new URL with console.log('Page changed to:', url);.

This example demonstrates how you can use the history.pushState() method to handle navigation within a single-page application by updating the browser's history and the URL shown in the address bar, without needing a page reload.

Example: Using replaceState

```
document.getElementById('updatePage').addEventListener('click', function() {
    const state = { page_id: 1, user_id: 'abc123' };
    const title = 'Updated Page';
    const url = '/updated-page';
```

```
      history.replaceState(state, title, url);
      document.title = title; // Update the document title
      // Update the current page content here
      console.log('Page URL updated to:', url);
});
```

The code snippet initiates by listening for a click event on an HTML element with the id 'updatePage'. This id presumably corresponds to a button or a link that, when clicked, triggers the function enclosed within the event listener.

Within the function, the first step is to create a new state object with const state = { page_id: 1, user_id: 'abc123' };. The state object is a JavaScript object that can hold any data relevant to the new history entry. In this example, it contains a page_id and user_id.

Following this, the title for the new page is defined with const title = 'Updated Page';. This title will be used to update the document's title later in the function.

The URL for the new page is also defined with const url = '/updated-page';. This URL will be shown in the address bar of the browser when the function is executed.

The heart of the function is the usage of the history.replaceState(state, title, url); method. The 'replaceState' method modifies the current history entry in the browser's history stack with the newly defined state object, title, and URL. It does this without adding a new entry to the history stack and without causing a page reload.

The document's title is then updated to match the new page title with document.title = title;. This helps maintain consistency between the document's title and the history entry.

At this point, it is assumed that the corresponding content of the new page would be loaded and displayed, however, this part is not shown in the code snippet.

Finally, a message is logged to the console indicating that the page has changed to the new URL with console.log('Page URL updated to:', url);.

This function demonstrates how the 'replaceState' method can be used in the History API to handle navigation within a single-page application. It shows how to update the browser's history and the URL shown in the address bar without needing a page reload.

7.3.3 Handling the popstate Event

When the user navigates to a new state, the browser fires a popstate event. Handling this event is crucial for restoring the state when the user navigates using the browser's back and forward buttons.

In web applications, the term 'state' often refers to the condition or the contents of the web page at a particular point of time. When a user navigates from one state to another in a web application, the browser fires an event known as the popstate event. This event is dispatched to the window each time the active history entry changes. It happens when the user clicks the browser's back or forward buttons, or when history.back(), history.forward(), or history.go() methods are programmatically invoked.

Handling this popstate event is crucial for a key aspect of user experience, which is restoring the state of the web application when the user navigates through it using the browser's back and forward buttons. This is particularly important for single page applications (SPAs), where multiple 'pages' or states of an application are managed within a single HTML document.

For instance, suppose a user is filling out a multi-step form on a single page application. They fill out the first step of the form and move to the second step. If they decide to use the browser's back button to review their information on the first step, the popstate event will be fired. A well-designed web application will have an event handler set up for this popstate event. The handler will take the state information provided by the popstate event, and use it to correctly display the first step of the form, as well as the data the user entered.

The popstate event plays a critical role in maintaining consistency and predictability in the user experience across web applications. Proper handling of this event allows web applications to respond to user navigation actions accurately, maintaining the correct state of the application as users navigate through it.

Example: Handling popstate

```
window.addEventListener('popstate', function(event) {
    if (event.state) {
        console.log('State:', event.state);
        // Restore the page using the state object
        document.title = event.state.title;
        // Load the content corresponding to event.state.page_id or other state
properties
    }
});
```

This example demonstrates how to respond to navigation actions that change the history state. The popstate event's state property contains the state object associated with the new history entry, which can be used to update the page content accordingly.

The 'popstate' event is fired by the browser when the user navigates through the session history. This can occur due to the user clicking the back or forward buttons, or when the history.back(), history.forward(), or history.go() methods are programmatically invoked.

The event listener is added to the 'popstate' event using the window.addEventListener() method. The first argument provided to this method is the string 'popstate', which specifies the event to listen for. The second argument is a function that defines what to do when the 'popstate' event is fired.

Inside the function, there is a conditional statement that checks if the 'state' property of the 'event' object exists. The 'state' property contains the state object associated with the current history entry. This state object is the same one that was specified when the history entry was created using the history.pushState() or history.replaceState() methods.

If the 'state' property does exist (i.e., it's truthy), several actions are taken. First, the state object is logged to the console using console.log(). This can be helpful for debugging purposes, allowing developers to see the contents of the state object when the 'popstate' event is fired.

Next, the title of the document is updated to match the 'title' property of the state object with document.title = event.state.title;. This helps maintain consistency between the document's title and the state of the application.

The comment in the code indicates that the next step would be to load and display the content corresponding to the 'page_id' or other properties of the state object. This could involve fetching data from a server and updating the DOM, or simply showing/hiding different elements on the page.

7.3.4 Synchronizing State with the UI

One of the challenges when using the History API is ensuring that the application's user interface remains in sync with the current state of the history. It's important to manage this synchronization carefully, especially in complex applications where the UI depends on multiple state variables.

Using the History API in web applications can sometimes present challenges, particularly when it comes to ensuring that the application's user interface (UI) accurately reflects the current state of the history. This synchronization between the UI and the state of the history is crucial for the consistency and coherence of the user experience.

The History API allows developers to manipulate the web browser's history stack. This is a particularly transformative feature for single-page applications (SPAs), where browser navigation can now be handled more efficiently, without the need for a full page refresh. However, as the state of the history changes - whether due to user actions such as clicking on links or buttons, or programmatically through methods like history.pushState() or history.replaceState() - it's important that these changes are appropriately mirrored in the application's UI.

In complex applications, where the UI depends on multiple state variables, managing this synchronization can become especially challenging. Changes in the application state need to be accurately and promptly reflected in the UI. For instance, if a user navigates from one page to another, not only should the URL reflect this change (which is handled by the History API), but the UI should also be updated to display the content of the new page.

Therefore, when working with the History API, developers need to carefully manage the synchronization between the state of the history and the UI, to ensure a seamless and intuitive user experience. This might involve setting up event listeners that respond to changes in the history state, and updating the UI accordingly. It could also involve using other features of the web development framework being used, such as React's state management features, to help manage this synchronization.

While the History API can significantly enhance the navigation experience for users, it's important to manage the synchronization between the state of the history and the UI carefully. By doing so, developers can ensure that their applications not only provide efficient and intuitive navigation, but also a consistent and accurate user interface.

Example: Syncing State with UI

```
function updateContent(state) {
    if (!state) return;

    // Update UI components based on state
    if (state.page_id === "home") {
        loadHomePage();
    } else if (state.page_id === "contact") {
        loadContactPage();
    }
    // Update other UI elements as necessary
}
```

```
window.addEventListener('popstate', function(event) {
    updateContent(event.state);
});
```

The updateContent function is defined as a way to update the User Interface (UI) components of the application based on the current state. The state is passed as a parameter to this function. If there is no state (i.e., if state is null or undefined), the function immediately returns and does nothing.

However, if the state does exist, the function will update the UI based on the page_id property of the state object. If the page_id is equal to "home", it calls a function named loadHomePage() which presumably loads and displays the home page content. If the page_id is "contact", it calls loadContactPage(), which would load and display the contact page content.

Additionally, the comment in the function indicates that there may be other UI elements that need to be updated based on the state. These updates aren't shown in this example, but would likely involve showing or hiding different elements on the page, updating the values of form fields, changing the active state of navigation links, or any other UI changes that need to happen when the application state changes.

After defining the updateContent function, an event listener is added to the 'popstate' event using the window.addEventListener() method. This means that whenever a 'popstate' event is fired, the function provided as the second argument to addEventListener() will be executed.

In this case, the function is an anonymous function that calls updateContent(), passing the state property of the 'popstate' event object as an argument. The 'state' property contains the state object associated with the current history entry. This state object is the same one that was specified when the history entry was created using the history.pushState() or history.replaceState() methods.

This setup allows the application to respond appropriately to user navigation actions, updating the UI to reflect the current state of the application whenever the active history entry changes.

7.3.5 Integrating with Frameworks

Many modern JavaScript frameworks and libraries, such as React, Vue.js, and Angular, have built-in support for managing history and routing, often integrating seamlessly with the History API. When working with these frameworks, it's typically better to use their routing solutions, which are designed to work naturally with the framework's reactive system.

One of the key features that these libraries offer is their built-in support for managing browser history and routing. This is a crucial aspect of web application development, especially when it comes to SPAs. In such applications, instead of loading a new page for each different view, the same page updates dynamically in response to user interaction, often needing to handle changes in the browser's history stack and URL to provide a seamless user experience.

The History API is a powerful tool that allows developers to manipulate the browser's history stack directly. However, frameworks like React, Vue.js, and Angular have gone a step further and have integrated this functionality into their systems, providing their own mechanisms for managing history and routing.

For instance, React has a library called React Router, Vue.js has vue-router, and Angular has @angular/router. These libraries provide high-level, abstracted interfaces for managing routing, which under the hood use the History API or fall back to other techniques for older browsers that do not support it.

When developers are working with these frameworks, it is typically more beneficial to use these routing solutions, as they are specifically designed to work smoothly and naturally with the respective framework's reactive system. Using these tools not only abstracts away the complexity of dealing with the History API directly, but it also ensures that the application's routing behavior is consistent and reliable, as it leverages the tried-and-tested solutions provided by the framework.

While the History API is a crucial part of modern web development, when working with modern JavaScript frameworks such as React, Vue.js, and Angular, it's usually better to leverage their built-in routing solutions. These solutions are designed to integrate seamlessly with the History API and the framework's architecture, providing a more powerful and developer-friendly interface for managing browser history and routing.

Example: Using React Router

```
// A basic example in a React application using React Router
import { BrowserRouter as Router, Route, Switch, Link } from 'react-router-dom';

function App() {
    return (
        <Router>
            <div>
                <nav>
                    <Link to="/">Home</Link>
                    <Link to="/about">About</Link>
                </nav>
                <Switch>
```

```
                    <Route path="/about">
                        <About />
                    </Route>
                    <Route path="/">
                        <Home />
                    </Route>
                </Switch>
            </div>
        </Router>
    );
}
```

This is a simple example of how routing is done in a React application using the React Router library.

The first line imports necessary components from the 'react-router-dom' library. 'BrowserRouter' is renamed to 'Router' for convenience, and 'Route', 'Switch', and 'Link' are also imported. These components are essential for configuring routing in a React application:

- 'BrowserRouter' or 'Router' is a component that uses the HTML5 history API (pushState, replaceState and the popstate event) to keep your UI in sync with the URL.
- 'Route' is a component that renders some UI when a location matches the route's path.
- 'Switch' is used to render only the first 'Route' or 'Redirect' that matches the current location.
- 'Link' is used to create links in your application. Clicking a 'Link' triggers a navigation and updates the URL.

The 'App' function is a functional component that returns a JSX (JavaScript XML) element. Inside this function, a 'Router' component is used to wrap the entire application.

Within the 'Router', there's a 'div' element that contains a 'nav' element and a 'Switch' component. The 'nav' element contains two 'Link' components that create links to the 'Home' and 'About' pages of the application.

The 'to' prop in the 'Link' component is used to specify the path to which the application will navigate when the link is clicked. Here, there are links to the root path ('/') and the '/about' path.

The 'Switch' component is used to group 'Route' components. It only renders the first 'Route' or 'Redirect' in its children that matches the location. Here, there are two 'Route' components - one for the '/about' path and one for the root path ('/').

When the path in the URL matches '/about', the 'About' component is rendered. When the path matches '/', the 'Home' component is rendered.

This React Router setup allows the application to navigate between the 'Home' and 'About' pages without a page refresh, which is a key advantage of single-page applications (SPAs).

7.3.6 Handling Edge Cases

When using the History API, consider edge cases such as what happens when a user directly modifies the URL or navigates to a URL manually. Ensure that your application can handle such scenarios gracefully, providing error pages or redirection as needed.

In practical terms, handling edge cases means considering scenarios that are not the most common, but can occur and can potentially lead to bugs or unexpected behavior if not handled properly. In the context of the History API, these edge cases might include situations where a user manually modifies the URL in the browser's address bar, or navigates to a URL directly by entering it in the address bar or clicking a bookmark, rather than reaching the page through the normal navigation flow of the application.

Such direct manipulations of the URL do not automatically update the state of the application, which can lead to a mismatch between the URL and the state of the application. This can be confusing for users and can lead to errors or unexpected behavior. For example, a user might manually navigate to a URL that corresponds to a specific state of the application that requires some preconditions to be met. If these preconditions are not met, the application might not work correctly.

To prevent such issues, the text advises developers to ensure that their applications can handle such scenarios gracefully. This could mean providing error pages that inform the user of an issue and guide them back to a valid state, or implementing redirection mechanisms that automatically navigate the user to a valid state of the application when they try to access an invalid state directly.

Overall, the handling of edge cases is an important aspect of robust application design. It ensures that the application can handle all potential user interactions gracefully and reliably, which improves the overall user experience and the robustness of the application.

Example: Validating State

```
window.addEventListener('popstate', function(event) {
    if (!event.state || !isValidState(event.state)) {
        console.error('Invalid state or direct navigation detected');
```

```
        loadDefaultPage();  // Load a default page or redirect
    } else {
        updateContent(event.state);
    }
});

function isValidState(state) {
    return state && state.page_id && isValidPageId(state.page_id);
}
```

This example code is written in the React JSX syntax and it shows how to handle a 'popstate' event in a web application. The 'popstate' event is fired by the browser when the user navigates through the browser's history using the back or forward buttons, or when history.back(), history.forward(), or history.go() methods are programmatically invoked.

In the context of a single-page application (SPA), the 'popstate' event is crucial for restoring the state of the application when the user navigates through it using the browser's back and forward buttons.

The code starts by adding an event listener to the 'popstate' event using the window.addEventListener() method. The first argument to this method is the string 'popstate', which specifies the event to listen for. The second argument is a callback function that defines what to do when the 'popstate' event is fired.

The callback function first checks if the state property of the event object exists and if it's valid using the isValidState() function. The state property of the event object contains the state object that was associated with the history entry when it was created using the history.pushState() or history.replaceState() methods.

If the state property doesn't exist or isn't valid (as determined by isValidState()), it logs an error message to the console and then calls the loadDefaultPage() function. This function presumably loads a default page or redirects the user to a default location. This is a way of handling edge cases where a user may manually navigate to a URL that doesn't correspond to a valid state of the application.

If the state property does exist and is valid, the callback function calls the updateContent() function, passing the state object as an argument. Presumably, the updateContent() function updates the content of the page based on the state.

The isValidState() function is a helper function that checks whether the state object is valid. It returns true if the state object exists, contains a page_id property, and if the page_id is valid (as determined by another function isValidPageId()), and false otherwise.

7.4 Web Storage

Web Storage is an integral and indispensable feature in contemporary web development. This powerful tool offers the ability to store data locally within a user's browser, eliminating the need for continuous server requests. By doing so, it significantly enhances the user experience by ensuring data persistence, allowing web applications to save, retrieve, and manipulate data across browser sessions. This capability is particularly crucial in scenarios where the user might need to temporarily step away from their computer or face intermittent connectivity issues.

In this comprehensive section, we will delve deep into the two primary mechanisms of Web Storage: localStorage and sessionStorage. We will discuss in detail their specific use cases, the key differences between them, and practical techniques to implement them effectively in your web applications.

By understanding and leveraging these mechanisms, developers can create more efficient, user-friendly web applications that remember user preferences, maintain application state, and even work offline. Through the course of this section, we will provide real-world examples and best practices, giving you the tools and knowledge necessary to implement robust and resilient storage solutions in your web development projects.

7.4.1 Understanding Web Storage

Web Storage, an important feature of modern web applications, provides two distinct types of storage:

- **localStorage**: This is a type of Web Storage that provides persistent storage across sessions. Unlike other types of storage, data stored in localStorage does not expire or get deleted when the browser is closed. Instead, it remains stored on the user's device until it is explicitly cleared, either by the user or the web application. This makes localStorage an excellent choice for storing data that needs to be accessed across multiple sessions or visits to the website, such as user preferences or saved game progress.
- **sessionStorage**: This storage type, on the other hand, offers storage that is strictly limited to the lifespan of the window or tab in which the web application is running. As soon as the window or tab is closed, any data stored in sessionStorage is immediately deleted. This makes sessionStorage a perfect choice for storing data that is relevant

only for the duration of a single session, such as form data that the user is currently entering.

Despite their differences in lifespan and use cases, both localStorage and sessionStorage provide a very similar interface for storing and retrieving data. Data is stored in a simple, easy-to-use key-value pair system, which allows web applications to quickly and easily save data and retrieve it later. This makes Web Storage a powerful tool for enhancing the user experience of a web application.

Key Features of Web Storage:

Web Storage is a powerful tool that has several unique features. One of its primary characteristics is its ability to store data in the form of key-value pairs. This means that each item of data (the value) is associated with a unique identifier (the key), which can be used to quickly retrieve the data when needed.

Another notable feature of Web Storage is its considerable data storage capacity. It allows for about 5MB of data to be stored per origin. This is a significant amount of space, which can be of great use in a variety of applications. Moreover, this large storage capacity does not affect the performance of the website, thereby ensuring that the user experience remains smooth and seamless.

Lastly, Web Storage is designed in such a way that it does not transmit data back to the server. This can help reduce the overall amount of data that is sent with each request, which can be beneficial in terms of improving the efficiency of data transmission and reducing the load on the server.

7.4.2 Using localStorage

localStorage is particularly useful for storing preferences, settings, and other data that needs to persist beyond the current session.

The Web Storage API provides mechanisms for web applications to store data in a user's web browser. Among its two storage types, localStorage is one that provides persistent storage of data. In other words, data stored in localStorage does not expire or get deleted when the browser session ends or when the browser is closed. Instead, this data remains stored on the user's device until it is explicitly deleted either by the user or by the web application.

localStorage is very useful when a web application needs to save certain types of data over the long term. For instance, a web application might use localStorage to save user preferences or settings. Since these are details that a user would likely want to keep the same over multiple visits to the website, saving them in localStorage means they can be easily retrieved each time the user returns to the site, enhancing the user experience.

Another common use case for localStorage is saving progress or state in a web application. For example, if a user is working on a task in a web application and needs to step away, the application could save the current state of the task in localStorage. When the user returns, even if it's from a different browser session, the application can retrieve the saved state from localStorage and restore it, allowing the user to pick up right where they left off.

In short, localStorage is a powerful tool for web developers, offering a simple, client-side method for storing and persisting data in a user's web browser. By leveraging localStorage, developers can enhance the functionality and user experience of their web applications by maintaining data across multiple sessions.

Example: Using localStorage to Store User Settings

```javascript
function saveSettings(settings) {
    for (const key in settings) {
        localStorage.setItem(key, settings[key]);
    }
    console.log('Settings saved:', settings);
}

function loadSettings() {
    return {
        theme: localStorage.getItem('theme') || 'light',
        notifications: localStorage.getItem('notifications') || 'enabled'
    };
}

// Example usage
saveSettings({ theme: 'dark', notifications: 'enabled' });
const settings = loadSettings();
console.log('Loaded settings:', settings);
```

This example code includes two functions: saveSettings() and loadSettings().

- The saveSettings(settings) function takes an object as its parameter, which should contain settings. It stores each setting in the web browser's local storage. After saving the settings, it logs a message to the console confirming that the settings were saved.

- The loadSettings() function retrieves the 'theme' and 'notifications' settings from local storage. If a setting is not found in local storage, it uses a default value ('light' for theme and 'enabled' for notifications). The function returns an object with these settings.

The example usage shows how to save a settings object with 'dark' theme and enabled notifications using saveSettings(), and then how to load these settings using loadSettings().

7.4.3 Using sessionStorage

sessionStorage is ideal for storing data that should not persist once the browser is closed, such as data related to a specific session. sessionStorage is unique in its functionality as its storage lifespan is limited to the duration of the page session. A page session lasts as long as the browser is open, and survives over page reloads and restores. However, it is designed to be much more transient and the data stored in the sessionStorage gets cleared when the page session ends - that is, when the user closes the specific browser tab.

This makes sessionStorage ideal for storing data that should not persist once the browser tab is closed. For instance, it can be used to store information related to a specific session, such as user inputs within a form before submission, or the state of a web application that needs to be preserved across multiple pages within a single session, but not beyond.

This feature provides developers with a powerful tool to enhance the user experience by making the web application more responsive and reducing the need for continuous server interactions. By storing data in the user's browser, the application can quickly and efficiently access and utilize this data to enhance the functionality and user experience of the web application during that specific session.

Example: Using sessionStorage for Session-specific Data

```javascript
function storeSessionData(key, data) {
    sessionStorage.setItem(key, data);
    console.log(`Session data stored [${key}]:`, data);
}

function getSessionData(key) {
    return sessionStorage.getItem(key);
}

// Example usage
storeSessionData('pageVisit', 'Homepage');
console.log('Session data loaded:', getSessionData('pageVisit'));
```

This example defines two functions for handling session-specific data: storeSessionData and getSessionData.

The storeSessionData function takes two parameters: key and data. The key is a string that acts as an identifier for the data you want to store. The data is the actual information you want to save in the user's session. This function uses the sessionStorage.setItem method to store the data in the user's browser for the duration of the session. This method takes two arguments: the key and the data, and stores the data under the specified key. After storing the data, a message is logged to the console to confirm the operation, showing the key and the data that was stored.

The getSessionData function, on the other hand, is used to retrieve data from the session storage. It takes one parameter: key, which is the identifier for the data you want to retrieve. This function uses the sessionStorage.getItem method, which takes a key as its argument and returns the data stored under that key. If no data is found under the specified key, getItem returns null.

At the end of the script, we have an example of how these functions can be used. First, the storeSessionData function is called with 'pageVisit' as the key and 'Homepage' as the data. This will store the string 'Homepage' in the session storage under the key 'pageVisit'. Then, the getSessionData function is called with 'pageVisit' as the key to retrieve the data that was just stored. The retrieved data is then logged to the console.

This example is particularly useful in scenarios where you need to store and retrieve data within a single session, and you want the data to be deleted as soon as the session ends (i.e., when the user closes the tab or the browser).

Best Practices for Using Web Storage

1. **Security Considerations**: Web Storage, while extremely convenient, does come with its own set of security considerations. Given that it isn't a secure storage solution, it's crucial to bear in mind that sensitive information should never be stored directly in localStorage or sessionStorage. The reason for this is that any scripts running on the page can easily access the data, potentially leading to security breaches.
2. **Data Size Limitations**: Another important factor to consider is the storage capacity, which is typically around 5MB. If you exceed this limit, exceptions can occur, disrupting the functionality of your application. Therefore, it's essential to monitor storage limits with tools such as localStorage.length or sessionStorage.length before attempting to add more data. This will help you to manage your storage effectively and avoid any potential pitfalls.

3. **Efficiency and Performance**: While Web Storage is generally regarded as a fast storage solution, overutilization can lead to a slowdown in your application's performance. This is especially true if large volumes of data are being read frequently. To optimize the usage of Web Storage, consider caching data in variables where feasible. This approach can help enhance performance and ensure that your application runs smoothly.

7.5 Practical Exercises for Chapter 7: Web APIs and Interfaces

To solidify your understanding of the web APIs discussed in Chapter 7, this section provides practical exercises that focus on implementing and using these technologies in real-world scenarios. Each exercise includes a detailed explanation and solution code, helping you apply what you've learned effectively.

Exercise 1: Using the Fetch API

Objective: Write a function using the Fetch API to retrieve user data from a public API and log the user names to the console.

Solution:

```javascript
function fetchUserData() {
    fetch('<https://jsonplaceholder.typicode.com/users>')
        .then(response => {
            if (!response.ok) {
                throw new Error('Network response was not ok');
            }
            return response.json();
        })
        .then(users => {
            users.forEach(user => {
                console.log(user.name);
            });
        })
        .catch(error => {
            console.error('Fetch error:', error);
        });
}

fetchUserData();
```

This function makes a GET request to a public API that returns a list of users. It processes the JSON response to extract and log each user's name.

Exercise 2: Implementing localStorage

Objective: Create functions to save, retrieve, and delete a theme setting in localStorage.

Solution:

```
function saveTheme(theme) {
    localStorage.setItem('theme', theme);
    console.log('Theme saved:', theme);
}

function getTheme() {
    return localStorage.getItem('theme') || 'default'; // Return 'default' if no theme
set
}

function removeTheme() {
    localStorage.removeItem('theme');
    console.log('Theme removed');
}

// Example usage
saveTheme('dark');
console.log('Current theme:', getTheme());
removeTheme();
```

This exercise demonstrates how to use localStorage to store, retrieve, and delete user settings, specifically a theme preference.

Exercise 3: Handling Session Data with sessionStorage

Objective: Write functions to store and retrieve session data about the user's current page visit count.

Solution:

```
function incrementPageVisit() {
    let visits = parseInt(sessionStorage.getItem('visitCount')) || 0;
    visits++;
    sessionStorage.setItem('visitCount', visits);
```

```
        console.log(`Visit count updated: ${visits}`);
}

function getPageVisits() {
    return sessionStorage.getItem('visitCount') || 0;
}

// Example usage
incrementPageVisit();
console.log(`Page visits: ${getPageVisits()}`);
```

This exercise showcases how to manage session-specific data using sessionStorage, tracking the number of times a user visits a page during a session.

Exercise 4: Manipulating History with the History API

Objective: Implement a function that navigates the user back to the homepage using the History API after modifying history entries.

Solution:

```
function navigateHome() {
    history.pushState({ page: 'homepage' }, 'homepage', '/home');
    history.go(); // Navigates to the new state
    console.log('Navigation to homepage triggered');
}

// Trigger the navigation
navigateHome();
```

This function uses the History API to programmatically add a new history entry for the homepage and then navigates to it. This is useful in single-page applications where you need to manage navigation history manually.

These exercises are designed to help you practice and understand the use of various web APIs and interfaces covered in Chapter 7. By completing these exercises, you should gain a practical understanding of how to integrate these APIs into web applications, enhancing their functionality and interactivity.

Chapter Summary 7: Web APIs and Interfaces

In Chapter 7 of "JavaScript from Scratch: Unlock your Web Development Superpowers," we delved into the essential web APIs and interfaces that empower developers to build sophisticated, interactive, and dynamic web applications. This chapter provided a comprehensive exploration of several key APIs provided by modern web browsers, including the Fetch API, Web Storage, the History API, and more. Each section not only introduced the functionalities and benefits of these APIs but also demonstrated practical applications through detailed examples and exercises.

The Fetch API

We began with the Fetch API, a modern tool for making HTTP requests. This API is crucial for communicating with servers and handling asynchronous data flows in web applications. The Fetch API offers a more powerful and flexible alternative to XMLHttpRequest, using Promises to make the handling of asynchronous operations simpler and more efficient. We explored how to make GET and POST requests, handle responses, and manage errors effectively, providing a foundation for developers to retrieve and send data seamlessly within their applications.

Web Storage

Next, we covered the Web Storage API, which includes localStorage and sessionStorage. This API enables the storage of data locally in the user's browser, allowing applications to save, retrieve, and manage user data across sessions. We discussed the differences between localStorage (which persists data across sessions) and sessionStorage (which retains data for the duration of the page session only), and provided examples of how to use these storage options to enhance user experience and maintain state within applications.

The History API

The History API was another focus of this chapter. It allows developers to interact with the browser's session history, providing methods to manipulate the history stack programmatically. This is particularly useful in single-page applications where managing the navigation stack is crucial for a seamless user experience. We examined how to use methods like pushState and replaceState to modify the browser's history without reloading the page, and how to handle the popstate event to update content when users navigate through their history.

Practical Exercises

Each section included practical exercises that challenged you to implement the discussed APIs in real-world scenarios. These exercises were designed to reinforce learning, enhance understanding, and provide hands-on experience with the APIs. From making network requests and managing local storage to manipulating browser history, these exercises prepared you to integrate these capabilities into your own projects effectively.

Conclusion

Understanding and utilizing web APIs and interfaces is essential for modern web development. These tools provide the mechanisms needed to create interactive, responsive, and user-friendly web applications. As we conclude this chapter, remember that mastering these APIs not only enhances the functionality of your applications but also significantly improves the overall user experience. The knowledge gained here sets a solid foundation for exploring more advanced features and integrations in future projects, ensuring you are well-equipped to tackle the challenges of dynamic web application development.

Chapter 8: Error Handling and Testing

Welcome to Chapter 8, titled "Error Handling and Testing," a topic of utmost importance when it comes to developing web applications that are robust and reliable. In the ever-evolving world of web development, ensuring the smooth operation of applications is paramount, and this chapter is dedicated to providing a comprehensive guide on the strategies, techniques, and tools that developers can leverage to identify, handle, and prevent errors effectively. Furthermore, it provides insights on how to ensure that the code behaves as expected through stringent testing procedures.

Effective error handling and thorough testing are two critical pillars in the development process. They not only improve the overall quality and reliability of applications but, equally important, enhance their maintainability. The user experience is significantly improved as well, as these processes work hand in hand to reduce bugs and curb unexpected behaviors that can disrupt the user's interaction with the application.

Throughout this chapter, we will embark on a journey exploring various JavaScript error handling mechanisms. These include the traditional try, catch, finally blocks, the essential concepts of error propagation, as well as the practice of custom error creation. These mechanisms serve as the first line of defense against runtime errors, ensuring that your application remains responsive and performant even in the face of unexpected issues.

In addition to error handling, we will also delve deep into the realm of testing strategies and frameworks. These tools are designed to help ensure that your codebase remains bug-free and operates at optimum performance. We will begin our exploration with an in-depth look at the foundational technique for managing runtime errors: the try, catch, finally statement. This statement forms the bedrock of error handling in JavaScript, and mastering it is key to creating resilient and reliable web applications.

8.1 Try, Catch, Finally

The try, catch, finally construct plays a crucial role in the world of JavaScript programming as a fundamental error handling mechanism. It provides developers with a structured pathway to gracefully handle exceptions, which are unforeseen errors that occur during program execution. Its beauty lies in the ability to not only capture these exceptions, but to also provide a means to respond to them in a controlled and orderly manner.

Furthermore, the finally clause within this construct holds significant importance. It ensures that specific cleanup actions get executed, regardless of whether an error occurred or not. This adds an additional layer of resilience to your code, making sure that important tasks (like closing connections or freeing up resources) always get done, thus maintaining the overall integrity of the program.

Understanding Try, Catch, Finally

- **try Block**: The try block encapsulates the code that might potentially result in an error or an exception. It serves as a protective shell, allowing the program to test a block of code for errors while it's being executed. If an exception occurs during the execution of this block, the normal flow of the code is disrupted, and the control is immediately passed over to the corresponding catch block.
- **catch Block**: The catch block is essentially a safety net for the try block. It is executed if and when an error occurs in the try block. It acts as an exception handler, which is a special block of code that defines what the program should do when a specific error or exception occurs. For example, if a file that does not exist is being opened within the try block, the catch block could define the action to create the file or to notify the user about the missing file.
- **finally Block**: The finally block serves a unique role in this construct. It contains the code that will be executed whether an error occurs in the try block or not. This block does not depend on the occurrence of an error, rather it guarantees that certain key parts of the code will run irrespective of an error. This is typically used for cleaning up resources or performing tasks that must be completed regardless of what happens in the try and catch blocks. For example, if a file was opened for reading in the try block, it must be closed in the finally block whether an error occurred or not. This ensures that resources like memory and file handles are properly managed, regardless of the outcome of the try and catch blocks.

In the larger context of programming, the try, catch, and finally blocks form the cornerstone of error handling, providing a structured and systematic approach to managing and responding to errors or exceptions that may occur during the execution of a program. Mastering the use of these blocks is crucial to developing robust and resilient software applications that can handle unexpected issues gracefully without disrupting the user experience.

Example: Basic Try, Catch, Finally Usage

```
function performCalculation() {
    try {
        const value = potentiallyFaultyFunction(); // This function may throw an error
        console.log('Calculation successful:', value);
    } catch (error) {
        console.error('An error occurred:', error.message);
    } finally {
        console.log('This always executes, error or no error.');
    }
}

function potentiallyFaultyFunction() {
    if (Math.random() < 0.5) {
        throw new Error('Fault occurred!');
    }
    return 'Success';
}

performCalculation();
```

In this example, potentiallyFaultyFunction might throw an error randomly. The try block attempts to execute this function, the catch block handles any errors that occur, and the finally block executes code that runs no matter the result, ensuring that all necessary final actions are taken.

The performCalculation function uses a try block to attempt to execute potentiallyFaultyFunction. The try block serves as a protective shell around the code that might potentially result in an error or an exception. If an exception occurs during the execution of this block, the normal flow of the code is disrupted, and control is immediately passed over to the corresponding catch block.

The potentiallyFaultyFunction is designed to randomly throw an error. This function uses Math.random() to generate a random number between 0 and 1. If the generated number is less than 0.5, it throws an error with the message 'Fault occurred!'. If it doesn't throw an error, it returns the string 'Success'.

Back in the performCalculation function, if the potentiallyFaultyFunction executes successfully (i.e., it doesn't throw an error), the try block logs the message 'Calculation successful:' followed by the return value of the function ('Success').

If the potentiallyFaultyFunction throws an error, the catch block in performCalculation is engaged. The catch block serves as a safety net for the try block. It is executed if and when an error occurs in the try block. In this case, the catch block logs the message 'An error occurred:' followed by the error message from the exception ('Fault occurred!').

Finally, the performCalculation function includes a finally block. The finally block serves a unique role in the try, catch, finally construct. It contains code that will be executed whether an error occurs in the try block or not. In this example, the finally block logs the message 'This always executes, error or no error.' This demonstrates that certain parts of the code will run irrespective of an error, a crucial aspect of maintaining the overall integrity of a program.

The try, catch, finally construct in this example demonstrates a structured and systematic approach to managing and responding to errors or exceptions that may occur during the execution of a program. By handling unexpected issues gracefully, it helps in developing robust and resilient software applications that can continue to function without disrupting the user experience, even when errors occur.

8.1.2 Using Try, Catch for Graceful Error Handling

Using try, catch allows programs to continue executing even after an error occurs, preventing the entire application from crashing. This is particularly useful in user-facing applications where abrupt crashes can lead to poor user experiences.

"Using Try, Catch for Graceful Error Handling" is a concept in programming that emphasizes the use of "try" and "catch" constructs to manage errors in a program. This approach is particularly critical in ensuring that a program can continue its execution even when an error occurs, instead of crashing abruptly.

In JavaScript, the "try" block is used to wrap the code that might potentially lead to an error during its execution. If an error occurs within the "try" block, the flow of control is immediately passed to the corresponding "catch" block.

The "catch" block acts as a safety net for the "try" block. It is executed when an error or exception occurs in the "try" block. The "catch" block serves as an exception handler, which is a special block of code that defines what the program should do when a specific error or exception occurs.

The use of "try" and "catch" allows errors to be handled gracefully, meaning the program can react to the error in a controlled manner, perhaps correcting the issue, logging it for review, or informing the user, instead of allowing the entire application to crash. This error handling mechanism significantly enhances the user experience, as the program remains functional and responsive even when unforeseen issues occur.

Mastering the use of "try" and "catch" is vital for developing robust, resilient, and user-friendly applications that can manage and respond to errors in a structured and systematic way.

Example: Handling User Input Errors

```javascript
function processUserInput(input) {
    try {
        validateInput(input); // Throws error if input is invalid
        console.log('Input is valid:', input);
    } catch (error) {
        console.error('Invalid user input:', error.message);
        return; // Return early or handle error by asking for new input
    } finally {
        console.log('User input processing attempt completed.');
    }
}

function validateInput(input) {
    if (!input || input.trim() === '') {
        throw new Error('Input cannot be empty');
    }
}

processUserInput('');
```

This scenario demonstrates handling potentially invalid user input. If the input validation fails, an error is thrown, caught, and handled, preventing the application from terminating unexpectedly while providing feedback to the user.

processUserInput receives an input, attempts to validate it using validateInput, and logs a success message if the input is valid. If the input validation fails (i.e., if validateInput throws an error), processUserInput catches the error, logs an error message, and returns early. Regardless of whether an error occurs, a "User input processing attempt completed." message is logged due to the finally clause.

validateInput checks if the input is either non-existent or only contains white space. If either is true, it throws an error with the message 'Input cannot be empty'.

The last line of the code executes processUserInput with an empty string as an argument, which will throw an error and log 'Invalid user input: Input cannot be empty'.

The try, catch, finally structure is a powerful tool for managing errors in JavaScript, allowing developers to write more resilient and user-friendly applications. By understanding and implementing these constructs effectively, you can safeguard your applications against unexpected failures and ensure that essential cleanup tasks are always performed.

8.1.3 Custom Error Handling

Beyond handling built-in JavaScript errors, you can create custom error types that are specific to your application's needs. This allows for more granular error management and clearer code.

Custom error handling in programming refers to the process of defining and implementing specific responses or actions for various types of errors that can occur within a codebase. This practice goes beyond handling built-in errors and involves creating custom error types that are specific to the needs of your application.

In JavaScript, for instance, you can create a custom error class that extends the built-in Error class. This custom error class can then be used to throw errors that are specific to certain situations in your application. When these custom errors are thrown, they can be caught and handled in a way that aligns with the specific needs of your application.

This approach allows for more granular error management, clearer code, and the ability to handle specific types of errors differently, thus improving the clarity and maintainability of error handling in the application. It provides developers with greater control over the application's behavior during error scenarios, allowing the software to gracefully recover from errors or provide meaningful error messages to users.

In complex applications, it's not uncommon to find nested try-catch blocks or asynchronous error handling to manage errors across different layers of the application logic. This structured approach to error handling is vital for developing robust, resilient, and user-friendly applications that can effectively manage and respond to errors in a systematic way.

Example: Defining and Throwing Custom Errors

```javascript
class ValidationError extends Error {
    constructor(message) {
        super(message);
        this.name = "ValidationError";
    }
}

function validateUsername(username) {
    if (username.length < 4) {
        throw new ValidationError("Username must be at least 4 characters long.");
    }
}
```

```
try {
    validateUsername("abc");
} catch (error) {
    if (error instanceof ValidationError) {
        console.error('Invalid data:', error.message);
    } else {
        console.error('Unexpected error:', error);
    }
} finally {
    console.log('Validation attempt completed.');
}
```

In this example, a custom ValidationError class is defined. This makes it easier to handle specific types of errors differently, improving the clarity and maintainability of error handling.

The code begins by defining a custom error type named 'ValidationError', which extends the built-in 'Error' class in JavaScript. Through this extension, the 'ValidationError' class inherits all the standard properties and methods of an Error, while also allowing us to add custom properties or methods if required. In this case, the name of the error is set to "ValidationError".

Next, a function named 'validateUsername' is defined. This function is designed to validate a username based on a specific condition, that is, the username must be at least 4 characters long. The function takes a parameter 'username' and checks if its length is less than 4. If this condition is met, indicating that the username is invalid, the function throws a new 'ValidationError'. The error message specifies the reason for the error, in this case, "Username must be at least 4 characters long."

Following this, a try-catch-finally statement is implemented. This is a built-in error handling mechanism in JavaScript that allows the program to "try" to execute a block of code, and "catch" any errors that occur during its execution. In this scenario, the "try" block attempts to execute the 'validateUsername' function with "abc" as an argument. Since "abc" is less than 4 characters long, the function will throw a 'ValidationError'.

The "catch" block is designed to catch and handle any errors that occur in the "try" block. In this case, it checks if the caught error is an instance of 'ValidationError'. If it is, a specific error message is logged to the console: 'Invalid data:' followed by the error message. If the error is not a 'ValidationError', meaning it's an unexpected error, a different message is logged to the console: 'Unexpected error:' followed by the error itself. This differentiation in error handling provides clarity and aids in debugging by providing specific and meaningful error messages.

Finally, the "finally" block executes code that will run regardless of whether an error occurred or not. This block does not depend on the occurrence of an error, but guarantees that certain key parts of the code will run irrespective of the outcome in the "try" and "catch" blocks. In this case, it logs the message 'Validation attempt completed.' to the console, indicating that the validation process has finished, regardless of whether it was successful or not.

This example not only showcases how to define custom errors and throw them under certain conditions but also how to catch and handle these errors in a meaningful and controlled way, thereby enhancing the robustness and reliability of the software.

8.1.4 Nested Try-Catch Blocks

In complex applications, you might encounter situations where a try-catch block is nested within another. This can be useful for handling errors in different layers of your application logic.

Nested try-catch blocks are used in programming when you have a situation where a try-catch block is enclosed within another try-catch block. In such a situation, you are essentially creating multiple layers of error handling in your code.

The outer try block contains a section of code that could potentially throw an exception. If an exception occurs, the control is passed to the associated catch block. However, within this outer try block, we may have another try block - this is what we call a nested try block. This nested try block is used to handle a different section of the code that could also potentially throw an exception. If an exception does occur within this nested try block, it has its own associated catch block that will handle the exception.

This structure can be particularly useful in complex applications where different parts of the code may throw different exceptions, and each exception might need to be handled in a specific way. By using nested try-catch blocks, developers can handle errors at different layers of application logic, providing multiple layers of fallback and ensuring that all possible recovery options are attempted.

An example of this would be a situation where a high-level operation (handled by the outer try-catch block) involves several sub-operations, each of which could potentially fail (handled by the nested try-catch blocks). By nesting the try-catch blocks, you can handle errors at the level of each sub-operation, while also providing a catch-all safety net at the high level.

In summary, nested try-catch blocks provide a powerful tool for managing and responding to errors at various levels of complexity within an application, enabling developers to build more robust and resilient software.

Example: Using Nested Try-Catch

```
try {
    performTask();
} catch (error) {
    console.error('High-level error handler:', error);
    try {
        recoverFromError();
    } catch (recoveryError) {
        console.error('Failed to recover:', recoveryError);
    }
}

function performTask() {
    throw new Error("Something went wrong!");
}

function recoverFromError() {
    throw new Error("Recovery attempt failed!");
}
```

This structure allows handling errors and recovery attempts distinctly, providing multiple layers of fallback and ensuring that all possible recovery options are attempted.

The performTask function is called inside the outer try block. This function, when invoked, is intentionally designed to throw an error with the message "Something went wrong!". The throw statement in JavaScript is used to create custom errors. When an error is thrown, the JavaScript runtime immediately stops execution of the current function and jumps to the catch block of the nearest enclosing try-catch structure. In this case, the catch block logs the error message to the console using console.error.

The console.error function is similar to console.log, but it also includes the stack trace in the browser console and is styled differently (usually in red) to stand out as an error. The error message 'High-level error handler:' is logged along with the error caught.

Within this catch block, there is a nested try-catch block. This nested try block calls the recoverFromError function. This function is a hypothetical recovery mechanism that is triggered when performTask fails. But as with performTask, recoverFromError is also designed to throw an error saying "Recovery attempt failed!".

The purpose of this is to simulate a scenario where the recovery mechanism itself fails. In real-world applications, the recovery mechanism might involve actions like retrying the failed operation, switching to a backup service, or prompting the user to provide valid input, and it's possible that these actions might fail as well.

If the recovery fails and throws an error, the nested catch block catches this error and logs it to the console with the message 'Failed to recover:'.

This script is a simplified representation of how you might handle errors and recovery attempts in JavaScript. In a real application, both performTask and recoverFromError would have more complex logic, and there might be additional error handling and recovery attempts at various levels of the application.

8.1.5 Asynchronous Error Handling

Handling errors from asynchronous operations within try, catch, finally blocks requires special consideration, especially when using Promises or async/await.

Asynchronous error handling refers to a programming method used to manage and resolve errors that occur during asynchronous operations. Asynchronous operations are tasks that can occur independently of the main program flow, meaning they do not need to wait for other tasks to complete before they can begin.

In JavaScript, asynchronous tasks are often represented by Promises or can be handled using async/await syntax. Asynchronous operations can be resources fetched from a network, file system operations, or any operation that relies on some sort of waiting time.

When asynchronous operations are used within a try-catch-finally block, special consideration is needed to handle potential errors. This is because the try block will complete before the Promise resolves or the async function completes its execution, so any errors that occur within the Promise or async function will not be caught by the catch block.

One way to handle asynchronous errors is by attaching .catch handlers to the Promise. Alternatively, if you're using async/await, you can use a try-catch block inside an async function. When an error occurs in the try block of an async function, it can be caught in the catch block just like synchronous errors.

Example 1:

```
async function fetchData() {
    try {
        const response = await fetch('<https://api.example.com/data>');
        const data = await response.json();
```

```
        console.log('Fetched data:', data);
    } catch (error) {
        console.error('Failed to fetch data:', error);
    } finally {
        console.log('Fetch attempt completed.');
    }
}

fetchData();
```

In this example, the fetchData async function attempts to fetch data from an API and convert the response to JSON format. If either of these operations fails, the error is caught in the catch block and logged to the console. Regardless of whether an error occurs, the finally block logs 'Fetch attempt completed.' to the console. This asynchronous error handling can make asynchronous code easier to read and manage, similar to how synchronous code is handled.

The function uses the fetch API, a built-in browser function for making HTTP requests. The fetch API returns a Promise that resolves to the Response object representing the response to the request. This promise can either be fulfilled (if the operation was successful) or rejected (if the operation failed).

Inside the fetchData function, the try block is used to encapsulate the code that may potentially throw an error. In this case, two operations are contained within the try block. First, the function makes a fetch request to the URL 'https://api.example.com/data'. This operation is prefixed with the await keyword, which makes JavaScript wait until the Promise settles and returns its result.

If the fetch operation is successful, the function then attempts to parse the response data into JSON format using the response.json() method. This method also returns a promise that resolves with the result of parsing the body text as JSON, hence the await keyword is used again.

If both operations are successful, the function logs the fetched data to the console using console.log.

In the event of an error during the fetch operation or while converting the response into JSON, the catch block will be executed. The catch block acts as a fallback mechanism, allowing the program to handle errors or exceptions gracefully without crashing entirely. If an error occurs, the function logs the error message to the console using console.error.

The finally block contains code that will be executed regardless of whether an error occurred or not. This is useful for performing cleanup operations or logging that aren't contingent on the success of the operations in the try block. In this case, it logs 'Fetch attempt completed.' to the console.

After defining the fetchData function, it is then called and executed using fetchData(). This triggers the function's operations, starting the asynchronous fetch operation.

Best Practices for Using Try, Catch, Finally

- **Minimize Code in Try Blocks**: It's a good practice to only include the code that could potentially throw an exception within try blocks. This way, you can avoid catching unintended exceptions which might be difficult to debug and could lead to misleading error information. By isolating the code that might fail, you can manage exceptions more effectively.
- **Be Specific with Error Types in Catch Blocks**: When catching errors, it's advisable to be as specific as possible regarding the types of errors you're handling. This precision helps prevent masking unrelated issues that could be occurring in your code. By specifying the types of exceptions, you can have more control over error handling and provide more accurate feedback to users.
- **Clean Up Resources in Finally Blocks**: Always use the finally block to ensure that all necessary cleanup operations are performed. This could include closing files or releasing network connections, among other tasks. This is crucial regardless of whether an error occurred or not. Ensuring that resources are properly released or closed can prevent memory leaks and other related issues, improving the robustness of your code.

Mastering the use of try, catch, finally in JavaScript is crucial for writing robust, reliable, and user-friendly applications. By employing advanced techniques and adhering to best practices, you can effectively manage a wide range of error conditions and ensure your applications behave predictably even under adverse conditions.

8.2 Throwing Errors

In software development, the strategic employment of error throwing is an integral facet of developing a robust error handling mechanism. This critical technique empowers developers to enforce specific conditions, ensure data validation, and manage the flow of execution in a methodical, controlled manner, thereby enhancing the overall reliability and performance of the application.

In the subsequent section of this document, we will delve deeper into the nuanced use of the 'throw' statement, a powerful tool in JavaScript, to devise and implement custom error conditions. This exploration will include a step-by-step guide on how to effectively manage and address these intentionally induced errors.

By mastering these techniques, you can uphold the integrity of your applications, whilst simultaneously improving their reliability and robustness, even in the face of unexpected circumstances or data inputs.

8.2.1 Understanding throw in JavaScript

The throw statement in JavaScript is used to create a custom error. When an error is thrown, the normal flow of the program is halted, and control is passed to the nearest exception handler, typically a catch block.

The throw statement in JavaScript is a powerful tool used to create and throw custom errors. The primary function of throw is to halt the normal execution of code and pass control over to the nearest exception handler, which is typically a catch block within a try...catch statement. This is particularly useful for enforcing rules and conditions in your code, such as input validation, and for signaling that something unexpected or erroneous has occurred that the program cannot handle or recover from.

For instance, you might use a throw statement when a function receives an argument that is outside of an acceptable range, or when a required resource (like a network connection or file) is unavailable. When a throw statement is encountered, the JavaScript interpreter immediately stops normal execution and looks for the closest catch block to handle the exception.

Here's a basic syntax of a throw statement:

throw expression;

In this syntax, expression can be a string, number, boolean, or more commonly, an Error object. The Error object is typically used because it automatically includes a stack trace that can be extremely helpful for debugging.

Here's an example of throwing a simple error:

```javascript
function checkAge(age) {
    if (age < 18) {
        throw new Error("Access denied - you are too young!");
    }
    console.log("Access granted.");
}
```

```
try {
    checkAge(16);
} catch (error) {
    console.error(error.message);
}
```

In this example, the function checkAge throws an error if the age is below 18. This error is then caught in the catch block, where an appropriate message is displayed.

In addition to the standard Error object provided by JavaScript, you can also define custom error types by extending the Error class. This allows for more nuanced error handling and better distinguishes different types of error conditions in your code.

For instance, you might define a ValidationError class for handling input validation errors, providing additional clarity and granularity in your error handling strategy.

As a best practice, it's important to use meaningful error messages, consider error types, throw errors early, and document any errors your functions can throw.

In conclusion, understanding how to use the throw statement in JavaScript is crucial for effective error handling, as it allows you to control the program flow, enforce specific conditions, and manage errors in a methodical and controlled manner.

8.2.2 Custom Error Types

While JavaScript provides a standard Error object, often it is beneficial to define custom error types. This can be achieved by extending the Error class. Custom errors are helpful for more detailed error handling and for distinguishing different kinds of error conditions in your code.

Custom Error Types are user-defined errors in programming that extend the built-in error types. They are particularly beneficial when the error you need to throw is specific to the business logic or problem domain of your application, and the standard error types provided by the programming language are not sufficient.

In the context of JavaScript, as in the provided example, you can define a custom error type by extending the built-in Error class. This allows you to create a named error with a specific message. The custom error can then be thrown when a certain condition is met.

In the given example, a custom error called ValidationError is defined. This error is thrown by the validateUsername function if the provided username does not meet the required condition, which is being at least 4 characters long.

This custom error type can then be specifically handled in a try-catch block. In the catch block, it checks if the caught error is an instance of ValidationError. If it is, a specific error message is logged to the console. If it's not, a different generic error message is logged.

Defining custom error types allows for more detailed and specific error handling. It enables developers to distinguish between different kinds of error conditions in their code, and handle each error in a way that is appropriate and specific to that error. This can greatly improve debugging, error reporting, and the overall robustness of an application.

Example: Defining a Custom Error Type

```javascript
class ValidationError extends Error {
    constructor(message) {
        super(message); // Call the superclass constructor with the message
        this.name = "ValidationError";
        this.date = new Date();
    }
}

function validateUsername(username) {
    if (username.length < 4) {
        throw new ValidationError("Username must be at least 4 characters long.");
    }
}

try {
    validateUsername("abc");
} catch (error) {
    if (error instanceof ValidationError) {
        console.error(`${error.name} on ${error.date}: ${error.message}`);
    } else {
        console.error('Unexpected error:', error);
    }
}
```

This example introduces a ValidationError class for handling validation errors. It provides a clear indication that the error is specifically related to validation, adding an additional layer of clarity to the error handling process.

In the 'ValidationError' class, which extends JavaScript's built-in Error class, the constructor method is used to create a new instance of a ValidationError. The constructor accepts a 'message' parameter and passes it to the superclass constructor (Error). It also sets the 'name' property to 'ValidationError' and the 'date' property to the current date.

In the 'validateUsername' function, the input username is evaluated. If the length of the username is less than 4 characters, a new 'ValidationError' is thrown with a specific error message.

The 'try-catch' mechanism is used to handle potential errors thrown by the 'validateUsername' function. If the function throws a 'ValidationError' (which it will do when the username is less than 4 characters long), the error is caught and logged to the console with a specific error message. If the error is not a 'ValidationError', it is considered an unexpected error and is logged as such.

The example also discusses the use of nested try-catch blocks, which can be useful in complex applications for handling errors at different layers of logic. An example is provided where a high-level operation involves several sub-operations, each of which could potentially fail. By nesting the try-catch blocks, you can handle errors at the level of each sub-operation while also providing a catch-all safety net at the high level.

It further discusses asynchronous error handling, especially when using Promises or async/await. It explains that special consideration is needed because the try block will complete before the Promise resolves or the async function completes its execution, so any errors that occur within the Promise or async function will not be caught by the catch block. An example is provided to illustrate this.

Finally, it covers best practices for using try-catch-finally blocks, including minimizing code in try blocks, being specific with error types in catch blocks, and cleaning up resources in finally blocks. It then goes into detail about throwing errors and creating custom error types, explaining why these techniques are crucial for effective error handling in JavaScript applications.

Best Practices When Throwing Errors

- **Use meaningful error messages**: It's important to make sure that the error messages your code throws are both descriptive and useful when it comes to identifying and rectifying issues. They should include enough detail for anyone who reads them to fully understand the context in which the error occurred.
- **Consider error types**: Always use specific error types where it's appropriate to do so. By doing this, you can greatly assist with error handling strategies because it becomes

easier to implement different responses for different kinds of errors. This can streamline the debugging process and help to prevent further issues.

- **Throw early**: It's vital to throw errors as soon as possible, ideally the moment something wrong is detected. This helps to prevent the further execution of any operations that could potentially be corrupted, thereby minimising the risk of more serious issues developing later on.
- **Document thrown errors**: Make sure to document any errors that your functions may throw in the function's documentation or comments. This is particularly crucial when it comes to public APIs and libraries, as it ensures that others who use your code can understand what the potential issues are and how they can be solved.

Throwing and handling errors effectively are fundamental skills in JavaScript programming. By using the throw statement responsibly and defining custom error types, you can greatly enhance the robustness and usability of your applications. Understanding these concepts allows you to prevent erroneous states, guide application execution, and provide meaningful feedback to users and other developers, contributing to overall application stability and reliability.

8.2.3 Contextual Error Information

When throwing errors, including contextual information can significantly aid in debugging and error resolution. This involves not just stating what went wrong, but where and why it went wrong, which can be crucial for quickly identifying and fixing issues.

In the context of programming, an error message typically includes a description of the problem. However, just having this description may not be enough to diagnose and fix the problem. Therefore, it's important to provide additional context about the state of the system or application when the error occurred.

For example, if an error happens while processing a payment in an online store, the error message might state that the payment has failed. But to identify the cause of the problem, additional information would be helpful, such as the user's account details, the payment method used, the time the error occurred, and any error codes returned by the payment gateway.

Including contextual information in error messages can significantly aid in debugging and error resolution. This involves not just stating what went wrong, but where and why it went wrong, which can be crucial for quickly identifying and fixing issues. This information can then be used to improve the robustness of the application and prevent such errors from happening in the future.

For instance, if certain errors always occur with specific types of payment methods, then the payment processing code for those methods can be reviewed and improved. Or if certain errors always occur at specific times, this could indicate a problem with server load, leading to improvements in server capacity or performance.

In conclusion, contextual error information is a crucial part of error handling and resolution in software development, aiding developers in diagnosing issues, improving application robustness, and providing better user experiences.

Example: Including Context in Errors

```javascript
function processPayment(amount, account) {
    if (amount <= 0) {
        throw new Error(`Invalid amount: ${amount}. Amount must be greater than zero.`);
    }
    if (!account.isActive) {
        throw new Error(`Account ${account.id} is inactive. Cannot process payment.`);
    }
    // Process the payment
}

try {
    processPayment(0, { id: 123, isActive: true });
} catch (error) {
    console.error(`Payment processing error: ${error.message}`);
}
```

This code snippet includes specific details in the error messages, such as the amount that caused the failure and the account status, which can be immensely helpful during troubleshooting.

It features a function named processPayment which is designed to process payments. The function takes two parameters: amount, which refers to the amount to be paid, and account, which refers to the account from which the payment will be made.

Inside the processPayment function, there are two conditional statements that check for specific conditions and throw errors if the conditions are not met.

The first if statement checks if the amount is less than or equal to zero. This is a basic validation to ensure that the payment amount is a positive number. If the amount is less than or equal to zero, the function throws an error with a message stating that the amount is invalid and that it must be greater than zero.

The second if statement checks if the account is active by checking the isActive attribute of the account object. If the account is not active, the function throws an error indicating that the account is inactive and cannot process the payment.

These error messages are useful because they provide contextual information about what went wrong, which can aid in debugging and error resolution.

Following the definition of the processPayment function, a try-catch block is used to test the function. The try-catch mechanism in JavaScript is used to handle exceptions (errors) that are thrown during the execution of code inside the try block.

In this case, the processPayment function is called inside the try block with an amount of 0 and an active account. Because the amount is 0, this will trigger the error in the first if statement of the processPayment function.

When this error is thrown, the execution of the try block is halted, and control is passed to the catch block. The catch block catches the error and executes its own block of code, which in this case, is to log the error message to the console.

This is a common pattern in JavaScript for handling errors and exceptions gracefully, preventing them from crashing the entire program and allowing for more informative error messages to be displayed or logged.

8.2.4 Error Chaining

Error chaining is a programming concept that occurs in complex applications where errors often result from other errors. In such situations, JavaScript allows you to chain errors by including an original error as part of a new error. This provides a trail of what went wrong at each step, allowing developers to trace the progression of errors through the chain.

This method of error handling is particularly useful in scenarios where low-level errors need to be transformed into more meaningful, high-level errors for the calling code. It helps to maintain the original error information, which can be crucial for debugging, while also providing additional context about the high-level operation that failed.

For example, consider a case where a low-level database operation fails. This low-level error can be caught and wrapped in a new, higher-level error, such as DatabaseError. The new error includes the original error as a cause, preserving the original error information and providing more context on the higher-level operation that failed.

Here's a code example that illustrates this:

```javascript
class DatabaseError extends Error {
    constructor(message, cause) {
        super(message);
        this.name = 'DatabaseError';
        this.cause = cause;
    }
}

function updateDatabase(entry) {
    try {
        // Simulate a database operation that fails
        throw new Error('Low-level database error');
    } catch (err) {
        throw new DatabaseError('Failed to update database', err);
    }
}

try {
    updateDatabase({ data: 'some data' });
} catch (error) {
    console.error(`${error.name}: ${error.message}`);
    if (error.cause) {
        console.error(`Caused by: ${error.cause.message}`);
    }
}
```

In this example, a DatabaseError wraps a lower-level error, preserving the original error information and providing more context on the higher-level operation that failed. When the error is logged, both the high-level and low-level error messages are displayed, giving a clear picture of what went wrong at each step.

At the beginning, a custom error class called 'DatabaseError' is defined. This class extends the built-in 'Error' class in JavaScript, forming a subclass that inherits all the properties and methods of the 'Error' class but also adds some custom ones. In the 'DatabaseError' class, a constructor function is defined which accepts two parameters: 'message' and 'cause'. The 'message' parameter is passed to the superclass's (Error's) constructor using the 'super' keyword, while 'cause' is stored in a property of the same name. The 'name' property is also set to 'DatabaseError' to indicate the type of the error.

The 'updateDatabase' function is where a simulated database operation occurs. This operation is designed to fail and thus throws an error, indicated by the 'throw' statement. The error message here is 'Low-level database error', signifying a typical error that might occur at the database level. This error is immediately caught in the 'catch' block that follows the 'try' block.

In the 'catch' block, the caught error (denoted by 'err') is wrapped in a 'DatabaseError' and thrown again. This is an example of error chaining, where a low-level error is caught and wrapped in a higher-level error. The original error is passed as the cause of the 'DatabaseError', preserving the original error information while providing additional context about the operation that failed (in this case, updating the database).

Next, the 'updateDatabase' function is invoked within a 'try' block. This function call is expected to throw a 'DatabaseError' due to the simulated database failure. This error is then caught in the 'catch' block.

In the 'catch' block, the error message is logged to the console. If an additional cause is present (which will be the case here as the 'DatabaseError' includes a 'cause'), the message of the cause error is also logged to the console, preceded by the text 'Caused by: '.

This way, both the high-level error message ('Failed to update database') and the low-level error message ('Low-level database error') are displayed, providing a clear overview of what went wrong at each step.

This concept of creating and using custom error types is a powerful tool in error handling. It allows for more nuanced and detailed error reporting, making debugging and resolution easier and more efficient. The practice of error chaining demonstrated here is particularly useful in complex applications where low-level errors need to be transformed into more meaningful, high-level errors.

8.2.5 Conditional Error Throwing

Sometimes, whether to throw an error might depend on multiple conditions or application state. Strategically managing these conditions can prevent unnecessary error throwing and make your application logic clearer and more predictable.

In many programming languages, you can create a set of conditions that, when met, will trigger the system to throw an error. These conditions can be anything that the programmer defines - for instance, it could be when a function receives an argument that is outside of an acceptable range, when a required resource (like a network connection or file) is unavailable, or when an operation produces a result that is not as expected.

The purpose of throwing these errors is to prevent the program from continuing in an erroneous state, and to alert the developers or users about issues that the program cannot handle or recover from.

For example, consider a function that is supposed to read data from a file and perform some operations on it. If the file does not exist or is not accessible for some reason, the function can't perform its job. In such cases, instead of continuing execution and possibly producing incorrect results, the function can throw an error indicating that the required file is not available.

Once an error is thrown, the normal execution of the program is halted, and the control is passed to a special error handling routine, which can be designed to handle the error in a controlled manner and take appropriate action, such as logging the error, notifying the user or the developer, or attempting a recovery operation.

Conditional error throwing is a powerful tool for managing unexpected situations in software applications. By throwing errors under specific conditions, programmers can ensure that their applications behave predictably under error conditions, making them more robust and reliable.

Example: Conditional Error Throwing

```javascript
function loadData(data) {
    if (!data) {
        throw new Error('No data provided.');
    }

    if (data.isLoaded && !data.isDirty) {
        console.log('Data is already loaded and not dirty.');
        return;  // No need to throw an error if the data is already loaded and not
dirty
    }

    // Assume data needs reloading
    console.log('Reloading data...');
}

try {
    loadData(null);
} catch (error) {
    console.error(`Error loading data: ${error.message}`);
}
```

This example shows how conditions around data state influence whether an error is thrown, promoting efficient and error-free data handling.

Inside the loadData function, the first operation is a conditional check to see if the data argument exists. If the data argument is not provided or is null, the function throws an error with the message 'No data provided.'. This is an example of "fail-fast" error handling, where the function immediately stops execution when it encounters an error condition.

The function then checks two properties of the data argument: isLoaded and isDirty. If the data is already loaded (data.isLoaded is true) and the data is not dirty (data.isDirty is false), it simply logs a message 'Data is already loaded and not dirty.' and exits the function. In this case, the function considers that there's no need to proceed with loading the data because it's already loaded and hasn't changed since it was loaded.

If neither of the above conditions is met, the function makes an assumption that the data needs to be reloaded. It then logs a message 'Reloading data...'.

The loadData function is then called within a try block, passing null as an argument. Since null is not a valid argument for the loadData function (as it expects an object with isLoaded and isDirty properties), this results in throwing an error with the message 'No data provided.'.

The try block is paired with a catch block, which is designed to handle any errors thrown in the try block. When the loadData function throws an error, the catch block catches this error and executes its code. In this case, it logs an error message to the console, including the message from the caught error.

This code thus demonstrates a common pattern in JavaScript for working with potential errors - throwing errors when a function can't proceed correctly, and catching those errors to handle them appropriately and prevent them from crashing the whole program.

Best Practices for Throwing Errors

- **Consistency**: It's crucial to maintain consistency in the way and the timings of when you throw errors across your application. By doing so, you create a predictable environment, which in turn makes your code simpler to comprehend and maintain for both you and other developers.
- **Documentation**: In the API documentation, make sure to document the types of errors your functions are capable of throwing. This level of transparency is beneficial as it aids other developers in foreseeing and managing potential exceptions, thereby reducing the likelihood of unexpected issues.
- **Testing**: Don't forget to include tests specifically for your error handling logic. It's important to remember that ensuring your application behaves correctly under error conditions is just as vital as its normal operation. Robust testing under a variety of conditions helps ensure that unexpected errors won't derail your application's performance.

Effectively managing and throwing errors is essential for building resilient software. By incorporating advanced techniques such as contextual information, error chaining, and conditional throwing, along with adhering to best practices, you can enhance your application's stability and provide a better user and developer experience.

8.3 Unit Testing and Integration Testing

In the complex and intricate process of software development, one aspect stands paramount for the creation of robust, reliable, and maintainable software - that is rigorous testing. Delving deeper into this integral part of the software life cycle, we find two critical types of testing methods that bear significant importance: unit testing and integration testing.

Unit testing, as the name suggests, focuses on testing individual components or 'units' of the software to ensure they perform as expected under various conditions. On the other hand, integration testing takes a broader perspective, assessing how these individual units interact and work together as a cohesive whole, ensuring seamless functionality.

These testing methodologies, when understood and implemented correctly, form the backbone of efficient and effective software development. They serve a dual purpose: firstly, they significantly reduce the likelihood of bugs or errors slipping through the cracks and making their way into the final product; secondly, they facilitate maintenance by making it easier to identify and rectify issues within the system.

By promoting a culture of thorough testing, developers can not only enhance the quality of their software but also improve its reliability and longevity, ultimately leading to better user satisfaction.

8.3.1 Unit Testing

Unit testing is a crucial aspect of software testing where individual components or units of a software are tested. The main purpose of unit testing is to verify that each unit of the software performs as expected and designed, under a variety of conditions. A unit may be an individual function, method, module, or object in a programming language.

In unit testing, the units are tested in isolation from the rest of the system to ensure the test only covers the unit's functionality itself. This focus on a single unit helps to identify and fix bugs early in the development cycle, making it a key aspect of software development.

Unit testing is characterized by automation and repeatability. The tests are often automated to run with every build or through a continuous integration system to ensure that all tests are executed. This automation is crucial to promptly identify and fix any issues or bugs. Furthermore, unit tests can be run multiple times under the same conditions and should produce the same results each time, ensuring the consistency and reliability of the software unit.

For instance, in JavaScript, a simple unit test can be written for an add function using a testing framework like Jest or Mocha. The test would verify that the add function correctly sums two numbers.

Unit testing is a fundamental part of the software development process, contributing significantly towards producing robust, reliable, and high-quality software.

Characteristics of Unit Testing:

- **Isolation**: In this testing procedure, individual units within the system are examined in isolation from the rest of the integrated system. This is done to ensure that the test is solely focused on the functionality of the unit itself. This approach allows for a more precise identification of any potential errors or issues within each unit.
- **Automation**: The process of testing is automated, meaning that tests are programmed to run automatically with each new build. This can also be facilitated through a continuous integration system. The aim of this automation is to ensure that all tests are executed without fail, thus reducing the possibility of human error and increasing the overall efficiency of the testing procedure.
- **Repeatability**: A key feature of these tests is their repeatability. They can be run multiple times under the exact same conditions. This is crucial as it ensures that the tests should produce the same results each and every time they are executed. This aspect of repeatability allows for consistent tracking and detection of any issues or bugs within the system.

Example: Unit Testing a Simple Function

```
function add(a, b) {
    return a + b;
}

describe('add function', () => {
    it('adds two numbers correctly', () => {
        expect(add(2, 3)).toBe(5);
    });
});
```

In this example, a simple unit test is written for an add function using a JavaScript testing framework (such as Jest or Mocha). The test verifies that the add function correctly sums two numbers.

The first section of the code defines a function named 'add'. The purpose of this function is to perform a simple arithmetic operation which is the addition of two numbers. This function receives two arguments, namely 'a' and 'b'. It returns the result of the addition of these two arguments. The 'return' statement is used to specify the value that a function should return. In this case, it returns the sum of 'a' and 'b'.

Following the function definition, there's a test suite for the 'add' function. Testing is a crucial aspect of software development that ensures the code behaves as expected. The testing framework being used in this code isn't explicitly mentioned, but it resembles the syntax used by popular JavaScript testing libraries like Jest or Mocha.

The 'describe' function is used to group related tests in a test suite. Here, it groups the tests for the 'add' function. It takes two arguments: a string and a callback function. The string 'add function' is a description of the test suite that can be helpful when reading the test results. The callback function contains the actual tests.

Inside the 'describe' block, there's an 'it' function which defines a single test. The 'it' function also takes a string and a callback function as arguments. The string 'adds two numbers correctly' is a description of what the test is supposed to do. The callback function contains the test logic.

In this test, the 'expect' function is used to make an assertion about the value returned by the 'add' function when it's called with the arguments 2 and 3. The 'toBe' function is called on the result of the 'expect' function to assert that the returned value should be identical to 5.

If 'add(2, 3)' indeed returns 5, then this test will pass. If it returns any other value, the test will fail, indicating there's a problem with the 'add' function that needs to be fixed.

This piece of code is a simple yet clear demonstration of function definition and testing in JavaScript, showcasing how functions can be tested to ensure they work correctly under different scenarios.

8.3.2 Integration Testing

While unit tests cover individual components, integration testing focuses on the points of interaction between those components to ensure that their combinations produce the desired results. This type of testing is crucial for identifying problems that occur when individual modules are combined.

This type of testing is particularly important when multiple components, which may have been developed independently, are combined to create a larger system. It allows for the discovery of issues related to data communication among modules, function calls, or information shared by shared state or other resources.

For instance, consider a scenario where one function is supposed to pass its results to another function for further processing. Each function might work perfectly when tested independently (unit testing), but issues might arise when they are combined due to reasons such as mismatched data formats, incorrect assumptions about execution order, or other discrepancies. Integration testing is designed to catch such issues.

In addition, integration testing can help verify system-level functionality, performance, and reliability requirements. It can be conducted in a top-down, bottom-up, or sandwich manner.

- The top-down approach tests the high-level components first, using stubs for lower-level components that have not yet been integrated.
- The bottom-up approach tests the low-level components first, using drivers for high-level components that have not yet been integrated.
- The sandwich approach is a combination of top-down and bottom-up approaches.

Integration testing is typically carried out by a test team. It's done after unit testing and before system testing. Its main goal is to ensure that the integrated components work as expected and that any errors that arise due to module interactions are caught and rectified before the system goes into the final phases of testing or, worse, gets delivered to the end user.

Characteristics of Integration Testing:

- **Combination of Modules**: This process is aimed at testing the integration of two or more units. The main goal is to ensure that their combined operation and interaction lead to the production of the expected outcome. This is an essential step in maintaining the functionality and reliability of the system as a whole.
- **Data Flow and Control Flow**: This involves a thorough examination of both the data flow between modules and the control logic that seamlessly integrates the modules. By ensuring both the proper flow of data and the appropriate control logic, we can achieve a more efficient and error-free system operation.

Example: Integration Testing for a Web Application

```
// Assuming an application with a user module and a database module
function getUser(id) {
    return database.findUserById(id);   // This function interacts with the database
module
}

describe('getUser integration', () => {
    it('retrieves a user correctly from the database', () => {
        // Mock the database.findUserById to return a specific user
        const mockId = 1;
        const mockUser = { id: mockId, name: 'John Doe' };
        jest.spyOn(database, 'findUserById').mockReturnValue(mockUser);

        const user = getUser(mockId);
        expect(user).toEqual(mockUser);
        expect(database.findUserById).toHaveBeenCalledWith(mockId);
    });
});
```

This example demonstrates an integration test for a function that retrieves user data from a database. The database.findUserById function is mocked to ensure the test focuses on the integration points without relying on the actual database implementation.

The function getUser(id) communicates with a hypothetical database module in the system, specifically calling a function database.findUserById(id). This function interacts with the database to retrieve a user record associated with the given id.

The integration test is crafted within a describe block, a Jest testing construct that groups related tests together. In this case, it's grouping tests related to the 'getUser integration'. Nested within the describe block is a single unit test defined by the it function, another Jest construct that specifies a single test case. This test case is titled 'retrieves a user correctly from the database'.

In order to test the getUser function in isolation without making actual calls to the database, the database.findUserById function is mocked using jest.spyOn(database, 'findUserById').mockReturnValue(mockUser);. This line replaces the actual function with a mock function that always returns a predefined user object, mockUser, when called. This technique is known as mocking and is a powerful tool in testing because it allows for control over a function's behavior and output during a test.

The mock user object is defined as const mockUser = { id: mockId, name: 'John Doe' };, representing a user with ID 1 named 'John Doe'. This is the user object that's returned when database.findUserById is called during the test.

The actual testing takes place in the last two lines of the it block. The getUser function is called with mockId as its argument and the returned user is compared to mockUser. If getUser functions correctly, it should return a user object identical to mockUser. This is checked using the expect function from Jest along with the toEqual matcher.

The last line checks if the database.findUserById function was called with mockId as its argument. This helps verify that getUser is making the correct call to the database function with the correct argument.

This test ensures that getUser function is correctly integrating with the database.findUserById function to retrieve user data from the database. It demonstrates the use of mocking to isolate the function being tested and control the behavior of dependencies during a test.

8.3.3 Best Practices for Testing

- **Maintainability**: It's important to write tests that are not only easy to maintain, but also easy to understand. As your codebase evolves and undergoes changes, your tests should be simple to update. This ensures that they remain relevant and effective, providing the necessary checks for your code as it matures.
- **Coverage**: While it's beneficial to aim for high test coverage, it's essential to prioritize and focus on the critical paths. Not all code needs the same level of scrutiny or extensive testing. Instead, concentrate on areas that are crucial to your application's functionality or have a higher risk of causing significant issues.
- **Continuous Integration**: Incorporating testing into your continuous integration (CI) pipeline is a key step to catch potential problems early and often. This allows you to address issues promptly, ensuring your code is consistently of high quality and reducing the risk of problems persisting until the later stages of development.

Unit testing and integration testing are crucial for developing high-quality software. By ensuring that individual units perform correctly and that they integrate properly, developers can build more reliable and maintainable systems. Implementing these testing practices effectively not only catches errors early but also supports better design decisions, ultimately leading to more robust software solutions.

8.4 Tools and Libraries for Testing (Jest, Mocha)

In the complex and ever-evolving realm of software development, the selection of appropriate tools and libraries specifically for testing purposes can have a profound impact on the efficiency, effectiveness, and overall ease of your testing efforts. The choice of framework can either streamline your process or create unnecessary complexities, making this an important factor to consider for any project.

This particular section of the document is dedicated to shedding light on two of the most widely used and well-regarded JavaScript testing frameworks in the modern web development landscape: Jest and Mocha.

Each of these powerful frameworks comes with its unique suite of features, characteristics, and ecosystems, which collectively contribute to making them highly suitable for a variety of distinctive testing scenarios within the broader context of web development. Although they share some commonalities, the differences between Jest and Mocha can lead to one being a better fit than the other depending on specific project requirements and scenarios.

Understanding the intricacies of these tools, their strengths, and potential weaknesses, as well as how to utilize them in the most effective manner, is absolutely crucial for any software developer or team aiming to implement a robust, comprehensive, and reliable testing strategy. This understanding can optimize your workflow, ensure the quality of your code, and ultimately contribute to the successful completion of your software development project.

8.4.1 Jest

Jest, developed by Facebook, is a delightful JavaScript Testing Framework with a focus on simplicity and support for large web applications. It is often favored for its zero-configuration setup, which means you can start writing your tests with minimal setup.

Jest is a popular, robust, and feature-rich JavaScript testing framework developed by Facebook. It is equipped with an extensive set of features making it a go-to choice for testing JavaScript code, including ES6 syntax, and is particularly favored in the React and React Native communities.

Some of Jest's primary features include a zero-configuration setup, meaning it works right after installation without requiring any initial setup. This makes it very beginner-friendly and reduces the boilerplate code typically associated with setting up a testing environment.

Jest also offers a powerful and flexible mocking library. It allows you to replace JavaScript functionality with mock data or functions, isolating the code under test and ensuring that your tests run in a predictable manner. The mocking library can handle function mocking, manual mocks, and timer mocks, which is useful when testing code that relies on JavaScript's built-in timers like setTimeout or setInterval.

Another prominent feature is Jest's snapshot testing capability. Snapshot tests compare the output of your code (the "snapshot") against a stored version. If the output changes, the test fails. This is especially useful when testing React components, as it helps ensure the UI does not change unexpectedly.

Jest also runs tests in parallel, distributing the test load across the CPUs in your machine. This can significantly improve the speed of large test suites and provide faster feedback, especially in continuous integration (CI) environments.

One more notable feature is Jest's support for asynchronous testing. Asynchronous operations are common in JavaScript, and handling them correctly in tests can be tricky. Jest provides several methods to deal with this, making it straightforward to test asynchronous code.

In summary, Jest is a comprehensive testing solution for JavaScript applications. Its wide array of features, ease-of-use, and powerful capabilities make it an excellent choice for any JavaScript or React project. Whether you're a testing novice or an experienced tester, Jest has tools and functionalities that can streamline your testing process and help you create robust, error-free code.

Key Features of Jest:

- **Zero Configuration**: Jest distinguishes itself by working seamlessly with minimal setup right out of the box. This feature is particularly noticeable and beneficial in projects that have been created using Create React App, eliminating the need for time-consuming configuration.
- **Built-in Mocking and Spies**: Jest comes equipped with a comprehensive set of tools for mocking functions, modules, and timers. This feature simplifies the process of testing modules in isolation, saving developers time and enhancing the efficiency and reliability of the tests.
- **Snapshot Testing**: Jest supports snapshot testing, an important functionality for modern development. Snapshot tests are particularly useful for ensuring that the user interface does not change unexpectedly, thereby improving the stability and predictability of the application.
- **Parallel Test Runs**: Jest automatically executes tests in parallel, utilizing multiple CPUs. This feature drastically improves the speed of the test suite, leading to quicker iterations and more productive development cycles.

Example: A Simple Jest Test

```
// sum.test.js
function sum(a, b) {
```

```
    return a + b;
}

test('adds 1 + 2 to equal 3', () => {
    expect(sum(1, 2)).toBe(3);
});
```

To run this test with Jest, you simply need to install Jest (npm install --save-dev jest) and add a script to your package.json: "test": "jest"

In the function definition, we have function sum(a, b), where sum is the function name and a and b are parameters to this function. These parameters represent the two numbers that we will be adding together.

The body of the function contains the statement return a + b;. This is the operation that the function performs, which is adding together the parameters a and b. The return keyword specifies the result that the function produces, which in this case, is the sum of a and b.

Below the function definition, there's a Jest test defined using the test function. The test function is used to define a test in Jest. It takes two arguments, a string and a callback function. The string argument is a description of what the test is meant to do. In this case, the description is 'adds 1 + 2 to equal 3'.

The callback function argument contains the logic of the test. Inside this function, we have an expect function call expect(sum(1, 2)). The expect function is used in Jest to test values. It takes the actual value that your code produces as an argument, in this case, the return value of calling sum(1, 2).

The expect function call is followed by a matcher method .toBe(3);. Matcher methods are used in Jest to assert how the expected and actual values should compare. The .toBe method checks if the actual value is the same as the expected value. Here, it checks if the result of sum(1, 2) is 3.

In summary, this example is a simple yet clear demonstration of function definition and testing in JavaScript. It defines a function to add two numbers, and then writes a test to verify that this function works correctly.

8.4.2 Mocha

Mocha is a powerful JavaScript test framework that runs on both Node.js and in the browser, making it a versatile tool for testing in different environments. It simplifies asynchronous testing, making it straightforward and enjoyable for developers.

Mocha runs tests serially, which allows for flexible and accurate reporting. This feature is particularly helpful when debugging, as it maps uncaught exceptions to the correct test cases, making it easier to pinpoint the source of an error.

Key features of Mocha include its flexible and accurate reporting, a rich interface that supports different testing styles such as Behavior-Driven Development (BDD) and Test-Driven Development (TDD), and compatibility with both client-side and server-side JavaScript testing.

Furthermore, Mocha is highly customizable. It offers a wide variety of plugins, including reporters for different output formats, integrations with assertion libraries for more readable tests, and mocking utilities for isolating code under test. This makes Mocha an ideal choice for developers who need a flexible and feature-rich testing framework.

Key Features of Mocha:

- **Flexible and Accurate**: Our testing framework executes tests serially, providing the advantage of detailed reporting. This allows for more precise error tracking and easier debugging, enhancing the overall development process.
- **Rich Interface**: It supports various testing styles, including but not limited to, Behavior-Driven Development (BDD) and Test-Driven Development (TDD). This broad testing style support caters to diverse development methodologies and project requirements.
- **Browser and Node.js Support**: Our framework is a versatile tool that can be utilized for testing both client-side and server-side JavaScript. Its wide application range ensures comprehensive testing and consistent results regardless of the environment.
- **Customizable**: It is highly customizable, offering a wide variety of plugins. These include reporters that provide detailed test result information, test frameworks for structured testing, and mocking utilities that simulate function behaviors. This adaptability allows for a tailored testing environment that can meet unique project needs.

Example: A Simple Mocha Test with Chai Assertion Library

```
// test.js
const assert = require('chai').assert;
const sum = require('./sum');
```

```
describe('Sum Function', () => {
    it('adds 1 + 2 to equal 3', () => {
        assert.equal(sum(1, 2), 3);
    });
});

// sum.js
function sum(a, b) {
    return a + b;
}
module.exports = sum;
```

To run Mocha tests, you need to install Mocha and Chai (npm install --save-dev mocha chai), then add a test script to your package.json: "test": "mocha"

The sum.js file contains a function named sum which takes two arguments a and b, representing two numbers. The function performs a basic arithmetic operation of addition on these two numbers and returns the result.

The test.js file, on the other hand, is where the test suite for the sum function is defined. The suite is structured using Mocha's describe and it functions, which are used to organize and define the tests.

The describe function groups related tests in a test suite. Here, it's used to group the tests for the sum function. It takes two arguments: a string describing the suite and a callback function containing the tests.

Nested inside the describe block is Mocha's it function, which is used to define a single test. It also takes a string and a callback function as arguments. The string describes what the test is meant to do, in this case, it verifies the sum of 1 and 2 equals 3. The callback function contains the logic of the test.

The actual test is performed using Chai's assert function, which is used to make assertions in tests. Here, it's used to assert the equality of the result of sum(1, 2) and 3. If the sum function works correctly and returns 3, the test will pass. If it returns any other value, the test will fail.

The use of Mocha and Chai in this code provides a structured and descriptive way of defining a suite of tests for a function, asserting the function's correctness, and handling the pass or fail outcomes of the tests.

Conclusion

Choosing the right testing tool is essential for effective software testing. Jest offers a comprehensive, all-in-one solution with a focus on simplicity and performance, suitable for projects needing out-of-the-box functionality with minimal setup.

Mocha, with its flexible and accurate testing capabilities, is ideal for developers who need a highly customizable framework compatible with both Node.js and browser environments. By understanding and leveraging these tools, developers can ensure their applications are robust, maintainable, and free of bugs.

Practical Exercises for Chapter 8: Error Handling and Testing

To reinforce your understanding of error handling and testing in JavaScript, this section provides practical exercises centered around these concepts. These exercises are designed to help you apply the theories discussed in Chapter 8 through hands-on implementation using popular testing frameworks and error handling techniques.

Exercise 1: Handling Exceptions with Try, Catch, Finally

Objective: Write a function that attempts to parse JSON data and uses try, catch, finally to handle any errors that might occur during parsing, logging the error and ensuring that a cleanup action is taken.

Solution:

```javascript
function safeJsonParse(jsonString) {
    let parsedData = null;
    try {
        parsedData = JSON.parse(jsonString);
        console.log("Parsing successful:", parsedData);
    } catch (error) {
        console.error("Failed to parse JSON:", error);
    } finally {
        console.log("Parse attempt finished.");
    }
    return parsedData;
}

// Example usage
const jsonData = '{"name": "John", "age": 30}';
const malformedJsonData = '{"name": "John", age: 30}';
safeJsonParse(jsonData);   // Should log the parsed data
safeJsonParse(malformedJsonData);   // Should log an error
```

Exercise 2: Testing with Jest

Objective: Create a Jest test for a simple function that adds two numbers. Ensure the test verifies the function's correctness.

Solution:

```javascript
// sum.js
function sum(a, b) {
    return a + b;
}
module.exports = sum;

// sum.test.js
const sum = require('./sum');

test('adds 1 + 2 to equal 3', () => {
    expect(sum(1, 2)).toBe(3);
});

// Run this test by adding `"test": "jest"` to your package.json scripts and running
`npm test` in your terminal.
```

Exercise 3: Integration Testing with Mocha and Chai

Objective: Write an integration test for a function that fetches user data from an API. Use Mocha for the test framework and Chai for assertions. Assume the API returns a JSON object.

Solution:

```javascript
// userFetcher.js
const fetch = require('node-fetch');

async function fetchUser(userId) {
    const                              response                    =                await
fetch(`https://jsonplaceholder.typicode.com/users/${userId}`);
    return response.json();
}
module.exports = fetchUser;

// userFetcher.test.js
const fetchUser = require('./userFetcher');
const chai = require('chai');
const expect = chai.expect;

describe('fetchUser', function() {
```

```
    it('should fetch user data', async function() {
        const user = await fetchUser(1);
        expect(user).to.have.property('id');
        expect(user.id).to.equal(1);
    });
});

// Ensure you have Mocha and Chai installed (`npm install --save-dev mocha chai`), and
set the test script in package.json: `"test": "mocha"`
```

These exercises are designed to solidify your understanding of error handling and testing practices covered in Chapter 8. By completing these tasks, you not only get to practice implementing error handling mechanisms but also gain hands-on experience with writing unit and integration tests using popular JavaScript testing frameworks. This hands-on approach helps you build a robust foundation in writing safer, cleaner, and more reliable JavaScript applications.

Chapter 8 Summary: Error Handling and Testing

In Chapter 8, we explored the critical aspects of error handling and testing in JavaScript. These components are essential for developing reliable, robust, and maintainable software. Through detailed discussions and practical exercises, this chapter aimed to provide you with the tools and knowledge necessary to implement effective error management strategies and ensure the integrity of your code through systematic testing.

Importance of Error Handling

Error handling is a fundamental part of software development that ensures your application behaves predictably under all circumstances, including when things go wrong. We began the chapter by discussing the try, catch, finally construct, which allows developers to gracefully manage and respond to errors in JavaScript. This mechanism not only helps in maintaining application stability but also enhances user experience by preventing abrupt application failures.

We delved into the nuances of throwing errors, where you learned how to deliberately generate errors with the throw keyword. This is particularly useful for enforcing certain conditions within your application, such as validating user inputs or ensuring that required resources are available. Custom error types were also discussed, which facilitate specific error handling strategies tailored to particular kinds of errors, making the debugging process more intuitive and focused.

The Role of Testing

Testing is the cornerstone of developing dependable software. It involves verifying that your code works as expected and continues to do so as it evolves. In this chapter, we covered two main types of testing:

- **Unit Testing**: Focused on individual components or "units" of code, ensuring that each part functions correctly in isolation. We explored how unit tests are crucial for validating the behavior of small, discrete pieces of functionality within your application.
- **Integration Testing**: This testing confirms that multiple units work together as expected. Integration tests are key to ensuring that the combination of individual parts of your application results in a coherent and functional whole.

Tools and Libraries for Testing

We reviewed two prominent JavaScript testing frameworks, Jest and Mocha, which are instrumental in facilitating both unit and integration testing. Jest is praised for its simplicity and out-of-the-box functionality, including built-in test runners and assertion libraries, making it ideal for projects where quick setup and ease of use are priorities. Mocha, known for its flexibility and extensive ecosystem, allows for more customized testing environments and integrates seamlessly with various assertion libraries and mocking tools.

Conclusion

Error handling and testing are not merely about preventing or fixing bugs; they are about proactively creating a robust foundation for your applications. By understanding and implementing the strategies discussed in this chapter, you can significantly enhance the reliability and quality of your software.

These practices not only safeguard your applications against unexpected failures but also foster confidence in your codebase, both for you and for others who rely on your software. As you continue to develop and refine your JavaScript skills, remember that thorough testing and diligent error handling are indispensable tools in your developer toolkit, essential for crafting professional-grade applications in today's dynamic software landscape.

Quiz for Part II: Intermediate JavaScript

This quiz is designed to test your understanding of the concepts discussed in Part II: Intermediate JavaScript, covering Advanced Functions, Object-Oriented JavaScript, Web APIs and Interfaces, and Error Handling and Testing. Each question reflects key topics and principles that are essential for mastering Intermediate JavaScript skills.

Question 1: Advanced Functions

What is the primary benefit of using arrow functions in JavaScript?

A) They have their own this context. B) They cannot contain asynchronous code. C) They do not have their own this context. D) They are faster than traditional functions.

Question 2: Object-Oriented JavaScript

Which statement about the prototype chain in JavaScript is true?

A) JavaScript objects directly contain all methods of their prototype. B) Changes to an object's prototype affect only that instance. C) An object's prototype defines methods that can be shared by all instances of that object. D) Prototypes are typically used in JavaScript to prevent inheritance.

Question 3: Web APIs and Interfaces

Which API is used to store data that should persist across sessions in a web application?

A) sessionStorage B) localStorage C) fetch() D) XMLHttpRequest

Question 4: Error Handling and Testing

What does the finally block in a try...catch...finally statement do?

A) It is executed if an error occurs in the try block. B) It is executed after the catch block, but only if no errors were caught. C) It is executed regardless of whether an error was thrown or caught. D) It contains the cleanup code that runs depending on the error type.

Question 5: Advanced Functions

In JavaScript, what is a closure?

A) A type of function that can execute asynchronously. B) A combination of a function bundled together with references to its surrounding state. C) The process of combining several functions into one. D) A function that is returned by another function.

Question 6: Object-Oriented JavaScript

Which keyword is used to create a class inheritance in JavaScript?

A) inherits B) extends C) prototype D) super

Question 7: Web APIs and Interfaces

What is the primary use of the fetch() API?

A) To manipulate the browser's history. B) To make network requests and handle responses. C) To store data in the browser that disappears after the session ends. D) To send data to local storage.

Question 8: Error Handling and Testing

What is the purpose of throwing custom errors in JavaScript?

A) To break the application deliberately. B) To enhance the debugging process by providing clearer errors. C) To slow down the execution of functions. D) To bypass the need for external error handling libraries.

Instructions for Completion

Choose the best answer for each question. This quiz is designed to reflect on your knowledge and understanding of the intermediate JavaScript concepts discussed throughout Part II of the book. Correct answers will provide insight into how well you have grasped the subjects, and reviewing any incorrect answers can help reinforce learning.

Project 2: Creating a Weather Application Using APIs

1. Project Overview: Creating a Weather Application Using APIs

1.1 Purpose

The primary goal of this weather application is to provide users with real-time weather information including temperature, humidity, wind speed, and forecasts. This application will serve as a reliable tool for planning daily activities, travel, or any event that could be affected by weather conditions.

1.2 Features

The weather application will include several key features to ensure it meets the needs of its users:

1. **Current Weather Display**: Shows the current weather conditions of a specified location, including temperature, humidity, cloudiness, and wind information.
2. **Weather Forecast**: Provides a short-term forecast (next 24 hours) and a long-term forecast (up to 7 days) to help users plan ahead.
3. **City Search**: Allows users to search for weather conditions in different cities worldwide.
4. **Location-Based Weather**: Automatically detects the user's current location to display the local weather upon application startup.
5. **Interactive Map**: (Optional) Integrates an interactive map showing weather conditions across different regions.
6. **Responsive Design**: Ensures that the application is accessible on various devices, including desktops, tablets, and smartphones.

1.3 API Choice

For this project, we will use the OpenWeatherMap API. This choice is based on several factors:

- **Comprehensive Data**: OpenWeatherMap provides a wide range of weather data, including current conditions, minute-by-minute precipitation forecasts, hourly forecasts, daily forecasts, and historical data.
- **Global Coverage**: It offers weather data for locations all over the world, which is crucial for a weather application intended for a global user base.
- **Ease of Use**: OpenWeatherMap's API has a well-documented, straightforward interface that simplifies the integration process.
- **Free Tier Availability**: The API offers a generous free tier, allowing up to 60 calls per minute, which is adequate for development and moderate use.

These features make OpenWeatherMap an excellent choice for developers looking to integrate reliable weather data into applications without significant cost or complexity.

This project aims not only to build a functional weather application but also to enhance understanding of working with APIs, handling asynchronous operations in JavaScript, and developing responsive user interfaces. By the end of this project, you will have gained valuable experience in API integration, data handling, and application design—a skill set that is highly relevant in today's web development landscape.

2. Setup and Configuration

Setting up and configuring your development environment properly is crucial for a smooth development process and successful application deployment. This section outlines the steps needed to prepare for building the weather application, including environment setup, obtaining an API key, and initializing a project structure.

2.1 Environment Setup

1. **Development Tools**:
    - **Code Editor**: Install a code editor suited for web development, such as Visual Studio Code, which offers excellent support for JavaScript, HTML, CSS, and various extensions.
    - **Node.js**: Install Node.js if you plan to use any Node-based tools or server-side scripting. It comes with npm (Node package manager), which is essential for managing JavaScript libraries.

      o   **Git**: Install Git for version control, allowing you to manage and track changes in your project.
2. **Browser**: Ensure you have a modern web browser such as Google Chrome or Firefox for testing. These browsers support the latest web technologies and provide powerful developer tools.

2.2 API Key Registration

To fetch weather data from OpenWeatherMap, you'll need to obtain an API key. Follow these steps to register and obtain your key:

- **Visit the OpenWeatherMap Website**: Go to OpenWeatherMap and sign up for an account.
- **Subscribe to an API Plan**: Navigate to the 'API' section and choose a plan. The free plan should suffice for development purposes.
- **Get Your API Key**: After subscribing, you can find your API key in your account dashboard under the 'API Keys' tab. Note that it may take a few minutes for the API key to become active.

2.3 Project Initialization

1. **Create a Project Directory**:
2. mkdir weather-app
3. cd weather-app
4. **Initialize the Project**:

- If using plain HTML, CSS, and JavaScript, set up your directory with the basic files:
- touch index.html style.css app.js
- If using a JavaScript framework like React: This command sets up a new React application with all necessary dependencies and build configurations.
- npx create-react-app .

1. **Version Control**:

- Initialize a Git repository in your project directory:
- git init
- Create a .gitignore file to exclude node_modules and other non-essential files:
- node_modules/
- .env

1. **Environment Variables**:

-   To securely store your API key, use an environment variable. If you're using Node.js or a framework like React, install dotenv:
-   npm install dotenv
-   Create a .env file in your project root and store your API key:
-   REACT_APP_OPEN_WEATHER_MAP_API_KEY=your_api_key_here

Proper setup and configuration are the foundation of your project's development environment. By completing these steps, you ensure that your development process is efficient and that your application is ready for further development and eventual deployment. With your environment set up, API key secured, and project structure initialized, you're now ready to start developing the weather application's core functionalities.

3. Designing the User Interface

Designing an intuitive and effective user interface (UI) is crucial for the success of any application. For the weather application, the UI should not only be aesthetically pleasing but also functional and easy to navigate. This section covers the key components of the UI design process, including layout planning, component breakdown, and styling approaches.

3.1 Layout Planning

1. **Wireframe Creation**:
    -   Begin by sketching a basic wireframe of the application's interface. Focus on the placement of the main elements such as the search bar, weather display area, and navigation links if needed.
    -   Tools like Balsamiq, Adobe XD, or even simple paper sketches can be used for this process.
2. **Responsive Design Considerations**:
    -   Ensure that the layout is responsive, adapting to various screen sizes and orientations. Use a mobile-first approach, which designs for smaller screens first before scaling up to larger screens.
    -   Incorporate media queries in your CSS to handle different screen sizes and maintain layout integrity across devices.

3.2 UI Components

1. **Search Bar**:

- o This is where users will enter the name of the city for which they want weather information. It should be prominently placed, typically at the top of the page.
  - o Include input validation to ensure that the input is not empty.
2. **Weather Display Cards**:
  - o Design cards to display weather data such as temperature, humidity, wind speed, and weather conditions (sunny, cloudy, etc.).
  - o Consider using icons or visuals to represent different weather conditions dynamically based on the data received from the API.
3. **Navigation Menu** (Optional):
  - o If your application has multiple views (e.g., current weather, detailed forecast, historical data), include a navigation menu to switch between these views.
  - o Ensure that the navigation elements are accessible and easily identifiable.

3.3 Styling

1. **CSS/SASS**:
  - o Decide whether to use plain CSS or a preprocessor like SASS. SASS offers advantages like nested rules, variables, and mixins which can simplify complex stylesheets.
  - o Organize your stylesheets logically, separating styles specific to components and those that are more general.
2. **Using a CSS Framework**:
  - o Consider using frameworks like Bootstrap, Material-UI (for React), or Tailwind CSS to speed up the development process and ensure consistency and responsiveness.
  - o These frameworks provide pre-built components and utility classes that can significantly reduce the time needed for styling.
3. **Theme and Colors**:
  - o Choose a color scheme that reflects the nature of the application—soft blues and whites can evoke a sense of calm appropriate for weather apps.
  - o Ensure text is readable on all backgrounds, and interactive elements are highlighted effectively to guide user interaction.

3.4 Accessibility Considerations

- Make sure that all parts of the UI are accessible, including keyboard navigation support and screen reader compatibility.
- Use semantic HTML to improve accessibility. For instance, use <button> for buttons instead of <div> and ensure images have alt text.

Designing the user interface is a critical step that affects how users interact with your application. By carefully considering the layout, components, and styling, and ensuring accessibility, you create a user-friendly environment that can effectively communicate weather information. This thoughtful design process not only enhances user engagement but also promotes a positive user experience.

4. Application Functionality

Developing the core functionality of your weather application involves handling data retrieval, processing, and management. This section outlines how to implement the primary functions of the weather application, including fetching weather data, handling API responses, and managing application state.

4.1 Fetching Weather Data

1. **Using the Fetch API**:
    o Utilize the JavaScript Fetch API to make asynchronous requests to the OpenWeatherMap API. This involves constructing a URL with the necessary query parameters, such as the city name and API key.

Example Fetch Request:

```javascript
function fetchWeather(city) {
    const apiKey = process.env.REACT_APP_OPEN_WEATHER_MAP_API_KEY;
    const                              url                              =
`https://api.openweathermap.org/data/2.5/weather?q=${city}&appid=${apiKey}&units=met
ric`;

    fetch(url)
        .then(response => {
            if (!response.ok) {
                throw new Error('Network response was not ok');
            }
            return response.json();
        })
        .then(data => updateWeatherDisplay(data))
        .catch(error => console.error('Failed to fetch weather:', error));
}
```

2. **Error Handling**:

o   Properly handle errors that may occur during the API request, such as network issues or data errors. Provide user-friendly error messages and fallback mechanisms.

4.2 Processing API Responses

1. **Data Parsing**:
    o   Once the data is retrieved from the API, parse it to extract and format the necessary information like temperature, wind speed, humidity, and weather conditions.
2. **Update UI**:
    o   Use the parsed data to update the UI components dynamically. This could involve displaying the current weather, updating weather icons, and populating forecast data.

Example Data Update Function:

```javascript
function updateWeatherDisplay(weatherData) {
    const temperature = weatherData.main.temp;
    const conditions = weatherData.weather[0].description;
    const humidity = weatherData.main.humidity;

    document.getElementById('temp').textContent = `${temperature} °C`;
    document.getElementById('conditions').textContent = conditions;
    document.getElementById('humidity').textContent = `Humidity: ${humidity}%`;
}
```

4.3 State Management

1. **Using State Hooks (React)**:

If using React, utilize state hooks (e.g., useState) to manage the application's state, such as the current city, weather data, and any loading or error states.

Example State Management in React:

```javascript
import React, { useState } from 'react';

function WeatherApp() {
    const [city, setCity] = useState('');
    const [weather, setWeather] = useState(null);
    const [loading, setLoading] = useState(false);
    const [error, setError] = useState(null);
```

```
const handleSearch = () => {
    setLoading(true);
    setError(null);
    fetchWeather(city).then(data => {
        setWeather(data);
        setLoading(false);
    }).catch(err => {
        setError(err.message);
        setLoading(false);
    });
};

return (
    // JSX for rendering the UI
);
}
```

2.   **Local Storage for Recent Searches**:

Optionally, use localStorage to remember recent searches or save user preferences like units of measurement (Celsius or Fahrenheit).

Implementing the application functionality involves setting up efficient data fetching, robust error handling, and dynamic UI updates. By effectively managing application state and integrating these functionalities, your weather app becomes a powerful tool for providing timely and accurate weather information. As you refine these processes, consider adding more advanced features like notifications for severe weather or integrating other data sources for a richer user experience.

5. Displaying Weather Data

Once you have successfully fetched and processed the weather data, the next critical step is to display this information effectively in your user interface. This section focuses on how to dynamically present weather data to users in a clear and engaging manner.

5.1 Structuring the Display

1.   **Weather Information Cards**:

Create distinct UI components or "cards" for different pieces of weather information. For example, have separate cards for displaying current weather, hourly forecasts, and weekly forecasts. This modular approach makes it easier to manage and update specific sections of the display independently.

Example HTML Structure for Weather Cards:

```
<div id="currentWeather" class="weather-card">
    <h2>Current Weather</h2>
    <div id="temp" class="weather-detail"></div>
    <div id="conditions" class="weather-detail"></div>
    <div id="humidity" class="weather-detail"></div>
</div>
```

2. **Dynamic Updates**:

Implement JavaScript functions to update these cards dynamically based on the data received from the weather API. Ensure that updates are performed without refreshing the page to provide a seamless user experience.

5.2 Implementing Data Visualization

1. **Weather Icons**:

Use weather condition codes returned by the API to display appropriate weather icons that visually represent the current weather conditions. Icons can be more effective than text at quickly conveying weather information.

Example of Updating Weather Icons:

```
function updateWeatherIcon(conditionCode) {

    const iconElement = document.getElementById('weatherIcon');

    iconElement.src = `/path/to/icons/${conditionCode}.png`;

}
```

2. **Graphs and Charts** (Optional):

For more detailed forecasts, consider using graphical representations such as charts or graphs. Libraries like Chart.js or D3.js can be used to create interactive charts that show temperature changes, precipitation probabilities, or wind patterns over time.

5.3 Accessibility Considerations

1. **Readable Fonts and Colors**:

Choose font sizes and colors that ensure readability across all devices and lighting conditions. High contrast between text and background colors is essential for readability and accessibility.

2. **Alt Text for Icons**:

Provide descriptive alt text for all icons used in the application. This text should convey the same information as the icon itself, ensuring that the content is accessible to users with visual impairments.

5.4 Responsive Design

1. **Fluid Layouts**:

Use CSS Grid or Flexbox to create fluid layouts that adapt to different screen sizes. This responsiveness ensures that the weather information is presented neatly and legibly on both large screens and mobile devices.

2. **Media Queries**:

Employ media queries to adjust styles and layout configurations based on the device's characteristics, such as width, height, or orientation.

Displaying weather data effectively is about more than just showing numbers and text; it's about creating an intuitive, informative, and accessible presentation that enhances user engagement. By carefully structuring the display elements, incorporating visual aids like icons and charts, and ensuring accessibility, your weather application will provide a user-friendly experience that delivers essential weather information efficiently. As you refine the display of weather data, continue to gather user feedback to make iterative improvements, ensuring that the application remains useful and relevant to your audience.

6. Additional Features

Enhancing the core functionality of your weather application with additional features can significantly improve user engagement and satisfaction. This section discusses potential enhancements that could make your weather app more comprehensive, interactive, and user-friendly.

6.1 Extended Forecast

Adding an extended forecast feature allows users to plan better for future events by viewing weather predictions over a longer period.

Implementation:

- Fetch and display a 5-day or 7-day weather forecast using the same API by adjusting the API endpoint or parameters.
- Display each day's forecast in a separate card or section, showing key information such as high and low temperatures, weather conditions, and chances of precipitation.

6.2 Geolocation Integration

Automatically detecting the user's current location to display local weather is a convenient feature for mobile users.

Implementation:

- Use the Geolocation API to get the user's current latitude and longitude.
- Pass these coordinates to the weather API to fetch and display the local weather.
- Ensure to handle permissions properly and provide fallback options if the user denies geolocation permissions.

Example Geolocation Usage:

```javascript
function fetchLocalWeather() {
    navigator.geolocation.getCurrentPosition(position => {
        const { latitude, longitude } = position.coords;
        fetchWeatherByCoords(latitude, longitude);
    }, showError);
}

function fetchWeatherByCoords(lat, lon) {
```

```
    const                               url                              =
`https://api.openweathermap.org/data/2.5/weather?lat=${lat}&lon=${lon}&appid=${apiKe
y}`;
    // fetch and update UI logic here
}
```

6.3 Weather Alerts

Displaying weather alerts can be crucial for user safety during extreme weather conditions.

Implementation:

- Use a dedicated endpoint from the weather API that provides weather alerts.
- Display alerts in a prominent area of the application with distinctive styling to catch the user's attention.

6.4 Unit Conversion Toggle

Allow users to switch between Celsius and Fahrenheit for temperature readings, catering to preferences based on geographical or personal choice.

Implementation:

- Implement a toggle switch in the UI.
- Convert temperatures between units according to the user's selection and update the display accordingly.

6.5 Interactive Weather Map

Incorporate an interactive map that visualizes various weather conditions across different regions, enhancing the application's interactivity and informational value.

Implementation:

- Integrate a mapping service like Google Maps or Leaflet.
- Overlay weather data on the map, such as temperature gradients, precipitation, or cloud cover.

Example Map Integration:

```
function initWeatherMap() {
    const map = L.map('weatherMap').setView([51.505, -0.09], 13);
    L.tileLayer('https://{s}.tile.openstreetmap.org/{z}/{x}/{y}.png', {
        attribution: '© OpenStreetMap contributors'
    }).addTo(map);
    // Additional logic to overlay weather data
}
```

6.6 Personalized Weather Dashboard

Allow users to create a personalized dashboard where they can add multiple locations to monitor, set up weather notifications, and customize the layout.

Implementation:

- Provide a user login or local storage-based system to save user preferences.
- Allow users to add and remove cities or regions for weather monitoring.

By implementing these additional features, your weather application can offer more than just basic weather updates, transforming it into a comprehensive tool for weather monitoring and planning. These features not only enhance user engagement but also provide practical value, making your application a go-to resource for weather-related information. As you develop these features, continue to test and gather user feedback to refine functionality and usability, ensuring the app remains relevant and useful in diverse scenarios.

7. Testing and Deployment

Testing and deployment are crucial phases in the development of your weather application. This section outlines strategies for thoroughly testing your application to ensure it is reliable and performs well under various conditions, as well as steps to deploy the application to a live environment where users can access it.

7.1 Testing

1. **Unit Testing**:
   - Focus on testing individual components of the application, such as data fetching functions, UI components, and utility functions.
   - Use Jest or Mocha/Chai to write unit tests that verify the functionality of these components in isolation.

Example of a Unit Test:

```
// Testing a function that formats weather data
describe('formatWeatherData', () => {
    it('correctly formats temperature data', () => {
        const rawWeather = { temp: 283.15 }; // Kelvin
        const expectedOutput = { temp: 10 }; // Celsius
        expect(formatTemperature(rawWeather.temp)).toEqual(expectedOutput.temp);
    });
});
```

2. **Integration Testing**:
   - Ensure that different parts of the application work together as expected. Test scenarios like user interactions with the search bar leading to proper API calls and correct updates to the UI.
   - Simulate user actions and check for proper response handling and error management.

3. **End-to-End Testing**:
   - Use tools like Cypress or Selenium to simulate user journeys from start to finish.
   - Validate the complete workflow of the application, including the integration of all components, from entering a city name, fetching data, to displaying the weather and updating UI elements.

7.2 Deployment

1. **Preparing for Deployment**:
   - Ensure all environmental variables, like API keys, are secured and not hard-coded in your source files. Use .env files or similar mechanisms to manage them.
   - Minimize and optimize your application's assets (HTML, CSS, JavaScript, images) for production.

2. **Choosing a Hosting Platform**:
   - Select a suitable hosting service based on your application's needs. For a simple weather application, platforms like Netlify, Vercel, or GitHub Pages offer free and easy hosting solutions.
   - For more dynamic applications that may require backend services, consider platforms like Heroku or AWS.

3. **Deployment Process**:
   - Configure your project repository on GitHub or a similar service.
   - Connect your repository to the hosting platform.
   - Set up continuous deployment from your repository to automatically deploy new versions of your application when changes are pushed to the main branch.

Example Deployment using Netlify:

```
# Assuming the project is set up with a GitHub repository
# Link your GitHub repository to Netlify
# Set up build commands and publish directory in Netlify
npm run build   # Build your application for production
# Netlify will handle the rest, deploying your site after each push to your repo
```

4. **Post-Deployment**:
   o   After deploying, conduct tests to ensure that the application is functioning as expected in the production environment.
   o   Monitor performance and user interactions to gather insights that can guide further development or improvements.

Testing ensures that your weather application is robust and bug-free, while effective deployment makes it accessible to your users reliably. By carefully planning and executing these stages, you can enhance the quality of your application and provide a smooth and engaging user experience. Continue to monitor the application post-deployment to handle any issues and to improve based on user feedback.

8. Challenges and Extensions

Developing a weather application offers a valuable opportunity to tackle a variety of technical challenges and to explore potential extensions that can enhance its functionality and user experience. This section will discuss some of the common challenges you might encounter and suggest possible extensions to improve and expand your application.

8.1 Challenges

1. **API Limitations**:
   o   **Challenge**: Free tiers of weather APIs often come with limitations on the number of requests per minute or day, which can restrict how frequently you can fetch updates.
   o   **Solution**: Implement caching mechanisms to store weather data temporarily and reduce the number of API calls. Provide user feedback when the limit is reached, explaining why updates may be delayed.
2. **Data Accuracy and Timeliness**:
   o   **Challenge**: Weather data might not always reflect real-time conditions due to delays in data updates from the API.

- **Solution**: Display the time of the data update to set the right expectations for users. Consider using APIs that offer more frequent updates if timeliness is critical.

3. **Complex User Interfaces**:
   - **Challenge**: Managing a complex UI, especially when incorporating features like interactive maps or extensive forecast data, can lead to performance issues.
   - **Solution**: Optimize front-end assets and consider lazy loading heavy components like maps only when needed.

4. **Handling Diverse Data Formats**:
   - **Challenge**: Weather APIs may return data in various formats, making it challenging to standardize the data handling across different sources.
   - **Solution**: Create a data normalization layer that converts all incoming data into a standard format before it is processed or displayed.

8.2 Extensions

1. **User Customization**:
   - Allow users to customize the interface, such as choosing between a dark and light mode or selecting which weather data points they wish to see by default.
   - Implement widgets or dashboards that users can personalize with their preferred information and layout.

2. **Social Features**:
   - Integrate social features where users can share the weather forecasts on social media or communicate with others about weather-related plans.
   - Allow users to submit local weather reports and photos, enhancing community engagement.

3. **Advanced Weather Analytics**:
   - Provide historical weather data comparisons to offer insights into weather trends and anomalies.
   - Integrate predictive weather modeling features that can forecast weather changes more accurately using machine learning algorithms.

4. **Multi-Source Weather Aggregation**:
   - Combine data from multiple weather APIs to enhance the reliability and accuracy of the forecasts provided.
   - Implement a system to compare and contrast forecasts from different sources, giving users a "confidence score" based on how closely these forecasts agree.

5. **Mobile Application**:
   - Develop a dedicated mobile app to provide more robust functionality, such as notifications for weather changes, widget features, or offline availability.

    ○   Optimize location-based services in the mobile app to offer more precise weather updates and alerts.

Navigating the challenges and exploring potential extensions are integral parts of developing a robust weather application. These efforts not only improve the application's reliability and user satisfaction but also encourage continuous learning and improvement. As you develop your weather app, consider these challenges and extensions as opportunities to innovate and enhance the value of your project. By addressing these areas, you can create a more comprehensive, engaging, and user-friendly application that stands out in a crowded market of weather apps.

9. Conclusion

The journey of creating a weather application from scratch is both challenging and rewarding. This project has not only enhanced your skills in using JavaScript and various web technologies but also deepened your understanding of application design, API integration, user interface creation, and data handling.

9.1 Key Takeaways

1. **API Integration**: You've learned how to effectively use the OpenWeatherMap API to fetch real-time weather data. Handling API requests and responses with JavaScript's Fetch API has improved your skills in asynchronous programming and error management.
2. **Responsive UI Design**: The design and implementation of a user-friendly, responsive interface using modern HTML and CSS practices (and potentially JavaScript frameworks like React) have prepared you for building a variety of web applications that are accessible on any device.
3. **Advanced JavaScript**: Through this project, you've applied advanced JavaScript concepts, including handling asynchronous data, working with environmental variables, and creating dynamic content based on user interactions. This has solidified your JavaScript knowledge and how it can be applied to real-world projects.
4. **Testing and Deployment**: You've tackled the essential practices of testing and deploying web applications, ensuring reliability and availability to end-users. These skills are critical for any software development project and will aid in your future career endeavors.

9.2 Reflecting on Challenges

Throughout this project, you faced numerous challenges, from dealing with API rate limits to ensuring the application performs efficiently across different platforms. Overcoming these challenges taught you problem-solving and optimization strategies that are vital for a successful programming career.

9.3 Future Enhancements

While the core functionality of the weather application is complete, the possibilities for enhancement and expansion are vast. Whether it's through integrating additional data sources for more accurate weather predictions, adding social features, or developing a companion mobile app, there's always room for improvement and innovation.

9.4 Final Thoughts

The completion of this weather application project marks a significant milestone in your journey as a web developer. It serves as a testament to your hard work and dedication to learning and applying new technologies and concepts. As you move forward, use this experience as a foundation for more complex projects and continue to explore new technologies and methodologies.

Keep coding, keep learning, and remember that every line of code you write not only builds applications but also builds your skills and shapes your future. Congratulations on completing this project, and best of luck in your future development endeavors!

Part III: JavaScript and Beyond

Chapter 9: Modern JavaScript Frameworks

Welcome to Chapter 9, "Modern JavaScript Frameworks." In this enlightening chapter, we will take an immersive deep dive into the transformative and innovative world of JavaScript's most influential, groundbreaking frameworks and libraries that have had a tremendous impact on the current state of web development.

The primary objective of this chapter is to meticulously explore and shed light on the key players that have actively and significantly shaped the landscape of modern web development. We aim to illustrate, with vivid clarity, how these powerful, intuitive tools can drastically enhance productivity, greatly improve maintainability, and substantially elevate the overall quality of web applications.

These tools have changed the dynamics of web development, making it more efficient and accessible, and contributing to a richer user experience.

9.1 Introduction to Frameworks and Libraries

In the rapidly changing and ever-evolving landscape of web development, JavaScript frameworks and libraries have emerged as integral components of the ecosystem. These indispensable tools provide structured, reusable, and maintainable code bases that can drastically enhance the efficiency and productivity of developers.

They facilitate the building of complex, scalable, and robust web applications that can meet the demands of today's dynamic online environments. This section offers an in-depth introduction to the concept of frameworks and libraries. It draws clear distinctions between these two types of resources, each of which has its own unique features and advantages.

Moreover, the section delves into their profound significance in modern web development, elucidating how they contribute to the creation of innovative, user-friendly, and high-performance web solutions.

9.1.1 Understanding Frameworks and Libraries

Frameworks

In the context of programming and web development, frameworks are essentially comprehensive tools that serve as a fundamental structure upon which to construct and shape software applications. They are comprised of pre-written, reusable code that is designed to aid developers in building applications or components more efficiently and effectively.

Frameworks dictate the architecture of your software, providing a complete scaffold that developers can fill with their own, unique code. They are opinionated in nature, meaning they set specific rules and guidelines that developers are expected to follow. This structure aids in the creation of scalable and maintainable codebases. In essence, they streamline the programming process, offering standardized ways to build and deploy different types of applications, hence increasing a developer's efficiency and productivity.

Moreover, frameworks often come with built-in tools and features for tasks such as input validation, session handling, database interaction and more, reducing the amount of manual coding required and allowing developers to focus more on application logic rather than routine elements.

In the world of web development, there are numerous popular frameworks, each with its own unique features, advantages, and use-cases. These include Angular, React, and Vue.js among others.

Frameworks play an indispensable role in modern software development, providing developers with a robust and efficient foundation for building high-quality, scalable applications. They not only hasten the development process but also enforce best practices, contributing significantly to the overall quality and maintainability of the software.

Libraries

In the realm of programming and web development, "Libraries" are collections of prewritten code snippets or routines that developers can leverage to perform specific tasks or functions within their applications. These libraries typically provide a set of well-defined interfaces, classes, methods, and functions, which can be invoked or reused as needed.

Libraries play a crucial role in software development as they enable developers to avoid reinventing the wheel for common functionalities, thus saving precious development time and resources. They consist of reusable functions and components that serve specific functionalities and assist in building applications. Developers call these functions and components when needed, providing them with more control over the application's architecture.

This is where libraries differ from frameworks - libraries are less opinionated and more flexible than frameworks, allowing developers more freedom to structure their applications as they see fit.

Libraries can serve a wide range of functionalities and can be general or specific in their scope. For instance, some libraries focus on user interface components, some help with networking tasks, others provide mathematical functions, and so on. They can be used in virtually any area of application development, from frontend user interfaces to backend server operations.

In essence, a library is like a toolkit for developers, providing them with ready-to-use tools that can help them build functional and efficient software applications. The use of libraries not only accelerates the development process but also enhances code readability and maintainability, as libraries adhere to standardized coding practices and conventions.

Libraries are an integral part of software development, significantly contributing to the efficiency, scalability, and quality of software applications. They help developers avoid code duplication, promote code reuse, and help in building more robust and reliable software.

9.1.2 Examples of Frameworks and Libraries

React

React, developed by Facebook, is a widely-used JavaScript library that specializes in helping developers build user interfaces, or UIs. Primarily used for single-page applications, it allows developers to create large web applications which can change data without refreshing the page.

React is known for its efficiency and flexibility. It operates on a virtual DOM (Document Object Model), which allows it to only refresh the portion of the page that needs to be updated rather than refreshing the entire page. This leads to a much smoother and faster performance.

Furthermore, React allows developers to create components, which are reusable pieces of code that return a React element to be rendered to the page. The use of components promotes reusability, making the code easier to maintain and debug, as each component has its own logic. This modular approach also makes it easier for teams to work cohesively on large projects.

React is also known for its rich ecosystem. Its libraries can be combined with a variety of other libraries or frameworks, such as Redux for state management or Jest for testing. In addition,

React has a large and active community that provides a wealth of resources, including tutorials, forums, and third-party libraries.

React is a powerful JavaScript library that offers efficient performance and flexibility, making it an excellent choice for building complex user interfaces.

Angular

Angular is a prominent, open-source web application framework developed and maintained by Google. It is widely used by developers around the globe to build dynamic, single-page web applications.

Angular is written in TypeScript, a statically-typed superset of JavaScript, which offers enhanced code readability and predictability. It adopts a modular development structure, where functionality is divided into separate modules, making the code easier to organize, manage, and reuse.

One of the defining features of Angular is its use of declarative templates, which are HTML plus additional custom (directives). The templates are parsed by Angular's template engine to produce the rendered live view. The dependency injection system, another key feature, helps to increase efficiency and modularity by allowing developers to reuse components and services across different parts of an application.

Moreover, Angular provides a wealth of built-in functionalities like data binding, form validation, routing, and HTTP implementation. Data binding reduces the need for writing a substantial amount of boilerplate code. The form validation feature ensures data correctness before it is sent to the server.

Angular's routing feature enables navigation among different views of an application. The HTTP implementation allows easy communication with a remote HTTP server for data retrieval, modification, and storage.

By providing a structured approach to web application development, Angular helps developers write more organized, reusable, and testable code, thereby increasing their productivity and efficiency. It's suitable for building large-scale applications due to its powerful features and strong community support.

Vue.js

Vue.js is a popular JavaScript framework that is used for building user interfaces and single-page applications. It's known as a progressive framework. This means that it is designed to be incrementally adoptable, meaning developers can adopt as much or as little of the framework as they need, adding more complexity only when it's necessary. This flexibility makes Vue.js a versatile tool for both simple and complex projects.

Developed by former Google engineer Evan You, Vue.js has gained popularity due to its simplicity and ease of use. It features an adaptable architecture that focuses on declarative rendering and component composition, enabling developers to write clean, maintainable code.

Vue.js is also highly performant and efficient, providing developers with features like lazy-loading, asynchronous rendering, and a host of other optimization options. Its component-based architecture allows for reusability of components, leading to code efficiency and consistency.

Moreover, Vue.js has a vibrant and supportive community that can provide valuable resources and support. The framework's extensive documentation, tutorials, and examples make it accessible for both beginners and experienced developers.

In conclusion, Vue.js is a powerful, flexible, and user-friendly framework that has made a significant impact in the world of web development. Its combination of robust features, performance optimizations, and community support make it an excellent choice for many developers.

9.1.3 Why Use Frameworks and Libraries?

1. **Efficiency**: One of the main advantages of using frameworks and libraries is the improved efficiency they offer. They expedite the development process by providing predefined templates and functions. This automation of repetitive tasks frees up valuable time for developers, allowing them to shift their focus towards the implementation of unique features and innovation. This means less time spent writing boilerplate code and more time crafting the unique aspects of an application.
2. **Quality**: Many of these frameworks and libraries are the product of skilled development teams and large, active communities. They are continually developing, refining, and maintaining these tools, ensuring they adhere to high-quality standards. The ongoing updates and improvements offer the assurance of a reliable foundation for building applications.
3. **Scalability**: As businesses grow, their software needs to adapt and grow with them. Frameworks are equipped with patterns and tools designed to make scaling up applications more manageable and less time-consuming. They provide the necessary

infrastructure to support the growth of applications, ensuring the software can handle increased loads and complexity.

4. **Community and Resources**: A significant benefit of popular frameworks and libraries is their extensive, often global, communities. These communities contribute to an abundance of resources such as tutorials, forums, and third-party plugins. These resources can help solve a wide range of problems, from common issues faced by many developers to more uncommon, specific challenges. This community support can be a lifeline for developers, providing access to a wealth of knowledge and experience.

Example: Setting Up a Simple React Application

React has become synonymous with modern web development due to its flexibility and the richness of its ecosystem. Here's a basic example of setting up a simple React application:

```
# Install Create React App globally
npm install -g create-react-app

# Create a new React application
create-react-app my-react-app

# Navigate into your new application folder
cd my-react-app

# Start the development server
npm start
```

By following these instructions, you'll have a new React application set up and ready for development on your local machine.

This setup will give you a boilerplate React application with support for hot reloading, modern JavaScript features, and a good structure for both small and large applications.

create-react-app is a command-line interface (CLI) tool maintained by Facebook's open-source community that allows you to generate a new React application and use a pre-configured webpack build for development. It sets up your development environment so that you can use the latest JavaScript features, provides a nice developer experience, and optimizes your app for production.

Here's a breakdown of each command:

- npm install -g create-react-app: This command installs create-react-app globally on your computer. npm is the package manager for Node.js, which is a runtime environment that allows you to run JavaScript on your computer. The g flag installs the package globally, making the create-react-app command available from any location in your command line.
- create-react-app my-react-app: This command creates a new React application with the name "my-react-app". When you run this command, create-react-app will set up a new directory with the name "my-react-app", and it will fill this directory with the boilerplate files needed to run a React application.
- cd my-react-app: This command navigates into the directory of your new application. cd stands for "change directory", which changes the current directory in your command line to the "my-react-app" directory.
- npm start: This command starts the development server. When you run a React application, it runs on a local server in your development environment (often referred to as a "development server"). This command starts that server, and makes your new React application available to view in your web browser.

Frameworks and libraries are indispensable in the toolkit of modern web developers. They not only provide the building blocks for creating advanced web applications but also enforce best practices and patterns that are essential for team collaboration and project scalability.

9.1.4 Advanced Integration Concepts

Modular Development

Modern JavaScript frameworks and libraries support modular development practices, which involve breaking the application down into smaller, interchangeable components. This approach enhances code reusability and makes it easier to manage large codebases.

Modular development is a software design technique that breaks down the system into smaller, independent, and interchangeable components known as modules. Each module is a separate unit of software that handles a specific piece of functionality in the larger system.

This approach to software development comes with numerous advantages.

Firstly, it makes the code more manageable. By dividing the codebase into smaller parts, it's easier for developers to understand and work on a specific module without getting lost in the entire system's complexity. This is especially beneficial in large projects with several developers, as it allows multiple people to work on different modules simultaneously without interfering with each other's code.

Secondly, it enhances code reusability. Once a module is developed, it can be used in various parts of the application, reducing the need to write the same code multiple times. This not only saves development time but also helps maintain consistency across the application.

Thirdly, modular development promotes scalability. As the application grows, new modules can be added without disrupting the existing system. This makes it easier to expand the application's functionality and adapt to changing requirements.

Lastly, it improves testing and maintenance. Since each module is a separate unit, it can be tested independently. This makes it easier to isolate and fix bugs. It also simplifies updates and modifications, as changes to a single module don't affect the entire system.

In modern JavaScript frameworks like React, Angular, and Vue.js, modular development is an integral part of their design. For instance, in React, the application is divided into components (modules) that are reusable and can manage their own state and props. Similarly, Angular uses a hierarchical structure of components for its applications.

In summary, modular development is an effective strategy for managing complexity in large software systems. It streamlines the development process, promotes code reuse and scalability, and improves maintainability, making it a popular approach in modern web development.

Example: In React, components are the building blocks of the application. Each component has its own state and props, making them reusable and independent.

State Management

Complex applications require efficient state management solutions to handle data across different components. Libraries like Redux for React or Vuex for Vue.js provide robust tools to manage state on a global scale.

State management in web development refers to the handling of data that can be manipulated by user interactions and system events within an application. It involves storing, manipulating, and deleting data over the lifecycle of an application or a component within an application.

In a web application, the state could represent any data that can change over time and affect the behavior or output of the application. This could be the user's login status, the contents of a shopping cart, form data, or any other data that the application needs to function.

State management can be local or global. Local state management refers to state that is specific to a single component and does not affect other parts of the application. For instance, the input value of a form field is local to that field and does not affect other components unless explicitly shared.

On the other hand, global state management involves data that is shared across multiple components. An example would be a user's login status, which could affect the application's entire behavior and needs to be accessible by multiple components.

Managing state efficiently is crucial in ensuring that an application behaves consistently and predictably. It helps track changes in data over time and can aid in debugging and testing the application.

Utilizing libraries like Redux for React, Vuex for Vue.js, or NgRx for Angular can greatly simplify the task of managing state, especially in larger applications with complex data requirements. These libraries provide a centralized store for state that can be accessed throughout the application, making it easier to track and manage changes to state. They also provide additional benefits like time-travel debugging, middleware support, and more.

Example: Using Redux in a React application to manage global state like user authentication status, which needs to be accessed across multiple components.

Server-Side Rendering (SSR)

Frameworks like Next.js (for React) and Nuxt.js (for Vue.js) enable server-side rendering of applications, which can significantly improve the performance and SEO of web pages by serving fully rendered pages from the server.

Server-Side Rendering (SSR) is a technique used in modern web development that optimizes a website's performance and makes it more compatible with search engine optimization (SEO).

In a typical single-page application (SPA), most of the rendering work is done on the client-side, meaning that the user's browser downloads a minimal HTML page, which then gets populated with JavaScript. This JavaScript is responsible for fetching data and marking up the HTML. While this approach provides a smooth user experience, especially for websites where the content changes dynamically, it has some drawbacks. The most significant is that it can lead to slower page load times, as the browser needs to wait for all the JavaScript to be downloaded and executed before it can fully render the page.

Moreover, this approach can also have a negative impact on SEO. This is because search engines' web crawlers may not fully render and understand the content added through JavaScript, leading to poorer visibility in search engine results.

Server-Side Rendering addresses these issues by doing most of the rendering work on the server. With SSR, when a user navigates to a webpage, the server generates the full HTML for the page on the server in response to the request. The server then sends this fully rendered HTML to the client's browser, allowing the page to be rendered faster than it would be with client-side rendering. This gives the user a fully populated page as soon as the HTML is fully downloaded, resulting in a faster load time.

Furthermore, because the server sends a fully rendered page to the client, all the content on the page is visible to search engines, which can lead to improved SEO.

In practice, SSR is implemented using specific frameworks for JavaScript libraries, such as Next.js for React and Nuxt.js for Vue.js. These frameworks provide a set of tools and features that simplify the process of setting up server-side rendering for your application.

In conclusion, Server-Side Rendering is a valuable technique for improving both the performance and the SEO of a web application. By rendering the page on the server rather than the client, SSR can provide a faster, more SEO-friendly user experience.

Example: Implementing Next.js in a React project to pre-render pages on the server, improving load times and SEO by delivering ready-to-view content to the user and search engines.

9.1.5 Performance Optimization

Lazy Loading

Implementing lazy loading can significantly enhance application performance by loading resources only when they are required.

Lazy loading is a design pattern commonly used in computer programming that defers the initialization of an object or the execution of a complex process until the point at which it is actually needed. This can significantly improve the performance of a software application by reducing its initial load time and conserving system resources such as memory and processing power.

In the context of web development, lazy loading is often used to defer the loading of resources like images, scripts, or even entire sections of a web page until they are required. For example, when a user visits a webpage, instead of loading all the images on the page at once, a lazy loading technique might only load the images that are immediately visible in the user's viewport. As the user scrolls down the page, additional images are loaded just in time as they become visible. This can significantly speed up the initial page load time, resulting in a faster, more responsive user experience.

Lazy loading can also be applied to other areas of software development. For example, in object-oriented programming, an object might be set up with a placeholder or proxy object until the full object is needed, at which point it is fully initialized. This can be particularly useful in situations where creating the full object is a resource-intensive process.

Lazy loading is a useful design pattern that can help improve software performance and efficiency by deferring resource-intensive operations until they are absolutely necessary. By using lazy loading, developers can create applications that are faster, more responsive, and more efficient in their use of system resources.

Example: Using React's React.lazy and Suspense to split the code at a component level, which allows you to only load user-facing features as needed.

Code Splitting

Most modern frameworks support code splitting out of the box, which divides the code into various bundles or chunks that can be loaded on demand.

Code splitting is a technique used in modern web development that allows developers to divide their code into separate bundles or chunks, which can be loaded on demand or in parallel.

This process has substantial benefits for application performance and load times. When a user initially visits a webpage, instead of loading the entire JavaScript bundle for the whole application, only the required chunks for the current view are loaded. This reduces the amount of data that needs to be transferred and parsed, resulting in faster load times and a more responsive user experience.

As the user navigates through the application, additional chunks of code are loaded as needed. This on-demand loading is especially beneficial for large applications with numerous routes and features, as it ensures that users only download the code necessary for the features they are using at any given time.

Moreover, code splitting can also improve caching efficiency. Since code is divided into several smaller bundles, any changes in one part of the application won't invalidate the entire cache but only the affected chunk. This means that users only need to download updated code, while unchanged parts remain cached from previous visits.

Most modern JavaScript frameworks and bundlers, like React with Webpack or Vue.js with Vue CLI, support code splitting out of the box. For instance, in a Vue.js project, you can configure Webpack to use dynamic imports, which splits out each route's components into separate chunks, so they are only loaded when the user accesses that route.

In conclusion, code splitting is a powerful technique in web development that enhances application performance and user experience by optimizing the loading and caching of JavaScript code.

Example: Configuring webpack in a Vue.js project to use dynamic imports, which splits out each route's components into separate chunks so they are only loaded when the user accesses that route.

9.1.6 Testing Frameworks and Tools

The process of writing unit and integration tests can be significantly streamlined through the integration of testing frameworks and tools. These include but are not limited to Jest, Mocha, Enzyme for applications built with React, or Vue Test Utils for Vue.js-based applications. These tools not only simplify the testing process but also ensure maximum coverage and efficient error detection.

In addition, the use of continuous integration (CI) and continuous deployment (CD) platforms is highly advantageous. These platforms, such as Jenkins, CircleCI, or GitHub Actions, offer a means to automate testing and deployment procedures. This automation ensures that every commit or change made to the codebase is tested and verified, minimizing the risk of errors or bugs in the production environment.

Furthermore, these platforms ensure that stable builds are deployed automatically, effectively streamlining the release process and ensuring that end-users always have access to the latest and most stable version of the application.

While frameworks and libraries provide a solid foundation for building applications, understanding how to use advanced features and integrations is crucial for developing high-performance, scalable, and maintainable web applications. As you delve deeper into specific frameworks in the subsequent sections, consider how the architecture and features of each framework can be optimized to meet the specific needs of your project.

9.2 React Basics

React is an extremely powerful JavaScript library that is specifically designed for the construction of user interfaces. Its primary function and strength lie in its capacity to build dynamic and highly responsive single-page applications where a swift reaction to user interactions is paramount. Originating from the innovative team at Facebook, React has carved out a space for itself due to its remarkably efficient rendering capabilities and its straightforward, intuitive architecture based on components.

In this section, we embark on a journey through the fundamental concepts and features of React, with the aim of providing you with a solid understanding of how to utilize this tool effectively. It is designed to assist you in getting started with crafting interactive user interfaces that are both highly functional and aesthetically pleasing. We will cover everything from the basic concepts to advanced techniques, giving you the knowledge and confidence to create React applications that are both powerful and user-friendly.

9.2.1 Understanding React Components

In React, applications are structured around components. These components can be understood as the fundamental building blocks of any application built using React. Each component in a React application functions as a distinct, self-contained module. They are responsible for managing their own content, presentation, and behavior, creating an easily manageable structure within the application itself.

Components in React encapsulate all the necessary logic required for their operation. This encompasses the rendering of the user interface (UI), handling of the state (the data that may change over time and impact how the application behaves), and responding to user interactions. By encapsulating this logic within each component, React facilitates the creation of a clean, efficient, and scalable structure for applications.

There are two types of components in React: Functional Components and Class Components. Functional components are JavaScript functions that accept properties (props) and return HTML elements describing what the UI should look like. Class components were the primary method for creating components that handle complex state logic and lifecycle methods before the introduction of Hooks.

React also uses a syntax extension for JavaScript called JSX (JavaScript XML) to describe what the UI should look like. JSX allows you to write HTML-like code within your JavaScript, making the code more readable and easier to understand.

In React, the state is an object that determines how a component renders and behaves. React components can have a local state, managed either by useState in functional components or this.state in class components. Lifecycle methods in class components and the useEffect Hook in functional components allow you to run code at specific times in the component's lifecycle.

Handling user inputs and actions is a critical part of any application. React simplifies event handling with its own synthetic event system, ensuring consistency across all browsers.

Overall, understanding React components is crucial for developing applications using React. The concept of components allows developers to create complex user interfaces with reusable pieces of code, leading to applications that are easier to develop and maintain.

Types of Components:

Functional Components

These are JavaScript functions that accept properties (props) and return HTML elements describing the UI. With the introduction of Hooks, functional components can also manage state and other React features.

Functional components are a specific type of component architecture in React, which is a popular JavaScript library for building interactive user interfaces. They are named as such because they're simply JavaScript functions. Unlike class components, they don't extend any base class but return HTML via a render function.

Functional components have gained popularity for their simplicity and conciseness. They are less verbose, easier to read and test, which leads to fewer bugs in code. Functional components just receive data and display them in some form; that is, they are mainly responsible for the UI.

In the early versions of React, functional components were also known as stateless components as they didn't have access to state or lifecycle methods. However, with the introduction of Hooks in React 16.8, functional components can now manage state and side effects, which were capabilities previously exclusive to class components.

A significant advantage of functional components is the ability to use React's built-in hooks. Hooks allow functional components to use state and other React features without writing a class. The useState and useEffect hooks are the most commonly used ones, enabling state management and the use of lifecycle events respectively within functional components.

An example of a simple functional component would be:

```
import React from 'react';

function Welcome(props) {
    return <h1>Hello, {props.name}!</h1>;
}

export default Welcome;
```

This is a simple example of a functional component in React. The component is written in a language called JSX (JavaScript XML), a syntax extension for JavaScript that allows you to write what looks like HTML in your JavaScript code.

The component is a function named 'Welcome'. As is common with functional components in React, this function takes an argument called 'props' which stands for properties. These properties are essentially inputs to the component that can be used to pass data into it. In this case, 'props' is expected to contain a property named 'name'.

Inside the function, a single HTML-like element is returned - an 'h1' header. Between the opening and closing tags of this header, the expression {props.name} is written. This is an example of JSX syntax, where JavaScript expressions can be embedded inside the HTML-like code by wrapping the expression in curly braces. Here, the expression is accessing the 'name' property of the 'props' object.

When this component is used in a React application, it will render an 'h1' header with the content "Hello, {name}!", where {name} will be replaced with whatever value is passed as the 'name' property to the 'Welcome' component.

Finally, the line 'export default Welcome' at the end of the code is using JavaScript's module system to export the 'Welcome' function from this file. The 'default' keyword indicates that 'Welcome' is the default export from this file, meaning it can be imported without needing to use curly braces in the import statement. This makes the 'Welcome' component available to be imported and used in other parts of the application.

So to summarize, this is a simple React functional component that takes a 'name' property and renders a greeting message with that name in an 'h1' header. This component can be reused anywhere a greeting message is needed in the application.

Class Components

Before Hooks, class components were the primary method for creating components that handle complex state logic and lifecycle methods.

In the context of React, class components are JavaScript ES6 classes that extend the React.Component class imported from the React library. The React.Component class is an abstract base class that provides the core functionality for React components, including the lifecycle methods and the ability to hold and manage state.

Class components have a render method that returns a React element (typically written in JSX, a syntax extension for JavaScript that resembles HTML). This React element describes what should appear on the screen when the component is rendered.

One of the defining features of class components is their ability to have local state. State in React is a data structure that holds and manages the data that can change over the course of the component's lifecycle and affects the component's behavior and rendering. In class components, the state is initialized in the constructor and can be updated using the setState method provided by React.Component.

Another important feature of class components is the lifecycle methods. These are special methods that get automatically called during different stages of a component's lifecycle, such as when it gets created, updated, or destroyed. These methods allow developers to control what happens when components mount, update, or unmount, providing a high degree of control over the component's behavior.

However, while class components are powerful, they can also be verbose and complex, especially for beginners. Moreover, the introduction of Hooks in React 16.8 has made it possible to use state and lifecycle features in functional components, making them equally powerful as class components, leading to a shift in the React community towards functional components.

Still, understanding class components is crucial, as many older and existing React codebases use class components extensively, and they remain a fundamental part of React's component model.

9.2.3 JSX - JavaScript XML

JSX, which stands for JavaScript XML, is a syntax extension for JavaScript. It was developed and is heavily used by React, a popular JavaScript library for building user interfaces. JSX is not a programming language, but it allows developers to write HTML-like syntax directly in their JavaScript code.

JSX makes it easier and more intuitive to create and manage complex, dynamic HTML in your JavaScript application. It provides a more readable and expressive syntax to structure your UI code and benefits from the power and flexibility of JavaScript.

One of the unique aspects of JSX is that it's not only used for HTML markup. It can also create user-defined components, enabling the composition of complex user interfaces from smaller, reusable components. This component-based architecture is at the heart of libraries like React, and JSX plays a crucial role in it.

A simple example of JSX code could look like this:

```
const element = <h1 className="greeting">Hello, world!</h1>;
```

In this example, the JSX translates into a JavaScript function that creates an HTML h1 element with the class "greeting" and the text "Hello, world!".

The key thing to remember about JSX is that it ultimately compiles down to regular JavaScript. Under the hood, JSX syntax is transformed into calls to React.createElement(), a method provided by the React library. This conversion is usually done using a JavaScript compiler like Babel.

Despite its HTML-like syntax, JSX comes with the full power of JavaScript. It allows you to embed any JavaScript expression within curly braces {} in your JSX code.

In conclusion, JSX is a powerful tool for writing declarative, component-based UI code in JavaScript. It combines the expressiveness of HTML with the power of JavaScript, resulting in a more intuitive and efficient way of building user interfaces in JavaScript.

9.2.4 State and Lifecycle

In React, the state is an object that determines how a component renders and behaves. React components can have local state, managed either by useState in functional components or this.state in class components.

'State' in React is a built-in object that holds property values that belong to a component. When the state object changes, the component re-renders. State is used for data that will change over time or affect the component's behavior or rendering. For example, user input, server responses, and more. The state is initialized in the constructor of a class component, or by using the useState Hook in functional components. State updates are done through the setState method or the setter function returned by useState.

The 'Lifecycle' of a React component refers to the different phases a component goes through from its creation to its removal from the DOM. Each phase comes with methods that React calls at particular moments, allowing you to control what happens when a component mounts, updates or unmounts. In class components, these are methods like componentDidMount, componentDidUpdate, and componentWillUnmount. With the introduction of hooks in React, similar effects can be achieved in functional components using the useEffect Hook.

Understanding these concepts is key to managing data and behavior in React applications. They allow developers to control the rendering process and react to changes in state or props, creating dynamic and interactive user interfaces.

Example of State in a Functional Component:

```
import React, { useState } from 'react';

function Counter() {
    const [count, setCount] = useState(0);

    const increment = () => {
        setCount(count + 1);
    };

    return (
        <div>
            <p>You clicked {count} times</p>
            <button onClick={increment}>Click me</button>
        </div>
    );
}

export default Counter;
```

This example uses the React library to create a simple counter component. The code demonstrates the use of React's functional components and the useState hook, which is a feature introduced in React 16.8 version that allows you to add state to your functional components.

Let's break down the code:

1. The import React, { useState } from 'react'; statement is used to import the React library and the useState hook into the file.
2. function Counter() { ... } defines a functional component named Counter. In React, a component can be defined as a function that returns a React element. This element describes what should appear on the screen when the component is rendered.

3. Inside the Counter component, const [count, setCount] = useState(0); is using the useState hook to create a new state variable called count. This variable will hold the current count. The useState hook also returns a function (setCount) that we can use to update the count state. The argument passed to useState (in this case, 0) is the initial value of the state.

4. const increment = () => { ... }; defines a function called increment. This function, when called, will update the count state by calling setCount(count + 1), effectively increasing the count by 1.

5. The return statement in the function describes the component's rendered output. It returns a div element containing a paragraph and a button. The paragraph displays the current count, which is dynamically inserted using curly braces. The button, when clicked, will call the increment function, thereby increasing the count.

6. The line export default Counter; exports the Counter component, making it available for use in other parts of the application.

The output of this component on the screen would be a text displaying "You clicked X times", where X is the current count, and a button saying "Click me". Every time the button is clicked, the count would increase by 1.

This code example demonstrates the basics of state management in React using the useState hook and functional components, both of which are central to modern React development.

Lifecycle methods in class components allow you to run code at particular times in the lifecycle, such as componentDidMount, componentDidUpdate, and componentWillUnmount. With Hooks in functional components, similar effects are achieved using useEffect.

9.2.5 Handling Events

"Handling Events" refers to the process of managing and responding to user interactions or system events in a software application. These interactions can include a wide variety of actions, such as mouse clicks, keyboard key presses, touch gestures, or even voice commands in some applications. System events can be anything from a timer running out, a system status changing, data being received from a server, and so on.

When a user interacts with an application, events are created and dispatched to be handled by the application. For example, when a user clicks a button, a click event is generated. The application must then decide how to respond to this event, which is where event handling comes in. This response could be anything from opening a new window, fetching data, changing the state of the application, and more.

In the context of JavaScript and web applications, event handling is often associated with specific HTML elements. For example, a button element might have a click event handler that triggers a function when the button is clicked.

In JavaScript frameworks like React, event handling is done using what's known as Synthetic Events. React's Synthetic Event system is a cross-browser wrapper around the browser's native event system, which ensures that the events have consistent properties across different browsers.

Here's an example of handling events in React:

```
function ActionLink() {
    const handleClick = (e) => {
        e.preventDefault();
        console.log('The link was clicked.');
    };

    return (
        <a href="#" onClick={handleClick}>
            Click me
        </a>
    );
}
```

This example demonstrates the creation of a functional component in React. The specific component detailed in the code is named 'ActionLink'. This is a type of functional component in React. Functional components are a simpler way to write components in React. They are just JavaScript functions that return React elements.

The ActionLink component is defined as a JavaScript function:

```
function ActionLink() {
    ...
}
Within the ActionLink function, another function named handleClick is defined:
const handleClick = (e) => {
    e.preventDefault();
    console.log('The link was clicked.');
};
```

This handleClick function is an event handler for click events. It takes an event object e as an argument. This e object represents the event that occurred. The preventDefault method is called on the event object to prevent the default action associated with the event from being performed. In this case, it prevents the default action of a link click, which is navigating to a new URL.

Instead of navigating to a new URL, the function logs 'The link was clicked.' to the console. This is achieved with the console.log method, which prints the provided message to the web browser's console.

Finally, the ActionLink component returns a JSX element:

```
return (
    <a href="#" onClick={handleClick}>
        Click me
    </a>
);
```

JSX is a syntax extension for JavaScript that is used with React to describe what the UI should look like. The returned JSX element is an anchor tag, which is typically used to create links.

The onClick attribute is a special prop in React that is used to handle click events. The handleClick function is passed to the onClick prop. This means that when the link is clicked, the handleClick function will be executed.

In summary, this ActionLink component, when used in a React application, will render a link that says 'Click me'. When this link is clicked, instead of navigating to a new URL (which is the default behavior of links), it will log 'The link was clicked.' to the console.

React provides a rich set of features that make it ideal for developing complex user interfaces with less code and higher reusability. Starting with these basics—components, JSX, state, lifecycle methods, and event handling—you now have the foundational knowledge to dive deeper into more advanced React features and patterns.

If you want to delve deeper into React, check out our other published books at: https://www.cuantum.tech/books. Consider our React-specific book, or follow our entire Web development learning path to become a master in web development.

9.3 Vue Basics

Vue.js, which is commonly referred to as Vue, is a progressive JavaScript framework that is primarily used for constructing user interfaces. This is not like other monolithic frameworks that can be overwhelming, Vue.js is carefully designed to be incrementally adoptable from the ground up.

The design philosophy behind Vue.js is simple: it focuses on the view layer only. This renders it highly flexible and easy to integrate with other libraries or even with already existing projects. It doesn't impose structure and allows developers to structure their code as they see fit, which can be a huge advantage for projects that need a certain level of customizability.

In addition to its simplicity and flexibility, Vue.js is also a powerful tool for building sophisticated Single-Page Applications (SPAs). When used in combination with modern tooling and supporting libraries, Vue.js is perfectly capable of taking on large-scale projects. It's a versatile and flexible framework that can handle a wide range of projects, from small, simple websites to large, complex web applications.

9.3.1 Understanding Vue Components

At its heart, Vue works on the component-based architecture, much like React. Components are reusable Vue instances with a name, and they encapsulate templates, logic, and styles in a fine-grained manner.

Vue.js, often shortened to Vue, is a progressive JavaScript framework primarily used for constructing user interfaces. Unlike other monolithic frameworks, Vue is designed to be incrementally adoptable and focuses solely on the view layer making it easy to integrate with other libraries or existing projects.

In the heart of Vue's architecture are components. These are reusable Vue instances with a name, and they play a crucial role in building Vue applications. Components in Vue encapsulate templates, logic, and styles in a self-contained, reusable manner. This encapsulation makes it easy to create complex user interfaces from smaller, manageable pieces.

A Vue component has three main parts:

1. The 'template' which contains the HTML markup with directives and bindings. These link the template to the underlying component data.
2. The 'script' that defines the component's logic. This includes its data properties, computed properties, methods, and more.
3. The 'style' that describes the visual appearance of the component.

Creating a Vue component involves defining these three parts in a .vue file. Once defined, the component can be reused throughout the application.

Understanding how to create and use Vue components is a fundamental part of mastering Vue.js. As you become more comfortable with Vue components, you'll be capable of building more complex and interactive Vue applications.

Example of a Simple Vue Component:

```
<template>
  <div>
    <h1>Hello, {{ name }}!</h1>
  </div>
</template>

<script>
export default {
  data() {
    return {
      name: 'Vue World'
    };
  }
}
</script>

<style>
h1 {
  color: blue;
}
</style>
```

This Vue component includes three sections:

- **template**: Contains the HTML markup with directives and bindings that link the template to the underlying component data.
- **script**: Defines the component's logic, including its data properties, computed properties, methods, and more.
- **style**: Describes the visual appearance of the component.

Now, let's break down the code:

1. <template>: This section contains the HTML structure of the Vue component. Inside the <template>, we have a div containing an h1 tag. Inside the h1 tag, we have "Hello, {{ name }}!". Here, {{ name }} is a placeholder for a variable named name. This is an example of Vue's declarative rendering, where the rendered result will be updated when the name data changes.

2. <script>: This section contains the JavaScript that controls the component's behavior. Inside the <script>, we define and export a JavaScript object, which is the definition of the Vue component. The data function returns the reactive data object of the component. In this case, it returns an object with one property: name, which has a value of 'Vue World'. This name value is what gets rendered into the {{ name }} placeholder in the template.
3. <style>: This section contains the CSS rules for the Vue component. Here, we have a rule that sets the text color of h1 elements to blue.

So, when this Vue component is rendered, it will produce a blue heading saying "Hello, Vue World!". The power of Vue.js components comes from their reusability - this component can be reused anywhere a greeting message is needed in the application, and the name can be easily changed to greet different entities.

9.3.2 The Vue Instance

Every Vue application starts by creating a new Vue instance with the Vue.createApp method, which serves as the root of a Vue application.

The Vue Instance, denoted as Vue.createApp in Vue 3, is a fundamental aspect of the Vue.js framework. It's the main building block of Vue applications and is the starting point when you're building an app with Vue.js.

When creating a Vue instance, you pass in an options object which includes declarative properties such as data, methods, computed, watch, components, and lifecycle hooks like created, mounted, updated, and destroyed.

The data option contains the data object of the Vue instance. Every property declared in the data object will be reactive, which means that if its value changes, the Vue instance will update to reflect the changes.

methods are functions that can be invoked from within the Vue instance or in the DOM part of the component. They are often used for event handling (like user input).

computed properties are functions that are used to compute derived state based on instance data. These properties are cached based on their dependencies, and only re-evaluate when some dependency changes.

watch option allows for asynchronous or expensive operations in response to changing data. This is most useful when you want to perform some operation when a particular piece of data changes.

The components option is where you declare the components that can be used within the Vue instance's template.

Lifecycle hooks are special methods that provide visibility into the life of a Vue instance from creation to destruction. They allow you to execute code at specific stages in the life cycle of a Vue Instance.

In a nutshell, a Vue Instance is the root of every Vue application and is created by instantiating Vue with the Vue.createApp() method. It provides the functionality necessary to build a reactive Vue application and serves as the glue that holds a Vue application together.

Example of Creating a Vue Instance:

```
const App = Vue.createApp({
  data() {
    return {
      greeting: 'Hello Vue!'
    };
  }
});

App.mount('#app');
```

This snippet initializes a new Vue application and mounts it to a DOM element with the id app.

Here's a step-by-step breakdown of what the code does:

1. const App = Vue.createApp({}): This line initializes a new Vue application. The createApp method creates a new application and returns an application instance, which is stored in the App constant.
2. Inside the createApp method, an object is passed as an argument. This object, often referred to as the "options object", defines the properties of the Vue instance.
3. In the options object, a data function is defined. This function returns an object that represents the local state of the component, i.e., the reactive data that the component will use.
4. In this case, the data object consists of a single property, greeting, which is initialized with the string 'Hello Vue!'. This greeting property can now be accessed and

manipulated by the Vue instance, and any changes to it will automatically cause the relevant parts of the DOM to update.

5. App.mount('#app'): This line of code tells Vue to mount the application to an HTML element. The '#app' argument is a CSS selector that selects the HTML element that will serve as the mount point for the Vue application. This element is the root of the Vue application. In this case, the application is being mounted to an element with the id 'app'.

In conclusion, this script creates a simple Vue application with a single piece of reactive data, 'greeting', and mounts it to a DOM element with the id 'app'. This basic pattern of creating an application instance, defining its data, and then mounting it to the DOM is common in Vue applications.

9.3.3 Directives and Data Binding

"Directives and Data Binding" is an important concept in modern JavaScript frameworks such as Vue.js and Angular.

Directives are special attributes with the "v-" prefix that you can include in your HTML tags. They are used to apply reactive behavior to the rendered DOM (Document Object Model). In other words, directives extend the functionality of HTML by allowing you to create dynamic content based on your application's data.

An example of a directive is v-if, which conditionally renders an element based on the truthiness of the data property it's bound to. Another is v-for, which renders a list of items based on an array in your data.

Data Binding, on the other hand, is a technique that establishes a connection between the application's user interface (UI) and its data. This connection ensures that any changes to the data automatically reflect on the UI, and vice versa.

One of the most commonly used directives for data binding in Vue.js is v-model, which creates a two-way data binding on form input and textarea elements. This means that not only does the UI update whenever the data changes, but the data also changes whenever the UI is updated.

These two concepts play a crucial role in the development of interactive web applications, as they allow developers to create dynamic and responsive user interfaces with less code.

Vue uses directives to provide functionality to HTML applications, and these directives offer a way to reactively apply side effects to the DOM when the state of the application changes.

- **v-bind**: This is a Vue.js directive which is used to bind an attribute or a component prop to an expression. The 'v-bind' directive creates a connection between the data in your Vue application and the attribute or prop you're binding to. This means that if the data changes, the attribute or prop will automatically update to reflect this change. It's a way of saying "keep this attribute or prop in sync with the current value of this expression." For example, if you wanted to bind an HTML element's 'title' attribute to a property in your Vue instance's data, you could use v-bind:title="myTitle". Then, whenever myTitle changes, the 'title' attribute on that element will be updated to reflect the new value.
- **v-model**: This directive in Vue.js creates a two-way data binding on form input, textarea, or select elements. This means that it not only updates the view whenever the model changes, but it also updates the model when the view is updated. In other words, 'v-model' provides a way for your data and your view to stay in sync in both directions. For example, if you have an input element and you want to keep its value in sync with a property in your Vue instance's data, you could use v-model="myInput". Then, whenever the user changes the input, myInput will be updated with the new value, and vice versa - if myInput changes, the input element's value will be updated to reflect the new value.

By using these directives, Vue.js allows you to create dynamic and responsive web applications where the view automatically updates to reflect changes in the data, and the data can be updated based on user interactions in the view.

Example of Data Binding:

```
<div id="app">
  <input v-model="message" placeholder="edit me">
  <p>The message is: {{ message }}</p>
</div>

<script>
Vue.createApp({
  data() {
    return {
      message: 'Hello Vue!'
    };
  }
}).mount('#app');
</script>
```

This example shows how v-model can be used to create a two-way data binding on an input element so that it not only displays the value of message but also updates it whenever the user modifies the input.

The application consists of a single div element with an id of 'app'. Inside this div, there are two elements: an input field and a p (paragraph) tag. The input field has a v-model directive that binds it to a 'message' data property. This means that any changes made in the input field will automatically update the 'message' data property, and vice versa. The placeholder text for the input field is 'edit me'.

The p tag contains a placeholder- {{ message }}. This syntax is used in Vue.js to output reactive data. In this case, it is outputting the value of the 'message' data property. As this 'message' data property is bound to the input field through v-model, any changes made in the input field will be reflected in the paragraph text.

The script section of the application is where the Vue instance is created and mounted. The Vue.createApp function is used to create a new Vue instance. Inside this function, a data function is defined, which returns the initial data state of the application. In this case, it returns an object with a single property- 'message', which is initialized with the string 'Hello Vue!'.

The mount method is then called on the Vue instance, with '#app' passed as an argument. This tells Vue to mount the application to the HTML element with the id 'app'. This element serves as the root element of the Vue application.

In summary, this is a straightforward Vue.js application that demonstrates the use of the v-model directive to create a two-way data binding between an input field and a data property. This allows any changes in the input field to automatically update the data property, and any changes in the data property to automatically update the content of the input field.

9.3.4 Handling Events

"Handling Events" refers to the process of managing and responding to user interactions or system events in a software application. These interactions can include a wide variety of actions, such as mouse clicks, keyboard key presses, touch gestures, or even voice commands in some applications. System events can be anything from a timer running out, a system status changing, data being received from a server, and so on.

When a user interacts with an application, events are created and dispatched to be handled by the application. For example, when a user clicks a button, a click event is generated. The application must then decide how to respond to this event, which is where event handling comes in. This response could be anything from opening a new window, fetching data, changing the state of the application, and more.

In the context of JavaScript and web applications, event handling is often associated with specific HTML elements. For example, a button element might have a click event handler that triggers a function when the button is clicked.

In JavaScript frameworks like React, event handling is done using what's known as Synthetic Events. React's Synthetic Event system is a cross-browser wrapper around the browser's native event system, which ensures that the events have consistent properties across different browsers.

Vue.js, offers a feature known as directives. One such directive is v-on, which serves the purpose of listening to DOM (Document Object Model) events. The primary function of this v-on directive is to execute specific JavaScript code whenever these aforementioned DOM events are triggered. This feature is particularly useful in dynamic and interactive web development, enabling developers to create more responsive experiences.

Example of Handling Click Events:

```
<div id="event-example">
  <button v-on:click="count++">Click me</button>
  <p>Times clicked: {{ count }}</p>
</div>

<script>
Vue.createApp({
  data() {
    return {
      count: 0
    };
  }
}).mount('#event-example');
</script>
```

In this example, the v-on:click directive tells Vue to increment the count data property whenever the button is clicked.

The script creates a webpage element consisting of a button and a paragraph of text. The button is labeled "Click me". This button is set up with an event listener through Vue's v-on:click directive. This directive tells Vue.js to listen for click events on the button, and each time a click event occurs, the count data property is incremented by one.

The count data property is initially set to zero when the Vue app is created. This property is reactive, meaning that whenever its value changes, Vue.js will automatically update the DOM to reflect the new value.

The paragraph of text displays the string "Times clicked: " followed by the current value of the count data property. Because count is a reactive property, the text in this paragraph will automatically update each time the button is clicked, displaying the updated click count.

In summary, this script demonstrates a simple but fundamental aspect of Vue.js and many other JavaScript frameworks: the ability to respond to user interactions with dynamic behavior. In this case, the user interaction is a button click, and the dynamic behavior is the incrementing of a click count and the automatic updating of the displayed click count.

Vue's design is focused on simplicity and flexibility. It offers a gentle learning curve and can be a perfect fit for both new developers and seasoned professionals. The core system is straightforward, but it is also incredibly adaptable and allows for powerful customizations with minimal overhead. As you continue to explore Vue, consider leveraging its extensive ecosystem of plugins and community libraries to extend your applications further.

9.4 Angular Basics

Angular is a highly dynamic platform and an extensive framework for constructing client-side applications. It employs HTML and TypeScript, which are versatile programming languages that further enhance its capabilities. Angular was developed and is continuously maintained by Google, making it a reliable and robust framework for developers.

As one of the most comprehensive front-end frameworks available today, Angular brings to the table a myriad of robust tools and impressive capabilities for building complex, high-performance applications. It is a sophisticated solution that caters to the needs of modern web development, offering a seamless experience for both developers and users alike.

One of the reasons Angular stands out from other frameworks is its strong architectural design. This solid structure allows developers to create scalable applications that can handle heavy loads while maintaining high performance. Its rich feature set also includes a range of functionalities and components that are designed to simplify complex tasks and enhance productivity.

Another significant aspect of Angular is its vibrant ecosystem. It has a large, active community of developers from around the world who contribute to its continuous development and improvement, providing valuable resources, insights, and support to other users.

This section aims to introduce the fundamental concepts that underpin Angular. By understanding these core principles, you can start building effective applications using this powerful framework. Whether you are a beginner looking to get started with front-end development or an experienced developer aiming to improve your skills, Angular is a versatile and powerful tool that can help you achieve your goals.

9.4.1 Angular Architecture Overview

Angular is built around a high-level architecture that uses a hierarchy of components as its primary architectural characteristic. It also leverages services that provide specific functionality not directly related to views and injectable into components as dependencies.

Some of the core concepts in Angular architecture include:

- **Modules**
- **Components**
- **Templates**
- **Services**
- **Dependency Injection (DI)**

Understanding these core principles helps in building effective applications using this powerful framework. Angular is a versatile and powerful tool that can help you achieve your goals, whether you are a beginner looking to get started with front-end development or an experienced developer aiming to improve your skills.

Core Concepts:

Modules

Angular applications are modular in nature. This means they are built up of several different modules, each responsible for a specific feature or functionality within the application. This modularity helps in organizing code, making it more maintainable, reusable, and easier to understand.

Angular has developed its own modularity system, known as NgModule. An NgModule is a way to consolidate components, directives, pipes and services that are related, in such a way that they can be combined with other NgModules to create an entire application.

Every Angular application has at least one NgModule, the root module, which is conventionally named AppModule. The root module provides the bootstrap mechanism that launches the application. It's the base module using which the Angular framework creates the application context or environment.

NgModules can import functionality from other NgModules just like JavaScript modules. They can also declare their own components, directives, pipes, and services. Components define views, which are sets of screen elements that Angular can choose among and modify according to your program logic and data.

Directives provide program logic, and services that your application needs can be added to components as dependencies, making your code modular, reusable, and efficient. Pipes transform displayed values within a template.

In a nutshell, modules in Angular act as the building blocks of the application and play a crucial role in structuring and bootstrapping the application.

Components

Every Angular application has at least one component—the root component. Components in Angular serve as controllers for templates (view layers) and manage the interaction of the view with various services and other components.

Components are the primary building blocks of Angular applications, and they play a critical role in defining the application's structure. Every Angular application consists of a tree of components, starting with at least one root component. This root component serves as the entry point for the application's logic, and it's the first component that Angular creates and inserts into the DOM (Document Object Model) when the application starts running.

Components in Angular are essentially classes that interact with the .html file of the component, which gets displayed on the browser. The primary responsibilities of an Angular component are to encapsulate the data, the HTML structure, and the logic for the section of the screen they control. They serve as controllers for the associated templates (view layers) and manage the interaction of these view layers with various services and other components.

Each component in Angular can be thought of as a specific piece of your application's UI (User Interface). Components can contain other components, thus creating a hierarchical structure that neatly organizes the application's functionality into manageable, modular pieces. This hierarchical structure also mirrors the DOM structure, making the application more intuitive and easier to understand.

Furthermore, components handle data and functionality and can react to user input and other events. They encapsulate the data and the behavior that the application needs to display in the view and respond to user interactions. This encapsulation makes components reusable, as they can be plugged into different parts of the application's UI, or even into different applications entirely, without needing to duplicate code.

Components in Angular are powerful and flexible tools for building dynamic, interactive web applications. They provide the means to define custom, reusable elements that encapsulate their own behavior and rendering logic, which can dramatically simplify the construction of complex user interfaces.

Templates

Templates in Angular are a crucial part of the application structure. They define the views of the application. Views are what the users see and interact with in the browser. They represent the user interface of an Angular application.

These templates are written in HTML. However, they are not just plain HTML. They incorporate Angular-specific elements and attributes which are parsed by Angular and then transformed into the Document Object Model (DOM), which represents the structure of the website. This process is what allows Angular to provide dynamic and interactive features in its applications.

Angular templates can include control flow statements, data bindings, user input handling, and many more features. Control flow statements like loops and conditionals allow developers to dynamically manipulate the structure of the DOM. Data bindings enable the synchronization of data between the model (JavaScript) and the view (HTML). User input handling allows templates to respond to user interactions.

Moreover, these templates can also leverage Angular's directives. Directives are functions that are executed whenever Angular compiler finds them in the DOM. These directives can manipulate the DOM, control the layout, create reusable components, or even extend the syntax of HTML.

In summary, templates in Angular are much more than static HTML. They are dynamic, responsive, and highly customizable, providing developers with a powerful tool to create intricate and interactive user interfaces.

Services

In the context of software development, especially within frameworks like Angular, "Services" refer to a critical architectural component. Services essentially encapsulate reusable business logic that is separate from the view, which is the part of the application that the user interacts with.

Business logic refers to the rules, workflows, and procedures that an application uses to manipulate and process data. This could include calculations, data transformations, database interactions, and other core functions of the application.

What makes Services unique is that they are designed to be independent of views. This independence allows them to be reusable, meaning they can be utilized across different parts of an application without having to rewrite the same logic multiple times.

Services are injected into components as dependencies, a process that provides components with functionalities they need to perform their tasks. In Angular, this is done using a mechanism known as Dependency Injection (DI). DI is a design pattern where a class receives its dependencies from an external source rather than creating them itself.

By injecting Services into components, you can make your application's code more modular. Modularity is a design principle where software is divided into separate, interchangeable components. Each of these components has a specific role and can function independently of the others. This separation makes the code easier to understand, maintain, and scale up.

Moreover, the use of Services promotes code reusability. Since Services encapsulate business logic that can be used across multiple components, you don't have to duplicate the same logic in different parts of your application. This not only makes your codebase cleaner and more organized but also easier to manage and debug.

Lastly, Services contribute to the efficiency of your application. With business logic neatly encapsulated in Services, components can focus on their primary role: controlling views and handling user interactions. This clear separation of concerns leads to a more efficient and performant application.

In conclusion, Services in Angular, and similar frameworks, offer a robust and efficient way to manage and reuse business logic across your application. By making your code more modular and reusable, Services play a crucial role in building scalable and maintainable software.

Dependency Injection (DI)

Dependency Injection (DI) is a software design pattern that is used to make the code more maintainable, testable, and modular. It involves a class receiving its dependencies, which are the objects that it works with, from an external source rather than creating them itself.

In the context of Angular, DI is a core feature that enables the framework to provide dependencies to a class upon instantiation. These dependencies can include various services, which are reusable pieces of code that can be shared across multiple parts of an application.

The use of DI in Angular helps to minimize the amount of hard-coding within the application and promotes loose coupling between classes. This means that the classes can operate independently of each other, which makes the code easier to modify and test.

Additionally, DI provides a way to manage the code's dependencies in a centralized place rather than scattering them throughout the application. This makes the codebase cleaner, easier to understand, and easier to maintain.

Another advantage of DI is that it allows for better code reusability and efficiency. When a service is injected into a class as a dependency, that service can be reused across multiple components, which eliminates the need to duplicate code. Moreover, by injecting dependencies, a class doesn't need to create and manage its own dependencies, which makes it more efficient and easier to manage.

To summarize, Dependency Injection (DI) in Angular is a powerful design pattern that helps to create more maintainable, testable, and modular code by providing a class with its dependencies from an external source rather than creating them itself. This approach results in a cleaner, more efficient, and more reusable codebase that is easier to understand and maintain.

9.4.2 Setting Up an Angular Project

When you are planning to begin with Angular, there are a few necessary steps that you need to take in order to set up your development environment properly. Angular operates on a robust system that requires some configuration before you can start creating applications.

A key component of this setup process is the Angular CLI, also known as the Command Line Interface. This tool is not only essential for initializing your Angular projects, but it's equally important for the development, scaffolding, and maintenance of your Angular applications.

By using Angular CLI, you can streamline your workflow and enhance your efficiency as you navigate through your Angular projects.

Installation:

npm install -g @angular/cli

Create a new project:

ng new my-angular-app

cd my-angular-app

Run the application:

ng serve

This command launches the server, watches your files, and rebuilds the app as you make changes to those files.

This example provides instructions on how to install Angular CLI (Command Line Interface), create a new Angular application, and run it.

npm install -g @angular/cli is a command to globally install Angular CLI using npm (Node Package Manager).

The ng new my-angular-app command creates a new Angular application named "my-angular-app".

cd my-angular-app changes the current directory to the newly created Angular application's directory.

Finally, ng serve is used to run the Angular application.

9.4.3 Basic Angular Example

Let's create a simple component that displays a message:

Generate a new component:

```
ng generate component hello-world
```

Edit the component (src/app/hello-world/hello-world.component.ts):

```
import { Component } from '@angular/core';

@Component({
  selector: 'app-hello-world',
  template: `<h1>Hello, {{ name }}!</h1>`,
  styleUrls: ['./hello-world.component.css']
})
```

```
export class HelloWorldComponent {
  name: string = 'Angular';
}
```

Using the component in your app (src/app/app.component.html):

```
<!-- Display the hello-world component -->
<app-hello-world></app-hello-world>
```

In the given code example, we start by generating a new component. This is done using Angular's Command Line Interface (CLI), which is a command line tool that helps with tasks like generating new components, services, and more. The command ng generate component hello-world is used to create a new component named 'hello-world'. This command, when run, creates a new directory named 'hello-world' within the 'app' directory, and within this new directory, four new files are created: a CSS file for styles, a HTML file for the template, a spec file for testing, and a TypeScript file for the component's logic.

Next, the example instructs on how to edit the newly created component. This is done in the component's TypeScript file (src/app/hello-world/hello-world.component.ts). A new class HelloWorldComponent is defined and decorated with @Component decorator. This decorator identifies the class immediately below it as a component and provides the template and related component-specific metadata.

In the component's metadata, we specify the component's CSS selector as 'app-hello-world', its HTML template as <h1>Hello, {{ name }}!</h1>, and an array of CSS files (in this case, just one) that styles this component.

The HelloWorldComponent class has a property name set to 'Angular'. This property is used in the component's HTML template. The curly braces ({{ }}) is Angular's interpolation binding syntax, which is used here to display the component's name property. So, the text "Hello, Angular!" will be displayed in the browser.

Finally, the example shows how to use the hello-world component in the application. This is done by adding its selector (<app-hello-world></app-hello-world>) to the application's main HTML file (src/app/app.component.html). When Angular sees this selector, it replaces it with the HTML from this component's template. So in this case, it will display "Hello, Angular!" in the browser where the <app-hello-world></app-hello-world> tag is placed.

In summary, this example provides a basic introduction to creating and using a new component in Angular. Components are a crucial part of Angular applications, and understanding how to create and use them is key to mastering Angular.

9.4.4 Handling Data and Events

Angular provides two-way data binding, event binding, and property binding, which are essential for handling dynamic data and user interactions.

Data handling in Angular involves managing the application's state, fetching data from external sources (like a server or an API), and updating the data in response to user actions or other events. Angular provides various tools and techniques for data handling, such as services, HTTP client, and Observables, which allow developers to manage data efficiently.

Event handling, on the other hand, involves responding to user actions, such as clicks, key presses, or mouse movements. Angular offers a robust event handling system that allows developers to define custom behavior in response to these user interactions. This can include updating the application's state, making API calls, validating form input, and much more.

Angular combines these two aspects — data handling and event handling — to create dynamic, interactive web applications. Through two-way data binding, for example, Angular allows developers to create a seamless connection between the application's data and the user interface. This means that changes in the application's state are immediately reflected in the user interface, and vice versa.

In summary, handling data and events in Angular is a fundamental part of creating web applications using this powerful framework. It involves managing the application's data, responding to user interactions, and tying the two together to create a responsive and intuitive user experience.

Example of Two-Way Data Binding:

```
<!-- Add FormsModule to your module imports -->
<input [(ngModel)]="name" placeholder="Enter your name">
<p>Hello, {{ name }}!</p>
```

This example code snippet demonstrates how to create an input field and bind it to a variable, allowing for real-time updates in both the variable and the input field.

Before diving into the code, it's important to note that the comment suggests adding FormsModule to your module imports. FormsModule is an Angular module that exports directive classes which can be used to create forms and manage form controls. FormsModule is needed in this context because it includes the ngModel directive, which is key to implementing two-way data binding.

The input tag creates an input field in an HTML form. The [(ngModel)]="name" attribute within the input tag is Angular's two-way data binding syntax. Here, ngModel is a built-in directive in Angular that sets up two-way data binding on a form input element. The name in the [(ngModel)] expression refers to a property name in the component's class.

Two-way data binding means that if the user changes the value in the input field, the name property in the component's class is updated. Conversely, if the name property changes for any reason, the value in the input field will also be updated to reflect the new value. This is what makes the data binding "two-way", it works in both directions - from the input field to the class property and vice versa.

The <p>Hello, {{ name }}!</p> line is a paragraph that displays a greeting message. The {{ name }} syntax is Angular's interpolation binding syntax. This means that the name property's value will be dynamically inserted in place of {{ name }}. As the value of name changes, the displayed greeting will automatically update to incorporate the new name value.

In summary, this code snippet demonstrates how Angular's two-way data binding can create a dynamic interaction between the user interface and the underlying data. By tying together an input field and a variable, Angular allows for real-time, bidirectional updates that create a responsive and intuitive user experience.

Conclusion

Angular provides a well-structured and robust platform that is ideally suited for the development of large-scale applications. It takes a comprehensive approach to application architecture, which makes it a powerful tool for managing the complexity typically associated with large software projects. Among its many features, Angular offers modules, components, services, and a dependency injection system, all of which contribute to its suitability for enterprise-level applications.

Modules allow developers to organize code into cohesive blocks, while components enable the building of reusable chunks of code that can dramatically improve efficiency and maintainability. Services provide a way to share common functionality across different parts of an application, and the dependency injection system simplifies the task of supplying instances of classes with their dependencies.

As you delve deeper into Angular, you will discover that it not only encourages good coding practices such as modularity and testability, but it also comes with a wide array of tools and utilities designed to assist in the development of sophisticated, modern web applications. From testing utilities to a powerful template engine, Angular is a complete solution for professional web development.

Practical Exercises for Chapter 9: Modern JavaScript Frameworks

These exercises are designed to reinforce your understanding of the JavaScript frameworks discussed in Chapter 9. By completing these tasks, you'll gain practical experience in applying the fundamental concepts of React, Vue, and Angular to build basic components and manage application state effectively.

Exercise 1: React Counter Component

Objective: Create a React component that implements a simple counter. The component should display a button and a number. Each click of the button should increment the number.

Solution:

```
import React, { useState } from 'react';

function Counter() {
    const [count, setCount] = useState(0);

    const increment = () => {
        setCount(count + 1);
    };

    return (
        <div>
            <p>Count: {count}</p>
            <button onClick={increment}>Increment</button>
        </div>
    );
}

export default Counter;
```

Exercise 2: Vue Todo List

Objective: Build a simple Vue application that allows users to add items to a todo list and display them.

Solution:

```
<template>
  <div>
    <input v-model="newTodo" @keyup.enter="addTodo" placeholder="Add a todo">
    <ul>
      <li v-for="todo in todos" :key="todo.id">
        {{ todo.text }}
      </li>
    </ul>
  </div>
</template>

<script>
export default {
  data() {
    return {
      newTodo: '',
      todos: [],
      nextTodoId: 1
    };
  },
  methods: {
    addTodo() {
      this.todos.push({ id: this.nextTodoId++, text: this.newTodo });
      this.newTodo = '';
    }
  }
}
</script>

<style>
/* Add style as necessary */
</style>
```

Exercise 3: Angular Data Binding

Objective: Create an Angular component that binds an input field to a paragraph element, updating the paragraph content in real-time as you type in the input.

Solution:

```
import { Component } from '@angular/core';

@Component({
  selector: 'app-data-binding',
  template: `
    <input [(ngModel)]="name" placeholder="Enter your name">
    <p>Hello, {{ name }}!</p>
  `,
  styleUrls: ['./data-binding.component.css']
})
export class DataBindingComponent {
  name: string = '';
}
```

Note: Ensure you have imported FormsModule in your Angular module to use ngModel.

These exercises serve as a starting point for understanding how to work with React, Vue, and Angular. They focus on fundamental features like state management in React, dynamic list rendering in Vue, and two-way data binding in Angular.

As you become comfortable with these frameworks, try extending these exercises with additional features such as editing and deleting todos or implementing more complex state logic in the counter app. These skills are essential for any modern web developer and provide a solid foundation for building complex applications.

Chapter 9 Summary: Modern JavaScript Frameworks

In Chapter 9, we explored the essentials of modern JavaScript frameworks, focusing on React, Vue, and Angular—three of the most influential tools in today's web development landscape. Each framework offers unique philosophies and technical approaches to building web applications, catering to different project needs and developer preferences. This chapter aimed to provide a foundational understanding of these frameworks, enabling you to appreciate their capabilities and how they can be utilized to enhance your web development projects.

React: A Library for Building User Interfaces

React, developed by Facebook, emphasizes declarative programming and efficient data handling through its virtual DOM. Its component-based architecture allows developers to build encapsulated components that manage their own state, leading to efficient updates and a predictable codebase.

We discussed how React uses JSX for templating, which combines the power of JavaScript with HTML-like syntax, making the code more readable and expressive. We also covered state management and the React Hooks feature, which allows for state and other React features to be used in functional components, simplifying code and increasing flexibility.

Vue: The Progressive JavaScript Framework

Vue.js is known for its simplicity and ease of integration. It is designed to be incrementally adoptable, making it as easy to integrate with existing projects as it is powerful in driving sophisticated Single-Page Applications and complex web interfaces.

We examined Vue's use of single-file components that encapsulate the template, logic, and style specific to that component. Vue's directive system, such as v-model for two-way binding and v-if for conditional rendering, provides developers with powerful and intuitive tools to build dynamic user interfaces. The core of Vue's reactivity system ensures that changes in data are efficiently reflected in the UI.

Angular: A Platform for Mobile and Desktop Web Applications

Angular, maintained by Google, is a full-fledged MVC (Model-View-Controller) framework that provides strong opinions on how applications should be structured. We looked into Angular's components, services, and modules, which help in organizing code and promoting reusability.

Angular's extensive feature set includes dependency injection, comprehensive routing, forms management, and more, making it suitable for enterprise-level applications that require scalability, maintainability, and testing capabilities. TypeScript, a core part of Angular, offers type safety, which can catch errors at compile time, improving the quality of code.

Conclusion

The chapter provided practical insights into each framework with examples and exercises to cement your understanding of key concepts such as component-based architecture, reactivity, and state management. As we concluded with exercises for building simple applications and interactive features, it's clear that learning these frameworks not only boosts your productivity but also expands your ability to solve complex development challenges.

Understanding and utilizing modern JavaScript frameworks and libraries are crucial for any developer looking to excel in the web development field. Each framework has its strengths and ideal use cases, and knowing which to use and when can significantly affect the success of your projects. Whether you choose React's flexibility, Vue's simplicity, or Angular's robustness, your journey into modern web development is well-supported by these powerful tools.

Chapter 10: Developing Single Page Applications

Welcome to the extensive journey of Chapter 10, wherein we shall submerge ourselves into the intriguing and progressive world of Single Page Applications (SPAs). This chapter is wholly dedicated to the comprehensive understanding of how SPAs function, the reasoning behind their skyrocketing popularity in the realm of modern web applications, and the most effective methodologies to develop them utilizing the cutting-edge technologies available today.

Single Page Applications, in their essence, provide a more seamless, fluid, and significantly faster user experience. This is achieved by loading content dynamically, thus substantially reducing the necessity for page reloads.

This dynamic content loading is one of the key elements that make SPAs an appealing choice. It offers the end-users a smooth, uninterrupted experience, similar to a desktop application, while they navigate through the website, ultimately enhancing user satisfaction and engagement.

10.1 The SPA Model

The SPA model fundamentally changes the way web applications interact with users by loading a single HTML page and dynamically updating that page as the user interacts with the application.

The SPA (Single Page Application) Model is a design approach used in web development where a single HTML page is loaded once and is dynamically updated as the user interacts with the application. Unlike traditional web applications where the browser initiates communication with a server to request and load new pages, the SPA model reduces the need for page reloads, thus providing a more fluid, seamless, and significantly faster user experience.

In the SPA model, most resources like HTML, CSS, and scripts are loaded only once during the initial page load. The server sends the necessary files needed to load the web application. As users interact with the application, JavaScript intercepts their actions, makes API calls to fetch data, and uses this data to dynamically update the Document Object Model (DOM) without refreshing the page.

One of the challenges in SPAs is managing the application state since the browser page does not reload. The state must be handled efficiently to keep the user interface in sync with the underlying data.

The SPA model is an appealing choice for modern web applications as it offers end-users a smooth, uninterrupted experience, similar to a desktop application, while they navigate through the website, ultimately enhancing user satisfaction and engagement. However, understanding and implementing SPAs require a good grasp of JavaScript, AJAX, and state management techniques.

This section explores the SPA model, outlining its architecture, benefits, and how it differs from traditional multi-page applications.

10.1.1 Understanding the SPA Model

Core Concept:

In a traditional web application, the browser, also known as the client, initiates communication with a server to request new pages. Upon receipt of the new HTML content, the browser reloads the page. This process is often slow and can interfere with the fluidity of the user experience, causing potential disruptions and delays.

Contrastingly, a Single-Page Application (SPA) adopts a different approach. An SPA loads a single HTML page once upon the initial visit and dynamically updates that page as the user continues to interact with the application. The resources required for the page, including the HTML content, CSS stylesheets, and JavaScript scripts, are loaded only once during the first page load.

As the user navigates through the SPA, any additional data required is retrieved as needed. This data retrieval is usually done in JSON format, using AJAX calls. AJAX, which stands for Asynchronous JavaScript and XML, allows for the updating of parts of a webpage without requiring the entire page to reload. This significantly enhances the user's experience by providing a smoother, uninterrupted interaction with the web page.

In essence, the SPA model mimics the behavior of a desktop application within the web browser, providing a fluid, seamless, and significantly faster user experience. The process of dynamically loading content and minimizing page reloads is one of the key features that makes SPAs an appealing choice in modern web development. However, it's worth noting that understanding and implementing SPAs effectively requires a solid grasp of JavaScript, AJAX, and techniques to manage application state.

Technical Flow:

Initial Page Load

The server sends the necessary HTML, CSS, and JavaScript files needed to load the web application. This is the only actual 'page load' in the traditional sense. During this initial page load, the server sends all the necessary resources, including HTML, CSS, and JavaScript files, to the client (the browser). This is the only time the webpage loads in the traditional sense in the SPA model.

The HTML provides the basic structure of the page, the CSS styles the page, and the JavaScript adds interactivity. The browser then interprets these files to construct and render the web application on the user's device.

Once the initial page load is complete, the SPA operates by dynamically updating the existing page as the user interacts with the application. This means that any additional data needed to update the content of the webpage is retrieved as needed, without requiring a full page reload.

The initial page load is a critical aspect of SPAs - it's the first point of contact the user has with the application, and the efficiency and speed of this process can have a significant impact on the user's perception of the application's performance.

In essence, during the initial page load, the server sends the necessary files to load the web application, and these files are loaded only once. After this, as users interact with the application, JavaScript intercepts their actions, fetches data as needed, and updates the Document Object Model (DOM) dynamically without refreshing the page. This leads to a smoother and faster user experience, enhancing user satisfaction and engagement.

Interaction and Dynamic Content

As users interact with the application, JavaScript intercepts browser behaviors and makes API calls to fetch data. This data is then used to update the DOM dynamically without refreshing the page. In the context of Single Page Applications (SPAs), this is especially critical.

When users interact with an SPA, their actions are intercepted by JavaScript. This could include actions such as clicking a button, submitting a form, or navigating through different sections of the application. Following these interactions, JavaScript makes API calls to fetch the necessary data.

Once the data is retrieved, it is used to dynamically update the content displayed on the page without requiring a full page reload. This dynamic update is accomplished by modifying the Document Object Model (DOM) - the structure that represents the webpage in the browser.

This process of interaction and dynamic content loading is part of what makes SPAs so efficient and user-friendly. It allows for a smooth, uninterrupted user experience, similar to what one might find in a desktop application. As the user interacts with the SPA, the content updates and changes in real-time based on their actions, without the need for disruptive and time-consuming page reloads.

However, this dynamic and interactive approach also brings with it some challenges, such as the need for effective state management. Since the page does not reload, the state of the application - which includes all the data currently displayed on the page and the status of any ongoing interactions - must be carefully managed to ensure the user interface remains in sync with the underlying data.

State Management

Since SPAs do not reload the browser page, managing the application state becomes crucial. The state must be handled efficiently to keep the UI in sync with the underlying data.

In the context of Single Page Applications (SPAs), state management takes a central role due to the dynamic nature of these applications. Since SPAs do not reload the entire page but rather update parts of it based on user interactions, managing the application's state becomes a crucial task.

State management involves keeping track of the status of an application, including the data being displayed and the status of any ongoing interactions. This state must be handled efficiently to ensure that the user interface (UI) remains in sync with the underlying data. If the state is not managed properly, it can lead to inconsistencies in the UI, causing a discrepancy between what the user sees and the actual data.

For example, consider an online shopping application. The state of the application could include the items currently in the user's shopping cart, the user's personal details, and the status of any ongoing transactions. If the user adds an item to their cart, this action should immediately be reflected in the cart's state. Similarly, if the user updates their delivery address, this new information should instantly update the state.

In a SPA, these state updates happen dynamically, without refreshing the entire page. Therefore, state management in SPAs involves methods and techniques to efficiently update the application's state in real-time, keeping the UI in sync with the underlying data.

Implementing effective state management is a technical challenge that requires a good understanding of JavaScript and, often, the use of libraries or frameworks dedicated to this purpose. Proper state management improves the user experience by ensuring a smooth, seamless interaction with the application, similar to a desktop application, enhancing user satisfaction and engagement.

Example: Basic SPA Structure

Let's look at a basic example of an SPA structure using vanilla JavaScript and HTML:

HTML (index.html):

```
<!DOCTYPE html>
<html lang="en">
<head>
    <meta charset="UTF-8">
    <meta name="viewport" content="width=device-width, initial-scale=1.0">
    <title>Simple SPA</title>
</head>
<body>
    <div id="app">
        <header>
            <h1>Simple SPA Example</h1>
            <nav>
                <a href="#" onclick="loadHome()">Home</a>
```

```
                    <a href="#" onclick="loadAbout()">About</a>
            </nav>
        </header>
        <main id="content">
            <!-- Content updated dynamically -->
        </main>
    </div>

    <script src="app.js"></script>
</body>
</html>
```

JavaScript (app.js):

```
function loadHome() {
    document.getElementById('content').innerHTML = '<h2>Home Page</h2><p>Welcome to
the home page!</p>';
}

function loadAbout() {
    document.getElementById('content').innerHTML = '<h2>About Page</h2><p>Welcome to
the about page!</p>';
}

// Load the default page
loadHome();
```

This simple SPA consists of two "pages", Home and About, which are loaded dynamically into the main container without causing a full page reload. The links in the navigation bar trigger JavaScript functions that update the content inside the main element.

SPAs, in essence, provide a seamless, fluid, and faster user experience by dynamically loading content and reducing the need for page reloads. This dynamic content loading is one of the key elements that make SPAs an attractive choice as it offers uninterrupted user experience while navigating through the website, enhancing user satisfaction and engagement.

The example then delves deeper into the SPA model, explaining how it fundamentally changes the interaction between web applications and users. Unlike traditional web applications that require the browser to initiate communication with a server to request and load new pages, the SPA model reduces this need. It loads a single HTML page and dynamically updates that page as the user interacts with the application. This model provides a fluid, seamless, and faster user experience.

The SPA model loads most resources, such as HTML, CSS, and scripts, only once during the initial page load. As users interact with the application, JavaScript intercepts their actions, makes API calls to fetch data, and uses this data to dynamically update the Document Object Model (DOM) without refreshing the page. This efficient handling of resources and user interaction enhances the user experience but also introduces challenges like managing the application state since the page does not reload.

To exemplify these concepts, the code example provides a basic example of an SPA structure using vanilla JavaScript and HTML. The HTML file includes a basic structure with a header containing navigation links and a main section where the content will be dynamically inserted.

The JavaScript file defines two functions, loadHome and loadAbout, which are responsible for loading content into the main section of the webpage when the corresponding navigation links are clicked. When the page initially loads, the loadHome function is called to populate the main section with the home page content. This illustrates how SPAs dynamically load content without causing a full page reload.

In summary, the SPA model offers significant advantages in terms of user experience and performance by reducing the need for full page reloads and providing a smooth, native-app-like interaction. Understanding and implementing SPAs requires a good grasp of JavaScript, AJAX, and state management techniques.

The SPA model offers significant advantages in terms of user experience and performance by reducing the need for full page reloads and providing a smooth, native-app-like interaction. Understanding and implementing SPAs requires a good grasp of JavaScript, AJAX, and state management techniques.

10.2 Routing in SPAs

Routing in Single Page Applications (SPAs) is a critical aspect that allows users to effortlessly navigate between different parts of the application without the need for reloading the entire page each time a new section is accessed. This effective method of routing significantly improves the overall user experience by creating a seamless and intuitive navigation process that mirrors what users have come to expect from using a traditional desktop application.

In this section, we are going to delve deeper into the workings of routing within the context of SPAs. We will particularly focus on the implementation of client-side routing, an aspect that forms the bedrock of SPA architecture. Client-side routing plays a pivotal role in ensuring the application is responsive and user-friendly, delivering content swiftly without the need for constant server requests which can slow the application's performance and disrupt the user's experience.

10.2.1 Understanding Client-Side Routing

Client-side routing involves manipulating the browser's history API to update the URL without sending a request to the server to load a new page. This approach allows different "views" or "components" of the SPA to be associated with specific URLs. When a user navigates to a different part of the application, the SPA intercepts the URL change, usually through a router, and loads the appropriate content dynamically.

In traditional web development, when a user clicks on a link to navigate to a different part of the website, a request is sent to the server. The server then responds with the relevant HTML page. This process can often lead to a noticeable delay as the new page loads, disrupting the user's experience.

However, in SPAs, all the necessary code (HTML, CSS, JavaScript) is loaded on the initial page load, or dynamically loaded as required, and added to the page. Therefore, when a user navigates to a different part of the SPA, no request to the server is needed to load a new page. Instead, JavaScript running on the client side manages the navigation, updates the URL in the browser and changes the view in the browser, all without a page refresh. This leads to a much smoother user experience, as there are no delays caused by page reloads.

Client-side routing is implemented using the HTML5 History API, which allows for changes to the URL without a page reload. Different views or components of the SPA are associated with specific URLs. When a user navigates to a different part of the application, the SPA intercepts the URL change and loads the appropriate content dynamically.

Understanding client-side routing involves learning about how SPAs manage navigation on the client side using JavaScript, how they utilize the HTML5 History API to change the URL without a page refresh, and how different views or components are associated with specific URLs.

Key Concepts:

History API

Modern SPAs use the HTML5 History API to interact with the browser history programmatically. This API allows changes to the URL without reloading the page, which is essential for client-side routing.

The History API is a powerful tool provided by modern web browsers, designed to enable developers to manipulate a website's URL and interface with the web browser's history without causing a page refresh or directing to a new page. This is particularly crucial for Single Page Applications (SPAs) where it's essential to provide a smooth and seamless user experience.

The History API comprises various methods and properties that allow developers to create more complex navigational structures. For instance, it includes methods such as pushState and replaceState which are used to add and modify history entries respectively. These methods don't reload the page but instead update the URL and can store state objects with any kind of data that needs to be preserved.

Another critical component of the History API is the popstate event. This event is fired whenever the active history entry changes, and if the history entry being activated was created by a pushState or replaceState call, the event will also contain the state object which was created with the call.

By using the History API, developers can take full control of the browser navigation and effectively manage how users interact with the application. This includes handling forward and back buttons, changing the URL according to the application's state, and even storing state information that can be used when the user navigates back to a previous state.

The History API is a key part of creating interactive and user-friendly web applications, providing the tools necessary to manage and manipulate the browser history and URL in a way that enhances the overall user experience.

Routes and Views

Routes define the URL patterns that are associated with different views in your application. A view is a representation of a particular part of the application (e.g., a profile page, settings page).

Routes define the URL patterns that users follow when navigating through a web application. Each route corresponds to a specific part of the application, such as a particular page or feature. For example, in a blogging website, you might have routes like /posts for the blog posts list, /posts/new for creating a new blog post, and /posts/:id for viewing a particular blog post.

On the other hand, views represent the visual templates or components that are displayed to the users when they navigate to a specific route. Views are responsible for defining what users see and interact with on the screen. For instance, the /posts route might display a list of blog posts, the /posts/new route might show a form to create a new blog post, and the /posts/:id route might display the full content of a specific blog post.

In the context of SPAs, routes and views are managed on the client-side. The SPA loads a single HTML page and uses JavaScript to update the views dynamically based on the user's interactions and the current route, without needing to load new pages from the server. This provides a smoother and faster user experience, similar to a desktop application.

However, managing routes and views in SPAs can be complex, as it involves manipulating the browser's history API to update the URL without sending a request to the server, and dynamically updating the Document Object Model (DOM) to change the views. Therefore, understanding and implementing routes and views effectively in SPAs requires a solid grasp of JavaScript and state management techniques.

10.2.2 Example: Implementing Basic Routing in Vanilla JavaScript

To illustrate client-side routing in an SPA, let's implement a simple router using vanilla JavaScript:

HTML Structure (index.html):

```
<div id="app">
    <nav>
        <ul>
            <li><a href="#/home">Home</a></li>
            <li><a href="#/about">About</a></li>
        </ul>
    </nav>
    <div id="view"></div>
</div>
```

```
<script src="router.js"></script>
```

JavaScript Router (router.js):

```
const routes = {
    '/home': '<h1>Home Page</h1><p>Welcome to the Home Page.</p>',
    '/about': '<h1>About Page</h1><p>Learn more about us here.</p>'
};

function router() {
    const path = window.location.hash.slice(1) || '/home';
    const route = routes[path];
    document.getElementById('view').innerHTML = route || '<h1>404 - Page Not
Found</h1>';
}

window.addEventListener('load', router);
window.addEventListener('hashchange', router);
```

In this setup:

- **HTML** defines a simple navigation menu with hash-based links for Home and About.

- **JavaScript** sets up a router function that handles the loading of different views based on the current hash value in the URL. It listens to both the load and hashchange events to handle initial page load and subsequent hash changes.

This is a simple but effective way of creating SPAs, which are web applications that load a single HTML page and dynamically update that page as the user interacts with the app.

In the HTML structure we have two links: Home and About. These links have href attributes that will change the URL's hash to #/home and #/about when clicked. The div with the id of view is the area where the content of the pages will be dynamically inserted.

In the JavaScript router, Routes are defined for '/home' and '/about'. Each route is associated with an HTML string that will be inserted into the view div when the route is activated.

The router function is responsible for managing the loading of different views based on the current hash value in the URL. It gets the current path by taking the hash of the current URL and removing the '#'. If the hash is empty, it defaults to '/home'. Then, it gets the HTML string associated with the current path from the routes object and inserts it into the view div. If no route matches the current path, an error message is displayed instead.

This router function is called when the page loads (load event) and when the URL's hash changes (hashchange event). This means that the correct view is displayed when the page first loads, and the view is updated whenever the user clicks a link.

In conclusion, this example illustrates how a basic SPA with client-side routing can be implemented using HTML and JavaScript. It demonstrates how different views can be associated with different URL hashes and loaded dynamically, providing a seamless user experience where page content changes without the whole page having to reload.

10.2.3 Advanced Client-Side Routing with Frameworks

While vanilla JavaScript is capable of handling the rudimentary aspects of routing, contemporary frameworks such as React, Vue, and Angular provide far more sophisticated routing solutions. These modern solutions come equipped with additional features such as lazy loading, nested routes, and route guards that significantly enhance the functionality and user experience of web applications.

React Router

This is a widely utilized library specifically designed for routing within React applications. It offers dynamic routing capabilities that are seamlessly synchronized with the state of your application, thereby providing an efficient and intuitive routing experience.

React Router is a powerful routing library built specifically for applications using React, a widely used JavaScript library for creating dynamic user interfaces. The fundamental role of React Router is to make it easy for developers to implement dynamic routing in their applications.

Routing refers to the ability of an application to transition between different states or views in response to user interactions. This is a crucial aspect of single-page applications (SPAs), where instead of fetching a new HTML page from the server every time the user navigates to a different part of the application, the necessary data is retrieved and the content of the current page is updated dynamically.

React Router facilitates this by allowing developers to associate different components (which represent different views or parts of the application) with different URL paths. When a user navigates to a certain path, React Router ensures that the associated component is rendered, effectively updating the visible content of the page.

What sets React Router apart is its dynamic nature. Traditional routing approaches define static routes that are only capable of rendering specific components when the application's path matches a specific URL. React Router, on the other hand, allows for dynamic routing, where the routes can be changed and configured at runtime, providing a more flexible and responsive user experience.

Moreover, React Router is designed to synchronize with the application's current state. This means that the application's UI and the URL are always in sync, allowing developers to create complex applications with nested views and routes, all while maintaining a straightforward and intuitive navigation experience for the user.

Using React Router, developers can also implement features like route protection (restricting access to certain parts of the application based on user authentication) and lazy loading (loading components only when they are needed), making it a versatile and robust solution for managing navigation in React applications.

React Router is a comprehensive routing solution for React applications that allows developers to create dynamic, responsive, and user-friendly navigation experiences.

Vue Router

Serving as the official router for Vue.js, Vue Router brings to the table support for nested routes, smooth transitions, and fine-grained navigation control. Its integrated features make it an ideal choice for developers seeking a robust and comprehensive routing solution for their Vue.js applications.

Vue Router is the official routing library designed specifically for Vue.js, a popular JavaScript framework for building user interfaces. It provides developers with the tools necessary to build Single Page Applications (SPAs) with dynamic and nested routing, as well as fine-grained navigation control.

In a Single Page Application, all necessary HTML, CSS, and JavaScript are loaded upon the initial page load, or they are dynamically loaded and added to the page as necessary. Rather than loading new pages from a server when the user navigates, SPAs update the current page in real-time in response to user interactions. This provides a more seamless user experience, similar to a desktop application.

Vue Router plays a critical role in managing this dynamic update process. It maps specific URL paths to components in the Vue.js application. When a user navigates to a particular URL, the associated Vue component is loaded and rendered, updating the visible content on the page without requiring a full page reload.

Moreover, Vue Router supports advanced routing features like nested routes and named views. Nested routes allow developers to build more complex user interfaces with nested view hierarchies, where certain components are nested within others. Named views allow developers to have multiple "views" at the same route, each with its own associated component.

In addition to these features, Vue Router also provides smooth transitions between routes with integrated transition system, and offers fine-grained navigation control by allowing developers to react to route changes and even prevent a route change if certain conditions are not met.

Vue Router is an essential tool for developing Single Page Applications with Vue.js. It provides robust and flexible routing capabilities that enhance the user experience by facilitating dynamic content loading and seamless navigation.

Angular Router

Incorporated directly into the Angular framework, the Angular Router provides support for advanced routing concepts. This includes route guards, data resolution, and multiple named router outlets, thus ensuring a versatile and secure routing environment for complex Angular applications.

The Angular Router is an integral feature of the Angular framework that provides advanced routing capabilities, enabling seamless navigation within Single Page Applications (SPAs). An SPA is a type of web application that loads a single HTML page and dynamically updates that page as the user interacts with the application.

Angular Router manages the transitions between different views or components that users see as they interact with the application. It can map different URL paths to specific components, ensuring that the correct content is loaded when a user navigates to a particular route. It also maintains the browser history for each view, enabling users to use the browser's forward and backward buttons as they would in a traditional multi-page web application.

Additionally, Angular Router supports advanced routing features, including route guards, data resolution, and multiple named router outlets. Route guards allow developers to add authentication and authorization checks before a route is activated or deactivated. Data resolution enables developers to fetch data before navigating to a particular route, ensuring that all necessary data is available when a route is activated. Multiple named router outlets allow developers to have multiple "views" at the same route, each with its own associated component.

The Angular Router is a powerful tool within the Angular framework that provides a range of capabilities for managing navigation in SPAs, from basic routing and navigation to more complex and advanced features, enhancing the overall user experience.

In conclusion, Routing is a fundamental aspect of developing SPAs that facilitates user navigation within an application without the need for page reloads. Implementing effective client-side routing ensures that your SPA behaves more like a traditional multi-page application from a user's perspective, but with much smoother transitions and improved performance. Whether you choose to implement routing from scratch or use a framework-specific router, understanding these concepts will greatly enhance your ability to develop dynamic and interactive SPAs.

10.3 State Management

State management is an integral aspect of Single Page Applications (SPAs). It plays a pivotal role in managing the state of the application in a manner that is predictable and consistent. As the complexity of Single Page Applications increases, the need for efficient state management becomes more pronounced. This is because it is vital to ensuring seamless interactions and maintaining data consistency throughout the application.

In this section, we will delve into the fundamentals of state management. We will look at what it entails and why it is so crucial in the development and maintenance of SPAs. We will also discuss common challenges that developers face when dealing with state management. These challenges can vary from maintaining the synchronization of state across multiple components to managing the memory usage of the application.

Furthermore, we will explore various strategies to handle state effectively in SPAs. Different applications may require different approaches depending on their complexity and the specific requirements of the project. By understanding these strategies, developers can make more informed decisions about the best way to manage state in their applications, thereby improving the user experience and overall performance of the SPAs.

10.3.1 Understanding State Management

State in an SPA represents the data or conditions of the UI at any given point in time. This can include user inputs, server responses, UI controls like buttons or sliders, or any other factors that might affect the output of the application.

SPAs are unique because they load a single HTML page when the application starts and dynamically update that page as the user interacts with the application. This requires careful management of the application's state, as any changes in the state directly impact what the user sees on the screen.

In the context of SPAs, the state represents data or conditions of the User Interface (UI) at any given point in time. This could include user inputs, server responses, UI controls like buttons or sliders, or any other factors that affect the output of the application. State management, therefore, involves tracking these changes and updating the UI to reflect them.

There are several key challenges when it comes to state management. As applications grow in complexity, the state can become deeply nested and difficult to manage. Without a structured approach, state changes can be unpredictable and hard to trace, leading to potential bugs. Inefficient state management can also lead to unnecessary re-renders or updates, which can impact the performance of the application.

Different strategies can be used to manage state effectively in SPAs. One common approach is to distinguish between local and global state. The local state is managed within a specific component and does not need to be shared across the application. On the other hand, the global state needs to be accessed and mutated by multiple components across the application.

Another strategy involves using state management libraries. These libraries provide tools and patterns to help developers manage application state more effectively. Examples of state management libraries include Redux for React applications, Vuex for Vue.js applications, and NgRx for Angular applications.

Advanced techniques for state management include using middleware to handle asynchronous actions or logging, using libraries like Immer or Immutable.js to handle data immutably, and leveraging the reactive state management capabilities of libraries like Angular's RxJS or Vue's reactivity system.

In conclusion, understanding state management is a critical part of developing SPAs. It involves tracking changes to the application's state and updating the UI to reflect these changes. By choosing the right approach and tools, developers can ensure that their application's state is manageable, predictable, and scalable, leading to more efficient, maintainable, and high-performing applications.

Key Challenges in State Management:

Complexity

As applications grow, the state can become deeply nested and difficult to manage. In the realm of Single Page Applications (SPAs) and state management, complexity often refers to the increased intricacy or complication that arises as applications grow and evolve. This complexity can manifest in several ways, particularly in how the state of an application is managed.

The state of an application represents the data or conditions of the User Interface (UI) at any given point in time. In SPAs, this could include user inputs, server responses, UI controls like buttons or sliders, or any other factors that affect the output of the application. As applications grow, the state can become deeply nested and difficult to manage. This is where the complexity comes into play.

A key challenge in state management is dealing with this complexity. As applications expand, introducing more features and functionalities, the state becomes increasingly intricate. This nested state can be difficult to manage without a structured approach.

State changes can also become unpredictable and hard to trace, leading to potential bugs and errors. Moreover, inefficient state management can lead to unnecessary re-renders or updates, impacting the performance of the application.

Complexity in state management refers to the increased intricacy that arises as applications grow and the state becomes more deeply nested and harder to manage. This complexity poses several challenges, including the maintainability and performance of the application, which developers must effectively address to build efficient and high-performing SPAs.

Maintainability

Maintainability, in the context of software development, refers to the measure of how easily a software system or component can be modified to correct faults, improve performance or other attributes, or adapt to a changed environment. It's a key attribute of software quality and is crucial for the long-term success and usability of software applications.

Maintainability involves several aspects:

1. **Corrective Maintenance**: This is perhaps the most common form of maintainability, and it involves fixing bugs, defects, or other issues that are identified after the software has been released. The easier it is to isolate the cause of a bug and fix it, the more maintainable the software is considered to be.

2. **Adaptive Maintenance**: As the software environment changes (for instance, if the software needs to be ported to a new operating system), the software needs to adapt. The easier it is to make these adaptations, the more maintainable the software.

3. **Perfective Maintenance**: This refers to improvements made to the software to increase its performance, maintainability, or other attributes. It could involve code optimizations, refactoring, or other techniques.

4. **Preventive Maintenance**: This involves making changes to prevent future problems. For example, a piece of code might be working fine now, but if it's anticipated that it could cause issues in the future, it might be rewritten.

Several factors can affect the maintainability of a software system or component:

- **Code Complexity**: More complex code is generally harder to maintain. Simpler, cleaner code is easier to understand and modify.

- **Documentation**: Well-documented code, with clear explanations of what different parts of the program do, can greatly improve maintainability.

- **Coding Standards**: Consistent use of coding standards can make the code easier to understand and maintain.

- **Modularity**: Software that's divided into separate, independent modules is generally easier to maintain.

Maintainability is a key characteristic of good software design and is essential for the long-term success of a software application.

In the context of Single Page Applications (SPAs), "Performance" is a crucial aspect and it refers to the speed and efficiency with which these applications render and respond to user interactions. Good performance is essential in SPAs to provide a smooth, efficient, and seamless user experience.

Performance can be affected by a variety of factors, including the efficiency of the underlying code, the size and complexity of the application, the load on the server, and the speed of the user's internet connection, among others.

One of the key factors affecting performance in SPAs is state management. Inefficient state management can lead to unnecessary re-renders or updates, which can slow down the application and degrade the user experience.

In the context of state management, there are several strategies that can be used to enhance performance. For example, using local state for data that is only needed within a specific component can reduce unnecessary updates to other parts of the application. On the other hand, global state can be used for data that needs to be shared across multiple components, but care must be taken to manage this global state efficiently to avoid performance issues.

State management libraries, such as Redux for React applications, Vuex for Vue.js applications, and NgRx for Angular applications, can also help to improve performance by providing efficient methods for managing state.

Advanced techniques for state management, such as using middleware to handle asynchronous actions or logging, using libraries like Immer or Immutable.js to handle data immutably, and leveraging the reactive state management capabilities of libraries like Angular's RxJS or Vue's reactivity system, can also enhance performance.

Performance is a critical aspect of developing SPAs and can greatly affect the user experience. Effective state management is key to ensuring good performance and building robust, efficient, and user-friendly applications.

10.3.2 Strategies for Effective State Management

Local vs. Global State:

Local State

In the realm of programming and software development, particularly in the context of designing and building Single Page Applications (SPAs), the term "Local State" is used to refer to the state of a specific component or function within the application.

Unlike global state, local state is not accessible or shared across the entire application. Instead, it is contained within the scope of the specific component or function where it has been defined and is only accessible within that particular context.

This concept is particularly important when working with modern JavaScript frameworks such as React, Vue, or Angular, where applications are typically constructed using a component-based architecture. Each of these individual components can have its own local state, which it uses to manage its internal data and operations.

For example, imagine a simple form component in a React application. This form might have its own local state to keep track of the data entered into form fields by the user. This state is only relevant and necessary within the scope of the form component, and doesn't need to be shared or made available to other components in the application. Hence, it is managed as local state.

Local state is a fundamental concept in the construction of SPAs, helping to manage and regulate the behavior of individual components or functions within the application, thereby contributing to the efficient and effective operation of the application as a whole.

Global State

Global State in the context of software development, particularly Single Page Applications (SPAs), refers to the state that is accessible and mutable from any part of the application. This concept is especially important in modern JavaScript frameworks such as React, Vue, or Angular, which adopt a component-based architecture for building applications.

Unlike the local state, which is confined within a specific component or function, the global state is shared across the entire application. It typically contains data that needs to be accessed by multiple components. For instance, user login status, theme settings, or locale settings are often stored in the global state as they are usually required by various parts of an application.

Managing global state efficiently is crucial for the performance and maintainability of an application. It requires careful handling to ensure data consistency and to avoid unnecessary re-renders or updates that can degrade the performance of an application. There are various strategies and libraries available to handle global state effectively, such as Redux for React applications, Vuex for Vue.js applications, and NgRx for Angular applications.

Global State is a fundamental aspect of state management in SPAs. It plays a pivotal role in sharing data across different parts of an application, thereby contributing to the efficient and effective operation of the application as a whole.

Example of Local State in a React Component:

```
import React, { useState } from 'react';

function LoginForm() {
    const [username, setUsername] = useState('');
    const [password, setPassword] = useState('');

    const handleSubmit = (event) => {
        event.preventDefault();
        console.log(username, password);
    };

    return (
        <form onSubmit={handleSubmit}>
            <input       type="text"      value={username}      onChange={e      =>
setUsername(e.target.value)} />
            <input      type="password"      value={password}      onChange={e      =>
setPassword(e.target.value)} />
            <button type="submit">Login</button>
        </form>
    );
}
```

This piece of code represents a functional component in React called "LoginForm". The function LoginForm() is the main component function.

In React, components are reusable pieces of code that return a React element to be rendered to the page. In this case, LoginForm is a functional component which returns a form element.

The useState hook is used to declare and manage state variables in functional components. Here, it's being used to declare and manage state for two variables: 'username' and 'password'.

useState('') creates a state variable and initializes it with an empty string. The first element of the array that useState returns is the current state value (username or password), and the second element is a function that lets you update it (setUsername or setPassword).

The handleSubmit function is an event handler that is triggered when the form is submitted. The event object is passed to it, and event.preventDefault() is called to prevent the form from being submitted in the default way, which would cause the page to reload. Instead, the current state of 'username' and 'password' is logged to the console.

The return statement of the LoginForm function returns JSX code, which is a syntax extension for JavaScript that produces React elements.

The form element has an onSubmit attribute, which is a JSX attribute that defines the event handler for the submit event of the form. It is set to the handleSubmit function.

Inside the form, there are two input elements and a button element. The input elements are of type text and password respectively. Each input has a value attribute that is set to the respective state variable, meaning the value of the input field is always the current state of 'username' or 'password'.

The onChange attribute of each input field is set to an arrow function that takes the event object and calls setUsername or setPassword with the current value of the input field. This means that every time the user types into the input field, the corresponding state variable is updated with that value.

The button element is of type submit, meaning clicking on it will submit the form and trigger the handleSubmit function.

In summary, this LoginForm component is a simple form with username and password input fields, and a submit button. The state of the input fields is managed using React's useState hook, and the form submission is handled with a custom function that logs the current state of the username and password to the console.

Using State Management Libraries:

React

In the context of designing and building Single Page Applications (SPAs) with React, managing the application's state effectively is crucial. React is a popular JavaScript library for building user interfaces, and its component-based structure necessitates efficient management of state.

State in an application refers to the data or conditions of the user interface (UI) at any given point in time. This could include user inputs, server responses, UI controls like buttons or sliders, or any other factors that affect the output of the application. State management, therefore, involves tracking these changes and updating the UI to reflect them.

Managing state in React applications can become complex when the state needs to be shared and manipulated across multiple components or when the structure of the state data is nested or complicated. In such cases, using state management libraries like Redux or Context API can be of immense help.

Redux provides a centralized store for the application state, allowing state to be managed in a predictable manner. It follows a strict unidirectional data flow and uses concepts like actions and reducers to handle state changes. This makes state updates predictable and transparent, making it easier to test and debug the application.

On the other hand, Context API is a feature provided by React itself for managing global state. It allows you to share state data and functions to manipulate that data, without having to pass props through multiple levels of components. It achieves this by using a Context object and providing it where needed using a Provider and Consumer or the useContext hook.

When building applications with React, managing state effectively is key to creating smooth, efficient, and robust applications. Tools like Redux and Context API can provide solutions for managing complex global states, improving the readability, maintainability, and performance of your React applications.

Vue

VueX is a state management pattern and library specifically designed for Vue.js applications. It provides a centralized store for all the components in an application, which means that the state information is managed in a unified place and can be accessed and manipulated from any component within the application.

The main objective of VueX is to provide a single source of truth for state data, ensuring that the state of the application remains consistent across all components. This makes it easier to track and debug state changes, enhancing the maintainability of the application.

In VueX, the state management pattern consists of four main parts: state, getters, mutations, and actions.

- The state holds the actual data.

- The getters are similar to computed properties in Vue and are used to retrieve data from the state.

- The mutations are used to modify the state and are the only way to change data in the state in a VueX store.

- The actions are functions where you put your business logic. Actions commit mutations and can contain arbitrary asynchronous operations.

By handling state changes in a predictable manner, VueX helps to manage the complexity that comes as Vue.js applications grow in size and functionality. This makes VueX an essential tool for developing large-scale Vue.js applications, where efficient state management is key to maintaining performance and user experience.

Angular

In the context of Angular, a highly popular open-source web application framework developed by Google, NgRx is an extremely effective library that provides reactive state management solutions.

State management is a crucial aspect of any web application. It refers to the handling of data or conditions of the User Interface (UI) at any given point in time, including user inputs, server responses, UI controls like buttons or sliders, or any other factors that affect the application's output.

NgRx aligns effectively with Angular's unidirectional data flow. This is a design where data flows in one direction from the source, through the application logic, and finally to the view. This approach ensures consistent and predictable behavior of the application, making it easier to debug and test.

NgRx uses a pattern inspired by Redux, another state management library, but with the power of reactive programming provided by the Observable streams from RxJS, a library for reactive programming. This means that NgRx can handle data that arrives asynchronously, and can create complex data flow structures in a more straightforward and easy-to-understand way.

In essence, NgRx provides a single source of truth for the state, which allows developers to write cleaner, more maintainable code, helps avoid state-related bugs, and makes it easier to trace the changes in the state of the application over time. It's a powerful tool for developers looking to build robust, high-performing Angular applications.

Example of Global State Management with Redux:

```
// Action Type
const SET_USER = 'SET_USER';

// Action Creator
function setUser(user) {
    return {
        type: SET_USER,
        payload: user
    };
}

// Reducer
function userReducer(state = {}, action) {
    switch (action.type) {
        case SET_USER:
            return {...state, ...action.payload};
        default:
            return state;
    }
}
```

```
// Store
import { createStore } from 'redux';
const store = createStore(userReducer);

// Dispatching an action
store.dispatch(setUser({ name: 'Jane Doe', isLoggedIn: true }));
```

This code is a basic example of managing global state with Redux in a JavaScript application, specifically a React application. Redux is a predictable state container designed to help you write JavaScript apps that behave consistently across different environments and are easy to test.

The code starts with the definition of an action type 'SET_USER'. In Redux, actions are plain JavaScript objects that have a 'type' field. This type field should typically be a string that gives this action a descriptive name, like 'SET_USER'. It's common practice to store these as constants to avoid bugs caused by typos.

Next, an action creator named 'setUser' is defined. An action creator is simply a function that creates an action. In Redux, action creators do not necessarily need to be pure functions and are often used with 'thunk' middleware for delayed actions, such as data fetching. However, in this case, 'setUser' is a simple action creator that returns an action. This action is an object that contains a 'type' field and a 'payload' field. The 'payload' field is the new data that we want to store in the Redux state.

Then, a reducer called 'userReducer' is defined. Reducers are functions that specify how the application's state changes in response to actions sent to the store. The purpose of a reducer is to return a new state object based on the type of action it receives. The reducer function switches over the action type; in the case of 'SET_USER', it returns a new state that is a combination of all existing state and the new data from the action's payload. If the reducer receives an action type it doesn't understand, it should return the existing state unchanged.

After defining the action type, action creator, and reducer, the Redux store is created. The Redux store is essentially a JavaScript object that holds the application state. The 'createStore' function from the Redux library is used to create the Redux store. The 'userReducer' is passed as an argument to 'createStore', connecting the reducer to the store.

Finally, an action is dispatched using the Redux store's 'dispatch' method. The 'setUser' action creator is called with an object containing the user's information as the argument. This object represents the payload of the 'SET_USER' action. The dispatch method takes this action object returned by 'setUser' and passes it to the 'userReducer'. The reducer then handles the action and updates the state in the Redux store.

In conclusion, this code provides a simple yet complete example of how Redux can be used to manage global state in a JavaScript application. It represents some of the fundamental concepts of Redux - actions, action creators, reducers, the store, and dispatching actions to the store.

10.3.3 Advanced Techniques

Middleware

In the context of state management in Single Page Applications (SPAs), middleware can offer additional functionality that enhances the way state is managed within the application. One of the ways it does this is by handling asynchronous actions. Asynchronous actions are tasks that start now but finish later, allowing other tasks to run in the meantime without being blocked. Handling these actions correctly is critical to maintaining the performance and user experience of the application.

For instance, suppose an SPA needs to fetch data from a server. This operation is typically asynchronous because it can take some time, and you wouldn't want the whole application to freeze whilst waiting for the server's response. Middleware can manage this operation, ensuring the rest of the application can continue to function whilst waiting for the data to be fetched.

Middleware can also provide logging functionality. Logging is a way of recording the activities within an application. This can be incredibly useful for debugging, as logs can provide a detailed overview of what happened leading up to a problem. In the context of state management, logging can help track how the state changes over time, what actions led to those state changes, and any errors that occurred during those state changes.

So, to sum up, middleware in state management can enhance the functionality of SPAs by providing tools and services for handling asynchronous actions and logging, which are key to building efficient, maintainable, and high-performing applications.

Immutable Data Patterns

Immutable Data Patterns are programming techniques that emphasize immutability in your application's data structures. These patterns involve the use of libraries like Immer or Immutable.js, which provide APIs to work with data structures in an immutable manner.

Immutability is a core principle in functional programming, meaning that once a data structure is created, it cannot be changed. Any modifications or updates to the data will result in a new copy of the data structure, leaving the original untouched. This principle has several benefits for application development.

Firstly, it can greatly enhance the predictability of your application. Since data cannot be changed once it's created, you can be confident that it won't be unexpectedly altered elsewhere in your application. This can make the code easier to reason about and reduce the likelihood of bugs.

Secondly, Immutable Data Patterns can improve the performance of your application. Libraries like Immer and Immutable.js use sophisticated techniques to avoid unnecessary data copying. For example, when an update is made, they'll only copy the part of the data structure that changed, while sharing the unmodified parts between the old and new version. This is known as structural sharing and can result in significant memory and performance optimizations.

Lastly, Immutable Data Patterns can make your application easier to work with when using certain tools or frameworks. For example, they work excellently with Redux, a popular state management library for React. Redux relies on immutability for features like time-travel debugging, where you can step forwards and backwards through your application's state to understand the sequence of state changes.

In conclusion, Immutable Data Patterns, facilitated by libraries like Immer and Immutable.js, provide a robust strategy for managing data in your application. By ensuring data immutability, they enhance both performance and predictability, leading to more robust and maintainable applications.

Reactive State Management

Reactive State Management is a concept in programming that refers to a model where changes in the state of an application are managed reactively. Application state refers to the stored information at any given point in time that can change over the lifecycle of an application. It's an integral part of interactive applications, whether web, desktop, or mobile. This state can include user inputs, server responses, UI controls, or any other factors that affect the output of the application.

Angular's RxJS and Vue's reactivity system are two examples of libraries that provide reactive state management capabilities. These libraries offer a way to manage state changes reactively, which brings about a number of benefits.

In a reactive system, when the state of the application changes, these changes are automatically propagated throughout the system to all interested parts of the application. This means that instead of components having to inquire about changes in state, they are informed about these changes. This can greatly simplify the coding paradigm because developers no longer need to write code to constantly check for state changes.

The reactive model also makes state changes easier to track and manage, making the application more efficient. Instead of manually having to manage when and where to update the UI or other parts of the application based on state changes, the reactive system handles these updates automatically. This can result in more responsive and performant applications, as updates are handled as soon as state changes occur, and only those parts of the application that depend on the changed state are updated.

Angular's RxJS (Reactive Extensions for JavaScript) is a library for reactive programming that uses Observables, making it easier to compose asynchronous or callback-based code. This aligns very well with the reactive state management model, by making state changes into a stream of events that can be observed and reacted to.

On the other hand, Vue's reactivity system is built into Vue's core. It uses a system of reactive dependencies that are automatically tracked and updated whenever state changes. This makes it incredibly easy to build dynamic user interfaces that react to state changes, as Vue handles all the complexity of tracking dependencies and updating the DOM.

In conclusion, Reactive State Management, as facilitated by libraries like Angular's RxJS or Vue's reactivity system, provides a powerful model for managing state changes in modern, interactive applications. By reacting to state changes automatically and efficiently, it simplifies the development process and results in more performant and maintainable applications.

In conclusion, effective state management is key to building robust SPAs. By choosing the right strategy and tools, you can ensure that your application's state is manageable, predictable, and scalable. Whether you opt for built-in capabilities like React's useState, use comprehensive libraries like Redux or VueX, or leverage the full reactive power of Angular with NgRx, understanding these concepts is crucial for any SPA developer looking to build efficient, maintainable, and high-performing applications.

Practical Exercises for Chapter 10: Developing Single Page Applications

To solidify your understanding of key concepts covered in Chapter 10, we present several practical exercises. These exercises are designed to help you gain hands-on experience with Single Page Application (SPA) development, focusing on routing, state management, and the SPA model.

Exercise 1: Simple SPA Routing

Objective: Implement simple client-side routing in a vanilla JavaScript SPA without using any frameworks.

Solution:

```html
<!-- index.html -->
<!DOCTYPE html>
<html lang="en">
<head>
    <meta charset="UTF-8">
    <title>Simple SPA Routing</title>
</head>
<body>
    <nav>
        <ul>
            <li><a href="#home">Home</a></li>
            <li><a href="#about">About</a></li>
        </ul>
    </nav>
    <div id="content"></div>

    <script src="router.js"></script>
</body>
</html>
// router.js
```

```
const routes = {
    'home': '<h1>Home Page</h1><p>Welcome to the home page.</p>',
    'about': '<h1>About Page</h1><p>Learn more about our SPA.</p>'
};

function handleRouting() {
    let hash = window.location.hash.substring(1);
    document.getElementById('content').innerHTML = routes[hash] || '<h1>404 Not
Found</h1><p>The requested page does not exist.</p>';
}

window.addEventListener('hashchange', handleRouting);
window.addEventListener('load', handleRouting);
```

Exercise 2: State Management with Redux

Objective: Create a simple React application that uses Redux for state management to handle a counter.

Solution:

```
# First, set up a new React app and install Redux
npx create-react-app redux-counter
cd redux-counter
npm install redux react-redux
// src/redux/store.js
import { createStore } from 'redux';

function counterReducer(state = { count: 0 }, action) {
    switch (action.type) {
        case 'INCREMENT':
            return { count: state.count + 1 };
        case 'DECREMENT':
            return { count: state.count - 1 };
        default:
            return state;
    }
}

const store = createStore(counterReducer);
export default store;
// src/App.js
import React from 'react';
import { useSelector, useDispatch } from 'react-redux';

function App() {
    const count = useSelector(state => state.count);
    const dispatch = useDispatch();

    return (
        <div>
```

```
            <h1>Count: {count}</h1>
            <button onClick={() => dispatch({ type: 'INCREMENT' })}>Increment</button>
            <button onClick={() => dispatch({ type: 'DECREMENT' })}>Decrement</button>
        </div>
    );
}

export default App;
// src/index.js
import React from 'react';
import ReactDOM from 'react-dom';
import { Provider } from 'react-redux';
import store from './redux/store';
import App from './App';

ReactDOM.render(
    <Provider store={store}>
        <App />
    </Provider>,
    document.getElementById('root')
);
```

Exercise 3: Vue.js Dynamic Component Loading

Objective: Implement a Vue.js application that dynamically loads components based on the route.

Solution:

```
<!-- App.vue -->
<template>
  <div id="app">
    <nav>
      <button @click="currentView = 'home'">Home</button>
      <button @click="currentView = 'about'">About</button>
    </nav>
    <component :is="currentView"></component>
  </div>
</template>

<script>
import Home from './components/Home.vue'
import About from './components/About.vue'

export default {
  data() {
    return {
      currentView: 'home'
    }
  },
  components: {
```

```
      Home,
      About
    }
  }
</script>
<!-- components/Home.vue -->
<template>
  <div>
    <h1>Home</h1>
    <p>This is the home page.</p>
  </div>
</template>

<script>
export default {
  name: 'Home'
}
</script>
<!-- components/About.vue -->
<template>
  <div>
    <h1>About</h1>
    <p>This is the about page.</p>
  </div>
</template>

<script>
export default {
  name: 'About'
}
</script>
```

These exercises are designed to enhance your skills in SPA development, focusing on implementing core functionalities such as routing and state management across different frameworks and libraries. By completing these tasks, you'll gain a deeper understanding of how SPAs function and how to effectively manage application states and routes, key components in building modern web applications.

Chapter 10 Summary: Developing Single Page Applications

This chapter delved into the world of Single Page Applications (SPAs), a modern approach to building dynamic and interactive web applications that offer a seamless user experience similar to desktop applications. Throughout this chapter, we explored the essential concepts, techniques, and best practices that are fundamental to designing, implementing, and optimizing SPAs.

Key Concepts and Techniques

We started by defining the SPA model, which revolves around loading a single HTML page and dynamically updating that page as the user interacts with the application. This approach minimizes page reloads, reduces web server load, and provides an instantaneous response to user actions, which are crucial for enhancing the user experience.

Routing in SPAs was a major focus, where we discussed how to manage navigation within an SPA without full page refreshes. We explored implementing client-side routing using both vanilla JavaScript and popular frameworks like React, Vue, and Angular. Each offers tools for defining navigable routes, handling route changes, and dynamically rendering content that corresponds to specific URLs. This allows SPAs to maintain bookmarkable URLs, improve SEO, and support browser history navigation, making them behave more like traditional multi-page websites from a user's perspective.

State Management emerged as a critical aspect of SPA development, given the complexity and interactivity of these applications. Efficient state management ensures that the UI remains consistent with the underlying data models and application logic. We examined different strategies for local and global state management, discussing how state can be handled using context, props, and advanced state management libraries such as Redux for React, VueX for Vue, and NgRx for Angular. These tools help manage an application's state in a predictable way, making the applications more scalable and maintainable.

Practical Application

Through practical exercises, you applied what you learned by building features such as simple routing mechanisms and state management solutions. These exercises were designed to provide hands-on experience with key SPA functionalities, enhancing your understanding and skills in real-world scenarios.

Challenges and Solutions

Developing SPAs is not without challenges. We addressed common issues such as managing complex states, optimizing performance to prevent UI jank, and configuring SPAs for improved search engine visibility. Solutions such as server-side rendering, code splitting, and dynamic data loading were discussed to mitigate these challenges.

Future Directions

Looking forward, the SPA architecture continues to evolve with advancements in web technologies. Progressive Web Apps (PWAs), for example, extend the concept of SPAs by offering offline capabilities, push notifications, and device hardware access, which blurs the lines between web and native applications.

Conclusion

Single Page Applications represent a significant shift in web development, focusing on user-centric experiences. As you continue to explore and build SPAs, keep abreast of the latest developments in JavaScript frameworks, performance optimization techniques, and new web standards that could impact how SPAs are designed and implemented. The skills and knowledge acquired in this chapter provide a solid foundation for creating sophisticated and efficient web applications that are well-suited to the needs of modern users and enterprises.

Chapter 11: JavaScript and the Server

Welcome to Chapter 11, aptly titled "JavaScript and the Server." In this enlightening chapter, we are going to take a deep dive into the versatile and powerful role of JavaScript that extends far beyond the traditional confines of in-browser operations.

We will delve into the nuances of how JavaScript, with the help of robust platforms like Node.js, manages to stretch its capabilities to encompass server-side programming. This unique characteristic of JavaScript paves the way for full-stack development capabilities, all achievable with just a single programming language.

This paradigm shift in the way we approach web development has had a profound and transformative impact. It has opened up a whole new world of possibilities, allowing developers to build web applications that are far more scalable, efficient, and integrated than ever before. By merging the front and back ends, it has enabled a seamless flow of data and logic, revolutionizing the process of web development.

11.1 Node.js Basics

Node.js is a powerful runtime environment that extends the capabilities of JavaScript beyond the confines of the browser and into server-side development. This innovative tool was masterfully developed by Ryan Dahl in 2009, in response to the growing need for a more unified approach to web development.

What makes Node.js unique is its utilization of the V8 JavaScript engine. This engine, which also powers the popular web browser Google Chrome, enables JavaScript to run outside the browser. This pivotal innovation significantly bridges the noticeable gap between front-end and back-end development, making it possible for developers to use JavaScript across the entirety of the development stack.

Therefore, with Node.js, web developers can now write server-side code using the same language they use for client-side coding, fostering a more integrated and seamless approach to web development. This has the added advantage of reducing the learning curve for developers and promoting code reusability and efficiency.

11.1.1 Core Features of Node.js

Event-Driven and Non-Blocking I/O Model

Node.js operates on a single-thread, using non-blocking I/O calls, allowing it to handle tens of thousands of concurrent connections, resulting in high scalability.

In computer programming, a thread is the smallest sequence of programmed instructions that can be managed independently by an operating system scheduler. In a traditional multi-threaded environment, new threads are spawned for each task. However, Node.js uses a different approach. Instead of creating a new thread for every client request (which can be highly memory-intensive), Node.js operates on a single thread, using what is referred to as an "event loop." This allows Node.js to handle multiple operations concurrently, without waiting for tasks to complete and without consuming high amounts of system resources.

The "non-blocking I/O calls" part of the description refers to how Node.js handles Input/Output (I/O) operations, which include tasks like reading from the network, accessing a database, or the filesystem. In a blocking I/O model, the execution thread is halted until the I/O operation completes, which can be inefficient. However, Node.js uses a non-blocking I/O model. This means that the system doesn't wait for an I/O operation to complete before moving on to handle other operations. As a result, it can continue to process incoming requests while I/O operations are being handled in the background.

The combination of these features allows Node.js to handle tens of thousands of concurrent connections. This is where the "high scalability" part comes in. Scalability, in the context of servers, refers to the ability of a system to handle an increasing amount of work by adding resources. Since Node.js can handle a high number of connections with a single thread and does not block I/O operations, it can serve a large number of client requests without degrading performance, making it highly scalable.

These attributes contribute to making Node.js a powerful tool for developing server-side applications, particularly for real-time applications, microservices, and other systems that require handling a large number of simultaneous connections with low latency.

NPM (Node Package Manager)

An integral part of Node.js, npm is a robust and dynamic package manager that is pivotal to the seamless functioning and advanced capabilities of Node.js. It serves as an accessible gateway to a vast array of libraries and tools, totaling well over 800,000. This vast collection is not just a testament to the diversity and reach of npm, but it also places npm among the largest software registries globally.

The libraries and tools accessible via npm encompass a wide array of functionalities, catering to virtually every aspect of programming and web development. They range from simple utility libraries that aid in everyday coding tasks, to complex frameworks that provide a backbone for entire applications. This multitude of resources facilitates the development of applications of all sizes and complexities, offering readily available solutions and tools for a vast array of programming needs and challenges.

In addition, npm also serves as a platform for developers to share and distribute their packages, fostering an open and collaborative programming community. Developers can publish their packages to the npm registry, making them available for others to use, thereby contributing to the exponential growth and diversity of the npm ecosystem.

Moreover, npm also includes features for version control and dependency management. It allows developers to specify the versions of packages that their project depends on, thus preventing potential conflicts and ensuring the smooth functioning of their applications. It also supports the installation of packages globally, making them available across multiple projects on the same system.

Thus, npm not only significantly enhances the utility and versatility of Node.js, but it also contributes to the broader programming and web development community. It brings together a diverse range of tools and libraries, facilitates code sharing and reuse, and provides robust mechanisms for package management, thereby streamlining the process of web development and making it more efficient and productive.

11.1.2 Getting Started with Node.js

Installation: To begin using Node.js, you need to install it on your system. You can download it from the official Node.js website.

For a comprehensive, step-by-step guide on how to install Node.js, please visit our blog post: https://www.cuantum.tech/post/how-to-install-nodejs-on-windows-mac-and-linux-a-stepbystep-guide

Hello World in Node.js: Once installed, you can write your first simple Node.js program, which traditionally starts with a "Hello World" example.

Create a file called app.js:

```
const http = require('http');

const server = http.createServer((req, res) => {
    res.statusCode = 200;
    res.setHeader('Content-Type', 'text/plain');
    res.end('Hello World\\\\n');
});

const port = 3000;
server.listen(port, () => {
    console.log(`Server running at <http://localhost>:${port}/`);
});
```

The script begins by requiring the 'http' module. This module is a built-in module within Node.js, used for creating HTTP servers and making HTTP requests.

The createServer method is then called on the http object which creates a new HTTP server and returns it. This method takes in a callback function which is executed whenever a request is received by the server. This callback function itself takes two arguments: req (the request object) and res (the response object).

The req object represents the HTTP request and has properties for the request query string, parameters, body, HTTP headers, and more. In contrast, the res object is used to send back the desired HTTP response to the client who made the request. In this script, the response status is set to 200 (which signifies a successful HTTP request), and the content type is set to 'text/plain'.

res.end('Hello World\\\\\\\\n'); is used to end the response process. This method signals to the server that all of the response headers and body have been sent, and that the server should consider this message complete. In this case, it sends the string 'Hello World\n' as the body of the response.

A constant, port, is then declared and assigned the value of 3000. This is the port at which our server will be listening for any incoming requests.

Finally, the listen method is called on the server object, which causes the server to wait for a request over the specified port - 3000. This method also takes a callback function which is run once the server starts listening successfully. Here, this callback simply logs a message to the console indicating that the server is running.

Run Your Node.js Application: Open your terminal, navigate to the directory containing app.js, and type:

node app.js

This command starts a server on localhost port 3000. When you visit http://localhost:3000 in your web browser, you will see "Hello World".

11.1.3 Understanding the Node.js Runtime

Node.js executes JavaScript code server-side, which means you can write your server logic using JavaScript. This capability is revolutionary for developers who are already familiar with JavaScript, as it eliminates the need to learn a separate language for back-end development.

One of the key characteristics of Node.js is its event-driven and non-blocking I/O model. This means that Node.js operates on a single thread, using what is called an "event loop." Instead of creating a new thread for every client request, which can be memory-intensive, Node.js can handle multiple operations concurrently without waiting for tasks to complete. In addition, its non-blocking I/O model allows the system to continue processing incoming requests while tasks like reading a file or accessing a database are being handled in the background. This architecture allows Node.js to handle tens of thousands of concurrent connections, making it highly scalable.

Another significant feature is the Node Package Manager (npm), a package manager that serves as an accessible gateway to a vast array of libraries and tools, and facilitates version control and dependency management.

Understanding the Node.js runtime also involves learning how to install Node.js and write simple programs in it. For example, the traditional 'Hello World' program in Node.js involves creating an HTTP server that responds to incoming requests with the message 'Hello World'.

Finally, Node.js excels in I/O-bound tasks, such as reading files asynchronously. This is demonstrated by the fs (file system) module in Node.js, which can read a file and print its contents, or log an error if the file doesn't exist.

Understanding the Node.js runtime is about knowing how Node.js executes JavaScript code server-side, its unique features, and how to use it to build server-side applications.

Example: Reading Files Asynchronously: Node.js excels in I/O-bound tasks. Here's an example of how Node.js handles file reading asynchronously, which is a common task in web applications.

```
const fs = require('fs');

fs.readFile('example.txt', 'utf8', (err, data) => {
    if (err) {
        console.error('Error reading file:', err);
        return;
    }
    console.log('File contents:', data);
});
```

At the start of the script, the 'fs' (file system) module is imported. This module provides various methods for interacting with the file system, making it possible to perform I/O operations, like reading and writing files, directly in JavaScript.

The fs.readFile function is then used to read the contents of a file named 'example.txt'. This function is asynchronous, meaning it returns immediately and does not block the rest of the program from executing while the file is being read. Instead, it takes a callback function that will be invoked once the file has been completely read.

The callback function provided to fs.readFile accepts two arguments: err and data. If an error occurs during the reading of the file, the err argument will contain an Error object describing what went wrong. In this case, the script logs the error message to the console using console.error.

On the other hand, if the file is successfully read, the data argument will contain the contents of the file as a string. The script then logs these contents to the console using console.log.

In summary, this script demonstrates a basic but fundamental aspect of Node.js - asynchronous file I/O. By using the 'fs' module and callback functions, it's possible to read files from the file system without blocking the execution of the rest of the program, making for efficient and responsive server-side code.

This example uses Node.js's fs (file system) module to read a file asynchronously. If there's an error (like the file doesn't exist), it logs the error; otherwise, it prints the contents of the file.

In conclusion, Node.js introduces JavaScript to the server environment, leveraging JavaScript's event-driven nature to provide a powerful tool for building fast and scalable server-side applications. This capability significantly simplifies the development process, allowing developers to use JavaScript throughout their full stack.

11.2 Building a REST API with Express

As you continue your journey in exploring the depths of server-side JavaScript, you will find that one of the most common and powerful applications of Node.js is in creating RESTful APIs. REST, which stands for Representational State Transfer, is a widely-accepted architectural style that capitalizes on standard HTTP methods such as GET, POST, PUT, and DELETE for communication. This style is employed in the development of web services, and it facilitates the interaction between client and server in a seamless manner.

On the other hand, Express.js, often simply referred to as Express, is a minimalistic and flexible web application framework for Node.js. It is designed with the concept of simplicity and flexibility in mind, allowing developers to build web and mobile applications with ease.

Its robust set of features allows for the creation of single, multi-page, and hybrid web applications, thus making it an incredibly efficient tool for building REST APIs. With Express.js, developers can write less code, avoid repetition, and ultimately, save time. Its flexibility and minimalism, coupled with the power of Node.js, make for a feature-rich environment that is conducive for the development of robust web and mobile applications.

11.2.1 Why Express?

Express simplifies the process of building server-side applications with Node.js. It is designed for building web applications and APIs. It has been called the de facto standard server framework for Node.js due to its simplicity and the vast middleware ecosystem available.

The primary reason behind Express's popularity is its simplicity. It provides a straightforward and intuitive way of defining routes and handlers for various HTTP requests and responses. This simplicity accelerates the development process and allows developers to build applications more efficiently.

Express also introduces the concept of middleware. Middleware functions are essentially pieces of code that have access to the request object, the response object, and the next middleware function in the application's request-response cycle. They can execute any code, modify the request and response objects, end the request-response cycle, or call the next middleware function in the stack. This architecture allows developers to perform a wide variety of tasks, from managing cookies, parsing request bodies, to logging and more, simply by plugging in the appropriate middleware.

Moreover, Express is known for its scalability. Its lightweight nature, combined with the ability to manage server-side logic efficiently and integrate seamlessly with databases and other tools, makes Express an excellent choice for scaling applications. As the application's requirements grow, Express can easily handle the increased load, ensuring the application remains robust and performant.

Express has a large and active community. This means that it's easy to find solutions to problems, learn from others' experiences, and access a vast array of middleware and tools developed by the community. This support network can be invaluable for both novice and experienced developers.

In summary, Express simplifies the development of server-side applications with Node.js by providing a simple, scalable, and flexible framework with a robust middleware ecosystem. Its active community also ensures support and continuous development, making it an excellent choice for building web applications and APIs.

Key Features of Express:

- **Simplicity**: Express.js offers an uncomplicated, straightforward way to set up routes that your API can use for effective communication with clients. The simplicity of Express.js allows developers to handle requests and responses without unnecessary complexity, thus enhancing productivity.
- **Middleware**: Express.js has a robust middleware framework which allows developers to use existing middleware to add functionality to Express applications. Alternatively, you can write your own middleware to perform an array of functions like parsing request bodies, handling cookies, managing sessions or logging. This flexibility empowers developers to extend the functionality of their applications as per their specific needs.
- **Scalability**: Express.js exhibits efficient handling of server-side logic and offers seamless integration with databases and other tools, making it an excellent choice for scaling applications. Its lightweight architecture and high performance make it the preferred choice for developing applications that can handle a large number of requests without sacrificing speed or performance.

11.2.2 Setting Up an Express Project

To start, you'll need Node.js installed on your system. Then, you can set up an Express project with some initial setup:

```
mkdir myapi
cd myapi
```

```
npm init -y
npm install express
```

The commands create a new directory called 'myapi', navigate into that directory, initialize a new Node.js project with default settings (because of the '-y' flag), and then install the Express.js library, which is a popular framework for building web applications in Node.js.

Create a file named app.js and add the following basic setup:

```
const express = require('express');
const app = express();
const PORT = process.env.PORT || 3000;

app.get('/', (req, res) => {
    res.send('Hello World from Express!');
});

app.listen(PORT, () => {
    console.log(`Server running on <http://localhost>:${PORT}`);
});
```

The example code snippet uses the Express.js framework, a popular and flexible Node.js web application framework, to set up a simple web server.

Firstly, it imports the 'express' module. This is achieved using the require() function, which is a built-in function in Node.js used for importing modules (libraries or files). The imported 'express' module is then stored in the constant variable 'app'.

Next, it sets up a constant variable 'PORT'. This variable is assigned the value from the environment variable 'PORT' if it exists, or defaults to 3000 if it doesn't. This is done using the '||' (logical OR) operator. Environment variables are a universal mechanism for conveying configuration information to Unix programs. They are part of the environment in which a process runs.

The app.get() function is then used to set up a route for HTTP GET requests. In this case, it specifies that when the server receives a GET request at the root URL ('/'), it should run the provided callback function. The callback function takes two arguments: 'req' (the request object) and 'res' (the response object). In this case, the function simply uses 'res.send()' to send the string 'Hello World from Express!' back to the client making the request.

Finally, app.listen() is called with the 'PORT' constant as an argument, which tells the server to start listening for incoming connections on that port. This method also takes a callback function as an argument, which will be run once the server starts listening successfully. In this case, it logs a message to the console, indicating that the server is running and on which port, using a template string and including the 'PORT' variable within it.

Run your application using node app.js and visit http://localhost:3000 to see it in action.

11.2.3 Building a Simple REST API

Let's expand our application to include a REST API for a simple resource, such as users.

Step 1: Define Data and Routes

First, create a simple array to serve as our database:

```
let users = [
    { id: 1, name: 'Alice' },
    { id: 2, name: 'Bob' },
    { id: 3, name: 'Charlie' }
];
Next, define routes to handle CRUD operations:
// Get all users
app.get('/users', (req, res) => {
    res.status(200).json(users);
});

// Get a single user by id
app.get('/users/:id', (req, res) => {
    const user = users.find(u => u.id === parseInt(req.params.id));
    if (!user) res.status(404).send('User not found');
    else res.status(200).json(user);
});

// Create a new user
app.use(express.json()); // Middleware to parse JSON bodies
app.post('/users', (req, res) => {
    const user = {
        id: users.length + 1,
        name: req.body.name
    };
    users.push(user);
    res.status(201).send(user);
});

// Update existing user
```

```javascript
app.put('/users/:id', (req, res) => {
    let user = users.find(u => u.id === parseInt(req.params.id));
    if (!user) res.status(404).send('User not found');
    else {
        user.name = req.body.name;
        res.status(200).send(user);
    }
});

// Delete a user
app.delete('/users/:id', (req, res) => {
    users = users.filter(u => u.id !== parseInt(req.params.id));
    res.status(204).send();
});
```

This example code defines several HTTP endpoints for a user resource:

- GET /users: This endpoint fetches all users. When a GET request is made to '/users', the function responds with the status 200 (OK) and sends back the 'users' array in JSON format.
- GET /users/:id: This endpoint fetches a single user by their ID. The ID is accessed through the route parameters in the request object. The function then finds the user in the 'users' array that matches this ID. If a user is found, the function responds with status 200 and sends back the user in JSON format. If a user is not found, it responds with status 404 (Not Found) and sends a 'User not found' message.
- POST /users: This endpoint creates a new user. The 'express.json()' middleware is used to parse incoming JSON request bodies, allowing the function to access the requested name through 'req.body.name'. A new user object is created with an ID of 'users.length + 1' and the requested name, and this user is then added to 'users' array. The function responds with status 201 (Created) and sends back the new user.
- PUT /users/:id: This endpoint updates an existing user's name by their ID. Similar to the GET '/users/:id' endpoint, the function finds the user with the matching ID. If a user is found, it updates the user's name with the requested name and responds with status 200, sending back the updated user. If a user is not found, it responds with status 404 and a 'User not found' message.
- DELETE /users/:id: This endpoint deletes a user by their ID. The function filters the 'users' array to remove the user with the matching ID, effectively deleting the user. The function then responds with status 204 (No Content) and does not send back any content.

This example provides a simple example of a RESTful API with Express.js, demonstrating how to handle various HTTP requests, manipulate data, and respond to clients effectively. It serves as a foundation for building more complex APIs with additional functionalities such as error handling, authentication, database integration, and more.

In conclusion, Express makes it straightforward to set up routes and middleware, creating a clean and maintainable structure for your API. By following these steps, you've built a basic REST API that can handle various HTTP requests, manipulate data, and respond to clients effectively. As you expand your Express applications, you can integrate more complex functionalities, such as connecting to databases, handling authentication, and more. This setup forms a foundation that you can build upon as your applications grow in complexity and scale.

11.3 Real-time Communication with WebSockets

In the realm of modern web applications, real-time communication is not just a luxury but a critical necessity. This essential feature brings to life dynamic and interactive experiences like live messaging, immersive gaming, and collaborative editing, all of which are expected in today's digital landscape.

One of the key technologies enabling this kind of real-time interactivity is WebSockets. WebSockets provide a method to establish a bi-directional communication session between the user's browser – the client – and a server. Unlike traditional HTTP requests, where the client must initiate communication, WebSockets allow both the client and the server to send messages to each other independently, thus breaking the conventional request-response cycle. This innovative approach paves the way for faster and more efficient communication, thereby facilitating the kind of instant, seamless interaction that users demand.

This section aims to familiarize you with the fundamentals of WebSockets. We will delve into the mechanics of how WebSockets work, elucidate the principles that underpin this technology, and provide a practical guide on how you can implement WebSockets in your own web applications. Whether you're building a chat application, a real-time gaming platform, or any other interactive web experience, understanding and leveraging WebSockets can dramatically enhance the responsiveness and user experience of your applications.

11.3.1 Understanding WebSockets

WebSockets are a significant improvement over traditional HTTP communications as they provide a full-duplex communication channel that operates over a single, long-lived connection. This means that both the client and the server can send data to each other independently and concurrently, without the need to establish new connections for each interaction. This is a dramatic shift from the conventional request-response cycle of HTTP where the client must initiate all communications.

This unique feature of WebSockets makes them particularly useful in scenarios where real-time data exchange is critical. For instance, they are heavily employed in applications like live chat systems, multiplayer online games, live sports updates, real-time market data updates, and collaborative editing tools, among others.

Understanding how WebSockets work, how they can be implemented in a web application, and how they differ from traditional HTTP communications is key to leveraging their full potential and creating interactive, real-time web experiences.

Key Features of WebSockets:

Persistent Connection

Unlike HTTP, which is stateless, WebSockets maintain a connection open, allowing for lower latencies and better management of real-time data. In a typical HTTP communication, the client establishes a new connection every time it needs to communicate with the server. This is because HTTP is stateless - it doesn't maintain any sort of connection or remember any information between different requests from the same client.

However, WebSockets operate differently. Once a WebSocket connection is established between a client and a server, that connection is kept alive, or "persistent", until it is explicitly closed by either the client or the server. This is what is referred to as a "Persistent Connection".

This persistent connection allows for lower latencies because the client and server do not need to constantly establish and close connections for each exchange of data. Instead, data can be sent back and forth on the open connection as long as it remains open, leading to a more efficient communication process.

Furthermore, this persistent connection enables better management of real-time data. Applications that require real-time data exchange, such as live chat systems, multiplayer online games, or live sports updates, can greatly benefit from this feature of WebSockets. By maintaining an open connection, these applications can provide instant, real-time updates and interactivity, improving user experience and responsiveness.

The persistent connection provided by WebSockets offers significant improvements in terms of efficiency and real-time data handling over traditional HTTP communications, making it a preferred choice for building real-time, interactive web applications.

Full-Duplex Communication

Full-Duplex Communication is a critical feature in modern web applications and refers to a communication system where data transmission can occur concurrently in two directions. In the context of web development, this means that both the client (usually a web browser) and the server can send and receive data at the same time, independently of one another.

This is a significant shift from the traditional request-response model of HTTP communication, where the client initiates a request and then waits for a response from the server. In a full-duplex system like WebSockets, once a connection is established, both the client and server can start communication independently, sending and receiving data without waiting for the other to respond.

This allows for real-time interaction and improves the efficiency of communication, making it particularly useful for applications that require instantaneous data exchange such as live chats, online gaming, and collaborative editing tools.

Efficiency

WebSockets are ideal for scenarios where the overhead of HTTP would be too high, such as frequent, small messages in chat applications or live sports updates. This becomes particularly significant in situations where the communication involves frequent exchange of small packets of data, such as chat applications or live sports updates.

In such scenarios, the overhead of HTTP, which includes establishing a connection, sending the request, waiting for the response, and then closing the connection, can be considerably high. Each of these steps takes time and resources, which can add up quickly when the communication involves frequent, small exchanges of data. This overhead can affect the performance of the application, making it slower and less responsive.

On the other hand, WebSockets maintain an open connection between the client and the server, allowing data to be sent back and forth without the need to constantly open and close connections. This persistent connection significantly reduces the overhead involved in the communication process, leading to more efficient data exchange.

Additionally, WebSockets support full-duplex communication, meaning that both the client and the server can send and receive data simultaneously. This is a significant improvement over the half-duplex communication of HTTP, where the client sends a request and then waits for a response from the server before it can send another request.

The efficiency of WebSockets comes from their ability to maintain a persistent, full-duplex connection, which reduces overhead and allows for more efficient data transmission. This makes them an ideal choice for applications that require frequent, small exchanges of data.

11.3.2 Setting Up a WebSocket Server with Node.js

When it comes to implementing WebSockets in Node.js, there are various libraries available to assist with the process. Often, developers turn to libraries such as ws or socket.io for this purpose.

These libraries offer a high degree of functionality and are well-suited to handle the complexities of WebSockets. For instance, socket.io provides additional features on top of the basic WebSocket framework. These added features include automatic reconnection, which ensures that your application remains running smoothly even when connection issues arise.

It also offers rooms, a feature that allows for more organized data flow and communication in your application. Lastly, socket.io provides events, a crucial aspect that allows for effective event-driven programming. By using these libraries, you can greatly enhance the performance and functionality of your Node.js application.

Here is an example:

Step 1: Install ws

```
npm install ws
```

Step 2: Create a WebSocket Server Create a file named websocket-server.js and add the following code:

```
const WebSocket = require('ws');
const server = new WebSocket.Server({ port: 8080 });

server.on('connection', socket => {
    console.log('A new client connected!');

    socket.on('message', message => {
        console.log('Received message: ' + message);
        server.clients.forEach(client => {
            if (client.readyState === WebSocket.OPEN) {
                client.send("Someone said: " + message);
            }
        });
    });

    socket.on('close', () => {
        console.log('Client has disconnected.');
    });
```

```
});
```

The example code is a simple server-side script written in Node.js using WebSocket for real-time communication. It uses the ws module, a popular WebSocket library for Node.js.

Let's break down the code:

```
const WebSocket = require('ws');
```

This line imports the WebSocket library, which is stored in the constant variable 'WebSocket'.

```
const server = new WebSocket.Server({ port: 8080 });
```

Here, a new WebSocket server instance is created. The server listens for WebSocket connections on port 8080.

```
server.on('connection', socket => {
    console.log('A new client connected!');
```

The server listens for any new connections from a client. When a client connects to the server, a 'connection' event is fired, and the server logs a message "A new client connected!".

```
  socket.on('message', message => {
      console.log('Received message: ' + message);
```

The server listens for a 'message' event on the connected socket. This event is triggered when a message is received from the client. The server then logs the received message.

```
    server.clients.forEach(client => {
        if (client.readyState === WebSocket.OPEN) {
            client.send("Someone said: " + message);
        }
      });
    });
```

Here, the server iterates over each client that is connected to it. If the client's readyState is WebSocket.OPEN, which means the connection is open, the server sends a message to the client. The message is prefixed with "Someone said: " for clarity.

```
socket.on('close', () => {
        console.log('Client has disconnected.');
    });
});
```

The server also listens for a 'close' event on the connected socket. This event is triggered when the client disconnects from the server. When this happens, the server logs "Client has disconnected.".

In summary, this example sets up a WebSocket server that accepts connections from clients, receives messages from clients, broadcasts those messages to all connected clients, and listens for disconnections from clients. It's a simple example of how WebSockets can be used for real-time communication in a server-side JavaScript application using Node.js.

This server listens for new connections, logs messages received from clients, and broadcasts these messages to all connected clients.

11.3.3 Implementing a Simple Client

A simple HTML client can be used to connect to this server and send messages.

HTML Client (index.html):

```
<!DOCTYPE html>
<html lang="en">
<head>
    <meta charset="UTF-8">
    <title>WebSocket Client</title>
</head>
<body>
    <input type="text" id="messageInput" placeholder="Type a message">
    <button onclick="sendMessage()">Send</button>
    <ul id="messages"></ul>

    <script>
        const socket = new WebSocket('ws://localhost:8080');

        socket.onmessage = function(event) {
            const messageList = document.getElementById('messages');
```

```javascript
            const msg = document.createElement('li');
            msg.textContent = event.data;
            messageList.appendChild(msg);
        };

        function sendMessage() {
            const input = document.getElementById('messageInput');
            if (input.value) {
                socket.send(input.value);
                input.value = '';
            }
        }
    </script>
</body>
</html>
```

The document structure begins with the <!DOCTYPE html> declaration, which is used to inform the web browser about the version of HTML the page is written in - in this case, HTML5.

Inside the <html> tags, there are two main sections: <head> and <body>. The <head> section contains meta-information about the document and may include the document title (which is displayed in the browser's title bar or tab), links to stylesheets, scripts, and more. In this case, it includes a character encoding declaration (<meta charset="UTF-8">), which specifies the character encoding for the HTML document, and the title of the document (<title>WebSocket Client</title>).

The <body> section contains the main content of the HTML document - what you see rendered in the browser. In this case, it includes an input field where users can type their messages, a 'Send' button to dispatch those messages, and an unordered list (<ul id="messages"></ul>) where incoming messages from the WebSocket server will be displayed.

The script block within the <body> section establishes a connection to the WebSocket server, sets up event listeners, and defines the sendMessage function.

The line const socket = new WebSocket('ws://localhost:8080'); creates a new WebSocket connection to the server located at 'ws://localhost:8080'.

The socket.onmessage event listener waits for messages from the server. When a message is received, a new list item (<li> element) is created, the incoming message is set as its content, and it is appended to the 'messages' list.

The sendMessage function is called when the 'Send' button is clicked. It first grabs the user's input from the text field. If the input isn't empty, it sends the message to the server using socket.send(input.value) and then clears the input field.

In essence, this document facilitates real-time communication with a WebSocket server, allowing users to send messages to the server and see responses from the server instantly.

This HTML page includes an input field for typing messages and a button to send them. It uses the WebSocket API to open a connection to the server, send messages, and display incoming messages.

In conclusion, WebSockets open up a plethora of possibilities for real-time data exchange in web applications, enhancing the interactivity and responsiveness of modern web experiences. By understanding and utilizing WebSockets, you can significantly improve the performance of applications that require real-time capabilities, such as chat applications, live notifications, or multiplayer games. This technology is a cornerstone for developers looking to build dynamic, engaging, and responsive web applications.

Practical Exercises for Chapter 11: JavaScript and the Server

These practical exercises are designed to reinforce your understanding of the concepts discussed in Chapter 11, focusing on Node.js, building REST APIs with Express, and implementing real-time communication using WebSockets. By completing these exercises, you will gain hands-on experience with server-side JavaScript, enhancing your ability to develop dynamic and interactive web applications.

Exercise 1: Basic Node.js Server

Objective: Create a simple Node.js server that responds with "Hello, Node.js!" for any requests.

Solution:

```
// Create a file named server.js
const http = require('http');

const server = http.createServer((req, res) => {
    res.statusCode = 200;
    res.setHeader('Content-Type', 'text/plain');
    res.end('Hello, Node.js!');
});

const port = 3000;
```

```
server.listen(port, () => {
    console.log(`Server running at <http://localhost>:${port}/`);
});
```

Run this server with node server.js and navigate to http://localhost:3000 in your browser to see the response.

Exercise 2: Building a Simple REST API with Express

Objective: Create an Express application that manages a list of tasks, supporting operations to create, read, update, and delete tasks.

Solution:

```
const express = require('express');
const app = express();
app.use(express.json()); // Middleware to parse JSON bodies

let tasks = [{ id: 1, task: 'Do laundry' }, { id: 2, task: 'Write code' }];

app.get('/tasks', (req, res) => {
    res.status(200).json(tasks);
});

app.post('/tasks', (req, res) => {
    const newTask = { id: tasks.length + 1, task: req.body.task };
    tasks.push(newTask);
    res.status(201).json(newTask);
});

app.put('/tasks/:id', (req, res) => {
    let task = tasks.find(t => t.id === parseInt(req.params.id));
    if (!task) res.status(404).send('Task not found');
    else {
        task.task = req.body.task;
        res.status(200).json(task);
    }
});

app.delete('/tasks/:id', (req, res) => {
    tasks = tasks.filter(t => t.id !== parseInt(req.params.id));
    res.status(204).send();
});

const port = 3000;
app.listen(port, () => {
    console.log(`Server running on <http://localhost>:${port}`);
```

```
});
```

Exercise 3: Real-time Chat Application with WebSockets

Objective: Implement a simple real-time chat application using WebSockets.

Solution:

```
// Server setup (server.js)
const WebSocket = require('ws');
const wss = new WebSocket.Server({ port: 8080 });

wss.on('connection', function connection(ws) {
    ws.on('message', function incoming(message) {
        console.log('received: %s', message);
        wss.clients.forEach(function each(client) {
            if (client !== ws && client.readyState === WebSocket.OPEN) {
                client.send(message);
            }
        });
    });
});
```

Client HTML (index.html):

```
<!DOCTYPE html>
<html>
<head>
    <title>WebSocket Chat</title>
</head>
<body>
    <textarea id="messages" cols="30" rows="10" readonly></textarea><br>
    <input        type="text"        id="messageBox"        autocomplete="off"><button
onclick="sendMessage()">Send</button>

    <script>
        const ws = new WebSocket('ws://localhost:8080');
        const messages = document.getElementById('messages');

        ws.onmessage = function (event) {
            messages.value += event.data + '\\\\n';
        };

        function sendMessage() {
            const messageBox = document.getElementById('messageBox');
            ws.send(messageBox.value);
```

```
            messageBox.value = '';
        }
    </script>
</body>
</html>
```

These exercises offer a practical way to apply the server-side JavaScript skills you've learned in this chapter. From setting up basic servers and creating RESTful services to implementing sophisticated real-time communication systems, you now have the tools to build robust, efficient, and interactive web applications.

Chapter 11 Summary: JavaScript and the Server

In Chapter 11, "JavaScript and the Server," we explored the powerful capabilities of JavaScript beyond the confines of the browser, focusing on server-side development with Node.js and other tools like Express and WebSockets. This journey into server-side JavaScript has provided comprehensive insights into how JavaScript can be leveraged to create robust, efficient, and scalable web servers and real-time applications.

Expanding JavaScript's Reach with Node.js

Node.js has revolutionized the way developers think about JavaScript. Traditionally confined to client-side scripting, JavaScript, with the help of Node.js, has become a major player in server-side application development. This transition allows developers to use a single programming language across both front-end and back-end, simplifying the development process and reducing the need to context switch between different languages for different parts of an application.

We began by introducing the basics of Node.js, emphasizing its non-blocking, event-driven architecture that makes it suitable for I/O-heavy operations. The ability to handle numerous simultaneous connections with a single server instance is a testament to its efficiency and has made Node.js a preferred environment for developing web applications and services.

Building REST APIs with Express

Express.js was highlighted as a minimalistic yet powerful framework for building web applications and APIs. Through detailed examples, we explored how to construct RESTful APIs with Express, allowing for the creation, retrieval, updating, and deletion of resources. This section provided practical knowledge on setting up routes, handling requests, and integrating middleware for extended functionalities, which are crucial for building modern web APIs.

Implementing Real-time Communication

The chapter also covered real-time communication using WebSockets, an essential feature for applications that require live interaction, such as chat applications, collaborative platforms, and live notifications. We delved into setting up a WebSocket server and clients, demonstrating how to facilitate bi-directional, low-latency communication. This capability is critical in the modern web landscape, where user expectations are shifting towards seamless and interactive experiences.

Practical Application and Exercises

The practical exercises reinforced the concepts discussed by guiding you through the creation of a basic Node.js server, developing a REST API with Express, and implementing a real-time chat application using WebSockets. These exercises were designed to provide hands-on experience, enhancing your understanding and skills in real-world scenarios.

Conclusion

This chapter has equipped you with the knowledge and tools to extend the functionality of JavaScript to the server side, opening up a world of possibilities for developing full-stack applications. As you continue to explore server-side JavaScript, remember that the principles of good software development—maintaining clean, efficient, and scalable code—are just as applicable here as they are in any other computing environment.

Moving forward, the skills acquired in this chapter will not only allow you to build more dynamic and responsive applications but also enable you to tackle complex problems with integrated solutions that span both client and server sides. As JavaScript continues to evolve, staying abreast of these developments will be crucial in advancing your capabilities as a developer and in meeting the challenges of modern web development head-on.

Chapter 12: Deploying JavaScript Applications

Welcome to Chapter 12, "Deploying JavaScript Applications," where we delve into the crucial stages of making your JavaScript applications available to the world. This chapter addresses the steps and tools essential for preparing and deploying your applications efficiently and securely. From version control to actual deployment on various platforms, this chapter provides a comprehensive guide to ensure your applications are robust, scalable, and ready for production.

12.1 Version Control with Git

Prior to exploring various deployment techniques, it is of paramount importance to grasp the role that version control systems play in managing and safeguarding the codebase of your application.

These systems serve as the backbone for any developer's toolkit, facilitating the process of tracking alterations made to the code, enabling the reversion to previous states when necessary, and providing a platform to collaborate effectively with other developers.

Among the plethora of version control systems available, Git stands out as the most widely used. It has garnered widespread adoption in the industry due to its inherent flexibility, immense power, and the ability to accommodate a variety of workflows. Git's popularity is further enhanced by its robust community support, providing resources and solutions for any challenges that might arise in the development process.

12.1.1 Understanding Git

Git is a distributed version control system designed to handle everything from small to very large projects with speed and efficiency. It allows multiple developers to work on the same project without interfering with each other's changes. Git operates on the concept of repositories, where your project's history is stored.

Git is designed to handle everything from small to very large projects with speed and efficiency. It allows multiple developers to work on the same project without interfering with each other's changes. Git operates on the concept of repositories, where your project's history is stored.

Understanding Git also involves setting it up on your machine. Once installed, you can initialize a new repository in your project's directory and start using Git's functionalities like adding files to the staging area, committing changes to the repository, and viewing commit history.

Moreover, there are best practices for using Git, which include making frequent commits with clear, descriptive messages, using branches for different features or fixes, and adopting consistent naming conventions for branches and commits.

Understanding Git is a crucial part of modern software development. It not only helps in tracking and managing changes to your code but also facilitates effective collaboration among developers.

Key Concepts of Git:

- **Commit**: A commit, in the context of Git, is essentially a snapshot of your project's current state at a specific moment in time. This state includes all the changes you have made to your files. Each commit possesses a unique ID, also known as a commit hash, which allows you to track specific changes made in the project. If there's ever a need to revert to a previous state of your project, these commit hashes come in handy for such instances.

- **Branch**: Branching in Git is a powerful feature that lets developers diverge from the main line of development and work independently without affecting other parts of the project. This is extremely useful when you want to add a new feature or experiment with something, but don't want to risk the stability of the main project. Once the work on this branch reaches a satisfactory level, it can then be merged back into the main line of the project.

- **Merge**: Merging is the method by which changes from different branches are brought together into a single branch. This process combines the divergent histories of these branches and potentially resolves any conflicts that may arise due to differences in these histories. It's a critical part of maintaining the coherent and unified progress of a project, ensuring that all beneficial changes and advancements are integrated into the main project.

12.1.2 Setting Up Git

Before you can start using Git, the first step is to install it on your computer. This open-source program is available for a variety of operating systems, including Windows, Mac OS, and Linux. You can find the necessary installation guides and download links for these different systems on the official website of Git.

Simply head over to Git's official website and follow the instructions provided for your specific operating system. This will ensure that you have the necessary software to begin managing and tracking changes in your source code projects.

Initialize a New Git Repository: Once Git is installed, you can initialize a new repository in your project's directory:

cd path/to/your/project

git init

These commands are used in a bash shell. cd path/to/your/project is used to change the current directory to the specified path where your project is located. git init is used to initialize a new Git repository in the current directory.

Basic Git Workflow: Here's a simple example of managing your project with Git:

1. **Add Files**: Add files to the staging area. This area holds the files you want to include in the next commit.

2. git add index.html app.js style.css

3. **Commit Changes**: Save the changes in the staging area to the repository.

4. git commit -m "Initial commit: Add main project files"

5. **View Commit History**: Check the history of commits to see what changes have been made.

6. git log

This example provides basic instructions for using Git, a version control system.

1. **Add Files**: This step describes how to add files to the staging area, which is a preparatory space for files that will be included in the next commit. The command 'git add' followed by the file names adds those files to the staging area.

2. **Commit Changes**: This step explains how to save the changes you've made to the repository. The 'git commit' command followed by '-m' and a message records the changes to the repository with a description of what was changed.

3. **View Commit History**: This step outlines how to view the history of commits, which is essentially a log of all the changes made in the repository. The 'git log' command displays this log.

12.1.3 Best Practices for Using Git

- **Frequent Commits**: It's highly recommended to commit changes often and to ensure each commit is accompanied by clear, descriptive messages. Not only does this practice make it easier to locate and understand the changes made, but it also assists in pinpointing the exact moment where potential issues might have been introduced, thereby simplifying the debugging process.

- **Branching Strategy**: One of the best practices in version control is the use of branches for different purposes such as features, fixes, or experiments. This strategy contributes to maintaining the main branch in a clean and deployment-ready state, preventing it from being cluttered with in-progress work or experimental code.

- **Consistent Naming Conventions**: To streamline the development process and collaboration among team members, it's important to adopt a consistent naming convention for branches and commits. This enhances the clarity and readability of your version control, making it easier for everyone in the team to understand what each commit and branch is for.

Version control with Git is an indispensable part of modern software development, particularly when preparing to deploy applications. Properly managing your codebase with Git not only safeguards your code but also enhances collaboration and efficiency.

12.2 Bundlers and Task Runners (Webpack, Gulp)

In the complex and nuanced journey of deploying JavaScript applications, there exists an absolutely essential step that cannot be overlooked: the optimization and meticulous organization of your code and resources. This particular stage is crucial in ensuring that your application runs smoothly and efficiently. This is the precise point in the process where the role of bundlers and task runners become significantly important.

These powerful tools greatly streamline and simplify the process of preparing your application for its final production stage. They achieve this by automating routine tasks that would otherwise be time-consuming, bundling together essential assets, and optimizing output to ensure the best performance. By using these tools, you can drastically reduce the resources and time spent on preparing your application, allowing you to focus on more important aspects of your project.

In this section, we will be focusing on two pivotal, industry-standard tools that have gained significant popularity due to their efficiency and ease of use: Webpack and Gulp. Webpack is primarily used for bundling together files and modules, while Gulp is renowned for its ability to automate tasks. Both of these tools play an integral role in the process of building efficient, scalable, and sustainable applications and are considered indispensable in modern JavaScript application development.

12.2.1 Understanding Bundlers: Webpack

Webpack is a powerful module bundler primarily used for JavaScript, but it can also transform front-end assets like HTML, CSS, and images if the corresponding loaders are included. It takes modules with dependencies and generates static assets representing those modules.

Notably, Webpack treats every piece of your application, including JavaScript, CSS, fonts, and images, as a module. This modular approach allows for better management and maintenance of code in large-scale projects.

Webpack also utilizes loaders and plugins to enhance its functionality. Loaders enable Webpack to process different types of files and convert them into modules which can be included in your output bundles. Plugins, on the other hand, extend Webpack's capabilities, allowing you to perform a wide range of tasks such as bundle optimization, asset management, and injection of environment variables.

In the context of deploying JavaScript applications, Webpack proves to be a crucial tool. Its ability to bundle together files and modules helps streamline the preparation of your application for the final production stage, ensuring the application runs smoothly and efficiently. It is therefore considered an indispensable tool in modern JavaScript application development.

Key Features of Webpack:

- **Modules**: Webpack, a powerful and flexible module bundler, treats every single component of your application as a module. This includes not just JavaScript files, but also CSS stylesheets, fonts, and image files. This approach allows for greater control and organization of your application structure.

- **Loaders**: Loaders are a key feature of Webpack. They provide a way for Webpack to process and transform different types of files before they are added to the dependency graph. This means they can convert files into modules that can then be included in your final output bundles. For example, a loader could transform a TypeScript file into JavaScript, or convert SASS into CSS.

- **Plugins**: Plugins are another core part of Webpack's architecture. They enhance Webpack's capabilities beyond the standard bundling and building. Plugins allow you to perform a wide range of tasks, including but not limited to bundle optimization, asset management, and injection of environment variables. With plugins, the possibilities are almost limitless, and you can tailor your build process to suit your unique requirements.

Example of Basic Webpack Configuration: Create a file named webpack.config.js in your project root:

```
const path = require('path');

module.exports = {
  // Entry point of your application
  entry: './src/index.js',

  // Output configuration
  output: {
    path: path.resolve(__dirname, 'dist'),
```

```
    filename: 'bundle.js'
  },

  // Loaders and rules
  module: {
    rules: [
      {
        test: /\\\\.css$/,
        use: ['style-loader', 'css-loader']
      },
      {
        test: /\\\\.js$/,
        exclude: /node_modules/,
        use: {
          loader: 'babel-loader',
          options: {
            presets: ['@babel/preset-env']
          }
        }
      }
    ]
  }
};
```

This example is a basic configuration for Webpack, a powerful and popular module bundler used predominantly in JavaScript application development.

The configuration begins by requiring the 'path' module, which provides utilities for working with file and directory paths. This module is used later in the configuration to resolve the absolute path of the 'dist' directory, where the output bundle will be placed.

The configuration object has three main sections: 'entry', 'output', and 'module'.

The 'entry' key specifies the entry point of your application, './src/index.js'. This is the JavaScript file that kicks off your application and where Webpack starts its bundling process. This file typically includes imports from other JavaScript modules. Webpack will then proceed to bundle this file along with all of the modules it depends upon.

The 'output' key is an object that defines where Webpack will output the bundles it creates and how it will name them. It includes 'path', which tells Webpack where to place the output files on your local machine, and 'filename', which specifies the name of the output bundle file. In this case, the output bundle will be placed in a 'dist' directory in the root of your project with the filename 'bundle.js'.

The 'module' key holds an object that defines different rules for different modules. In the context of Webpack, a module can be a JavaScript file, a CSS file, an image file, or any other asset that you might want to include in your application. The 'rules' key is an array of objects, each defining a rule for a certain type of module.

In this configuration, we can see two rules. The first rule tells Webpack to use the 'style-loader' and 'css-loader' for all files ending in '.css'. The 'style-loader' adds CSS to the DOM by injecting a 'style' tag, while the 'css-loader' interprets '@import' and 'url()' like 'import/require()' and resolves them.

The second rule targets '.js' files, excluding those in the 'node_modules' directory. For these files, the 'babel-loader' is used. This loader uses Babel, a tool for transpiling ES6 and beyond syntax to ES5, to ensure compatibility with older browsers. The 'options' key specifies that the '@babel/preset-env' preset should be used, which allows you to use the latest JavaScript without needing to micromanage which syntax transforms are needed based on your target environment.

12.2.2 Understanding Task Runners: Gulp

Gulp is a powerful task runner that utilizes Node.js as a platform. It plays a significant role in the development process by automating repetitive tasks, making your workflow faster and more efficient.

Key tasks automated by Gulp include minification, compilation, unit testing, linting, and more. Minification is a process that removes unnecessary characters from code to reduce its size, thereby improving load times. Compilation is the process of transforming source code written in one programming language into another language, often binary language. Unit testing involves testing individual components of the software to ensure that they are working as expected. Linting, on the other hand, is the process of running a program that analyses code for potential errors.

The popularity of Gulp stems from a few key features. Firstly, it champions simplicity by preferring code over configuration for defining tasks, making it straightforward and easier to use. Secondly, Gulp is stream-based and leverages Node.js streams, which allows you to perform multiple operations on files without the need to write intermediate files to the disk. This results in a faster, more efficient build process.

Lastly, just like Webpack, Gulp has a wide range of plugins available that can be harnessed to perform various tasks, thereby enhancing its functionality. This makes it a versatile tool that can be configured to suit a variety of project needs.

In a practical scenario, after installing Gulp in your project, you would create a gulpfile.js in your project root. This file is used to define tasks that Gulp will run. For example, you might define a task to minify JavaScript files, which involves specifying the source files, applying the minification process using a plugin like 'gulp-uglify', renaming the output file, and finally specifying the destination directory for the output file. You can also define a default task that runs when you simply use the 'gulp' command.

In conclusion, Gulp is a crucial tool in modern JavaScript application development. Its ability to automate numerous tasks saves developers a significant amount of time, thus speeding up the development process. By understanding and effectively utilizing Gulp, developers can focus more on the core aspects of their applications and less on repetitive tasks.

Key Features of Gulp:

- **Simplicity**: Gulp is designed keeping simplicity at the forefront. It uses a code-over-configuration approach to define tasks. This design philosophy makes it straightforward and easy to use, even for beginners. The developers aimed to create a tool that would not require excessive configuration, allowing more time for actual development work.

- **Stream-based**: Gulp utilizes the power of Node.js streams. This unique feature allows developers to perform multiple operations on the files in a pipeline fashion, eliminating the need to write intermediate files to disk. This approach not only makes processing faster but also significantly reduces the I/O overhead.

- **Plugins**: Similar to Webpack, Gulp is highly extensible and has a wide range of plugins available for various tasks. These plugins enhance its functionality, making it a powerful tool for any developer's toolkit. Whether you need to minify your code, compile your Sass files, or even optimize your images, there's likely a Gulp plugin that can do the job.

Example of a Gulp Task: First, install Gulp in your project:

```
npm install --save-dev gulp
Create a gulpfile.js in your project root:
const gulp = require('gulp');
const uglify = require('gulp-uglify');
const rename = require('gulp-rename');

// Define a task to minify JavaScript files
gulp.task('compress', function () {
  return gulp.src('src/*.js')
    .pipe(uglify())
    .pipe(rename({ suffix: '.min' }))
    .pipe(gulp.dest('dist'));
});

// Default task
gulp.task('default', gulp.series('compress'));
```

This example code demonstrates the use of Gulp, a powerful task runner that can automate repetitive tasks to make your workflow more efficient. In this specific code snippet, Gulp is being used to automate the task of minifying JavaScript files.

In the first three lines of the code, three packages are being required: 'gulp', 'gulp-uglify', and 'gulp-rename'. The 'gulp' package is the main Gulp library. 'gulp-uglify' is a Gulp plugin used for minifying JavaScript files, and 'gulp-rename' is a Gulp plugin used for renaming files.

Following this, a Gulp task named 'compress' is defined. This task is designed to minify JavaScript files. The function inside 'gulp.task' specifies what the task does. It returns a stream of files from the 'src' directory with a '.js' extension. These files are then piped into the 'uglify' function, which minifies the JavaScript files. The minified files are then piped into the 'rename' function, which adds a '.min' suffix to the file names. Finally, these renamed, minified files are piped into 'gulp.dest', which writes the files to the 'dist' directory.

The final line of the code defines a default task. Default tasks are tasks that are run when the 'gulp' command is run without specifying any tasks. In this case, the default task is set to run the 'compress' task. The 'gulp.series' method is used to define a series of tasks that should be executed one after the other. In this case, the only task in the series is 'compress'.

So, to summarize, this script defines a Gulp task that minifies all JavaScript files in the 'src' directory, renames them by adding a '.min' suffix, and then outputs them to the 'dist' directory. This task is also set as the default task, so it will run when the 'gulp' command is run without specifying any tasks.

This kind of automation can help developers save time and reduce the risk of errors that can occur when performing repetitive tasks manually. By understanding and using task runners like Gulp, developers can make their workflows more efficient and productive.

In conclusion, Webpack and Gulp are instrumental in preparing JavaScript applications for deployment. They optimize the process, reduce potential errors, and ensure that your applications are as efficient as possible. By understanding and utilizing these tools, you can automate many aspects of the build process, from bundling and minifying code to running predefined tasks, significantly easing the path from development to production.

12.3 Deployment and Hosting (Netlify, Vercel)

After your JavaScript application has been bundled effectively and optimized for production, the subsequent crucial phase in your development process is deployment and hosting. This pivotal stage requires you to make your application accessible to users through the internet, effectively taking your project from development to the hands of end-users.

In the landscape of web development, the last few years have seen a revolutionary shift in the way deployment and hosting processes are handled. Platforms like Netlify and Vercel have emerged at the forefront of this revolution, providing modern web applications with an unprecedented level of simplicity, speed, and a set of powerful features specifically designed to cater to front-end projects. These platforms have reshaped the deployment and hosting process, aligning it with the needs of the modern web.

In this section, we will delve deeper into these platforms, exploring their unique features and advantages. We will highlight how these platforms have been designed to cater to the unique requirements of modern deployment, providing a seamless and efficient process that integrates continuous integration and delivery. From automated build processes to instant cache invalidation, these platforms provide the tools necessary for a robust and efficient deployment process that meets the demands of modern web applications.

12.3.1 Overview of Modern Hosting Solutions

Netlify and **Vercel** represent two of the most popular cloud-based hosting services in the modern development world. Both of these services are known for their generous free basic plans, which have drawn a significant following of developers. These developers frequently rely on Netlify and Vercel for hosting a variety of digital properties, including static sites and serverless backends.

One of the key reasons for their popularity is the way these platforms integrate with your Git repositories. They provide seamless and efficient continuous deployment services that work in harmony with your development workflow.

This means that every time you make updates to your repository, perhaps by pushing a new set of changes, the platform jumps into action. It automatically deploys the new version of your site, saving you time and reducing the potential for human error. This functionality is a game-changer, making website updates and maintenance much more streamlined and manageable.

Key Features

- **Continuous Deployment**: Both platforms integrate seamlessly with your Git repositories, whether it's GitHub, GitLab, or Bitbucket, to automate the deployment process. This means that every time you push changes to your Git repository, a new deployment is triggered automatically, ensuring that your live application is always up-to-date with the latest changes.

- **Serverless Functions**: These platforms also support serverless functions. This powerful feature allows you to run backend code without having to manage an entire server setup, simplifying your development process and reducing overhead costs.

- **Instant Rollbacks**: Another standout feature is the ability to instantly revert to previous versions of your application. This eliminates the need to redeploy your application, saving you time and effort, especially when dealing with critical issues that require immediate fixes.

- **Custom Domains and SSL**: Lastly, you can easily configure custom domains on these platforms. They also offer automatic SSL certificate issuance and renewal, ensuring that your site is always secure and that your users' data is protected.

12.3.2 Deploying with Netlify

Step-by-Step Guide:

1. **Create a Netlify Account**: The first step is to create a Netlify account. You can do this by signing up for free at Netlify.

2. **New Site from Git**: Once you've signed up and logged in to your account, navigate to the Netlify dashboard. Here, you should choose to create a new site from Git. This will allow you to deploy directly from your Git repository, making updates and changes quick and easy.

3. **Connect Your Repository**: The next step is to connect your GitHub, GitLab, or Bitbucket account to Netlify. Follow the prompts provided by the platform to do this. Make sure to select the repository that contains the project you want to deploy.

4. **Build Settings**: Before you can deploy your site, you need to specify your build commands and publish directory. For instance, if you're working on a Webpack project, you might enter npm run build as your build command and dist/ as your publish directory.

5. **Deploy**: With everything set up, you can now deploy your site. Netlify will automatically take care of the deployment process and provide a URL where you can access your newly deployed site.

Example Build Settings for a React Application:

Build command: npm run build

Publish directory: build/

12.3.3 Deploying with Vercel

Step-by-Step Guide:Create a Vercel Account: Begin by signing up for a free account at Vercel. This platform will host your project, so creating an account is a necessary first step.

1. **Import Your Project**: Once you've created your account and logged in, navigate to the Vercel dashboard. Here, you'll click on the "New Project" button, which will lead you to the "Import Project" option. You can import your project directly from a Git repository.

2. **Configure Your Project**: Vercel has the capability to automatically detect build settings for a wide variety of frameworks, which can simplify the setup process. However, if you're using a custom setup, you'll need to specify the build command and the output directory manually.

3. **Environment Variables**: The next step involves configuring any necessary environment variables. This is an important step because these variables can affect the way your project runs.

4. **Deploy**: Finally, Vercel will take care of building and deploying your application. Upon completion, it will provide a live URL where you can access your deployed project.

Example Configuration for a Vue.js Application:

Build command: npm run build

Output directory: dist/

These instructions are for building a software project. "Build command: npm run build" is the command you run to start the build process using npm (Node Package Manager). "Output directory: dist/" indicates that the build results (compiled code or executable file) will be stored in a directory named 'dist/'.

In conclusion, deploying and hosting with platforms like Netlify and Vercel simplifies the process of making web applications available online. These platforms not only provide robust, scalable hosting solutions but also integrate modern development practices such as continuous integration and deployment, serverless functions, and automated HTTPS.

By utilizing these services, developers can focus more on building their applications and less on the intricacies of deployment and server management. As web development continues to evolve, the role of such platforms becomes increasingly crucial in the deployment pipeline, ensuring that developers have access to the best tools for delivering high-quality web experiences efficiently.

Practical Exercises for Chapter 12: Deploying JavaScript Applications

These practical exercises are designed to consolidate your understanding of deploying JavaScript applications, focusing on using version control, bundlers, task runners, and modern deployment platforms such as Netlify and Vercel. By completing these exercises, you will gain hands-on experience in preparing and deploying web applications efficiently.

Exercise 1: Version Control with Git

Objective: Initialize a new Git repository, add your project files, commit them, and push to a remote repository on GitHub.

Solution:

1. **Create a Local Repository:**

- Navigate to your project directory in the terminal.

- Initialize the repository:

- git init

- Add files to staging:

- git add .

- Commit the changes:

- git commit -m "Initial commit"

1. **Create a Remote Repository on GitHub**:
   - ○ Go to GitHub and create a new repository.
   - ○ Copy the remote repository URL provided by GitHub.

2. **Link Local Repository to Remote and Push**:
   - Add the remote repository:
   - git remote add origin YOUR_REPOSITORY_URL
   - Push your code to GitHub:
   - git push -u origin master

Exercise 2: Setting Up Webpack for a Simple Project

Objective: Configure Webpack to bundle JavaScript and CSS files for a simple project.

Solution:

1. **Install Webpack and Loaders**:
   - Install Webpack and necessary loaders:
   - npm install --save-dev webpack webpack-cli css-loader style-loader
1. **Create webpack.config.js**:
   - Set up configuration:
   - const path = require('path');
   -
   - module.exports = {
   - entry: './src/index.js',
   - output: {
   - filename: 'bundle.js',
   - path: path.resolve(__dirname, 'dist')
   - },
   - module: {
   - rules: [
   - {
   - test: /\\\\.css$/,

- use: ['style-loader', 'css-loader']
- }
-]
- }
- };
- Add a simple CSS file to your project and require it in your index.js.

1. **Run Webpack**:

- Add a build script in your package.json:
- "scripts": {
- "build": "webpack"
- }
- Build the project:
- npm run build

Exercise 3: Deploying a Static Site to Netlify

Objective: Deploy a simple static website to Netlify using continuous deployment from a Git repository.

Solution:

1. **Prepare Your Project**:
   - Ensure your project has an index.html and any associated CSS/JS files.
   - Push your project to GitHub if not already done.

2. **Set Up Netlify**:
   - Sign up for Netlify and log in.
   - Click "New site from Git" and select your GitHub repository.
   - Configure the build settings if necessary (for static sites, typically no build command is needed; just set the publish directory if your index.html is not in the root).

3. **Deploy**:
   - Follow the prompts to deploy your site.
   - Netlify will provide a URL to view your live site.

These exercises provide practical scenarios to apply the concepts learned in Chapter 12, from using Git for version control, configuring Webpack for asset bundling, to deploying a site using Netlify. Completing these tasks will enhance your ability to manage and deploy web applications effectively, ensuring they are accessible and performant for end-users.

Chapter 12 Summary: Deploying JavaScript Applications

In Chapter 12, "Deploying JavaScript Applications," we delved into the final stages of the development lifecycle, focusing on the crucial aspects of preparing and deploying JavaScript applications for production. This journey equipped you with the knowledge and tools needed to ensure your applications are not only ready for deployment but also optimized for performance, scalability, and maintainability.

Key Concepts and Technologies

We began by exploring **version control with Git**, emphasizing its critical role in any development project. Git serves as the backbone for managing changes, facilitating collaboration, and safeguarding your codebase against potential losses or errors. We discussed how to set up and manage a Git repository, including committing changes, branching, and merging, which are essential for maintaining a clean and efficient development history.

Following version control, we examined **bundlers and task runners**, specifically Webpack and Gulp. These tools streamline the development process by automating routine tasks such as minification, compilation, and transpilation. Webpack, a module bundler, focuses on assembling and optimizing your application's assets. It handles everything from JavaScript and CSS to images and fonts, ensuring that your project's files are efficiently packaged for deployment. Gulp, on the other hand, excels as a task runner, allowing you to automate repetitive tasks like CSS preprocessing and image optimization, which can significantly enhance your productivity.

The chapter then shifted to **deployment and hosting**, where we covered modern platforms such as Netlify and Vercel. These platforms revolutionize deployment by integrating directly with your version control system to automate the process of pushing your application live. We detailed the steps to deploy a web application using these services, highlighting their continuous deployment capabilities, which automatically update your live application with every commit to your repository. This integration of development and deployment processes underscores the modern approach to web hosting, where ease of use, scalability, and integration with development tools are paramount.

Practical Application and Exercises

The practical exercises provided hands-on experience with the tools and concepts discussed throughout the chapter. From initializing and managing a Git repository to configuring Webpack for asset bundling and deploying a static site with Netlify, these exercises aimed to solidify your understanding and enhance your skills in deploying web applications.

Conclusion

Deploying JavaScript applications involves more than just transferring files to a server; it requires a comprehensive approach that includes version control, code optimization, and automated deployments. The practices and tools we explored are fundamental to modern web development workflows, ensuring that your applications are delivered to users efficiently and reliably.

As you continue to develop and deploy applications, the insights gained from this chapter will serve as a foundation for adopting best practices and leveraging advanced tools to streamline your workflows and improve the quality of your deployments. The ability to effectively manage and deploy applications is crucial in a rapidly evolving digital landscape, and the skills you've acquired here will be invaluable as you tackle more complex projects and challenges in your development career.

Quiz Part III: JavaScript and Beyond

This quiz is designed to test your understanding of the key concepts discussed in Part III of the book, encompassing modern JavaScript frameworks, single-page application development, server-side JavaScript, and deployment strategies. Each question is crafted to help reinforce your knowledge and ensure you have grasped the essential elements from each chapter.

Question 1: Modern JavaScript Frameworks

Which statement best describes the use of Vue.js in developing user interfaces? A) Vue.js is exclusively used for server-side rendering. B) Vue.js uses a virtual DOM to optimize rendering. C) Vue.js treats everything as a component, including HTML, CSS, and JavaScript. D) Vue.js does not support the use of components.

Question 2: Developing Single Page Applications

What is the main benefit of using client-side routing in a Single Page Application (SPA)? A) It requires the server to render and return new HTML on navigation. B) It allows the application to load new pages without a full page refresh, enhancing user experience. C) It significantly increases the amount of data transferred between the server and client. D) It simplifies backend architecture by handling all rendering on the client side.

Question 3: JavaScript and the Server

What is Node.js primarily used for in web development? A) Creating animated web pages. B) Editing JavaScript code directly in the browser. C) Running JavaScript on the server to build scalable network applications. D) Enhancing CSS styling capabilities in web applications.

Question 4: Deploying JavaScript Applications

Which tool is described as a "module bundler" and is particularly effective in managing application assets like JavaScript, CSS, and images? A) Gulp B) Jenkins C) Webpack D) Git

Question 5: Real-time Communication Technologies

What technology allows for real-time, bi-directional communication between web clients and servers? A) HTTP/2 B) WebSockets C) AJAX D) REST API

Question 6: Continuous Deployment

Which platform provides a feature for continuous deployment that integrates directly with code repositories for automatic updates upon code commits? A) Apache B) Netlify C) FTP servers D) Localhost

Question 7: Task Runners

What is the primary use of Gulp in web development workflows? A) To create private branches in version control. B) To automate tasks like minification, compilation, and testing. C) To bundle modules and assets together. D) To deploy applications to production servers.

Answers:

1. C) Vue.js treats everything as a component, including HTML, CSS, and JavaScript.
2. B) It allows the application to load new pages without a full page refresh, enhancing user experience.
3. C) Running JavaScript on the server to build scalable network applications.
4. C) Webpack
5. B) WebSockets
6. B) Netlify
7. B) To automate tasks like minification, compilation, and testing.

This quiz should help solidify your understanding of the advanced JavaScript concepts covered in Part III of the book, preparing you for more complex projects and further learning in the field of modern web development.

Project 3: Full-Stack Note-Taking Application

1. Objective

The objective of this project is to develop a full-stack note-taking application that allows users to efficiently manage their notes with operations such as creating, reading, updating, and deleting (CRUD). The application will feature a user-friendly interface, secure and reliable storage, and seamless interaction between the front-end and back-end components.

1.1 Key Features

- **CRUD Operations**: Users will be able to create new notes, read existing notes, update their content, and delete them as needed.
- **Responsive Design**: The application will be responsive, ensuring a functional and attractive interface across various devices and screen sizes.
- **Real-time Updates**: Changes made to the notes will update in real-time, enhancing the user experience by providing immediate feedback.
- **Search Functionality**: Users can search through their notes using keywords to quickly find the information they need.
- **Data Persistence**: Notes will be stored in a MongoDB database, ensuring that user data is saved and persisted across sessions.

1.2 Technologies

- **Front-end**:
  - **React**: Utilized for its component-based architecture, which allows for modular, reusable code and an efficient rendering process.
  - **Redux** (optional): For managing and centralizing application state, facilitating easier communication between React components.
  - **Bootstrap** or **Material-UI**: To help with the styling and to speed up the development process with ready-to-use components that are also responsive.
- **Back-end**:

- o **Node.js**: As the runtime environment for executing JavaScript on the server.
- o **Express**: A minimal and flexible Node.js web application framework that provides a robust set of features to develop web and mobile applications.
- o **Mongoose**: An ODM (Object Data Modeling) library for MongoDB and Node.js that manages relationships between data, provides schema validation, and is used to translate between objects in code and their representation in MongoDB.
- **Database**:
  - o **MongoDB**: A NoSQL database known for its high performance, high availability, and easy scalability.

1.3 Development and Deployment Tools

- **Webpack**: For bundling JavaScript files and assets, including transpiling newer JavaScript and JSX code.
- **Babel**: Transpiler for writing next-generation JavaScript, especially JSX.
- **Git**: For version control, to manage and track changes in the source code.
- **Heroku** or **Netlify**: For hosting the application, offering easy deployment processes and integration with Git.
- **MongoDB Atlas**: For hosting the MongoDB database in the cloud, providing scalability and easy access.

1.4 Project Goals

The ultimate goal of this project is to provide a robust, intuitive, and full-featured note-taking platform that leverages modern web technologies and best practices. The application aims to offer users a seamless experience in managing their notes, whether for personal, educational, or professional purposes.

2. Setup and Configuration

Proper setup and configuration are foundational to a smooth development process for our full-stack note-taking application. This section will guide you through setting up the development environment, structuring the project, and installing necessary dependencies.

2.1 Environment Setup

1. **Node.js Installation**:
   - o Ensure Node.js is installed on your machine. You can download it from the official Node.js website.

o   Verify installation by running node -v in your command line to check the version.

2. **MongoDB Installation**:
   o   Install MongoDB locally for development purposes from the MongoDB website, or set up a free MongoDB Atlas cluster for cloud-based development.

3. **Text Editor**:
   o   Choose a text editor or an Integrated Development Environment (IDE) such as Visual Studio Code (VSCode), which supports JavaScript development and extensions for Node.js, React, and Git.

2.2 Project Directory Structure

Creating a well-organized directory structure is crucial for managing the complexities of a full-stack application efficiently. Here's a suggested structure:

```
note-taking-app/

├── client/              # Frontend React application
│   ├── public/
│   ├── src/
│   ├── package.json
│   └── webpack.config.js

├── server/              # Backend Node.js application
│   ├── config/
│   ├── models/
│   ├── routes/
│   ├── controllers/
│   ├── server.js
│   └── package.json

└── README.md            # Project documentation
```

2.3 Initializing the Project

1. **Create the Project Folders**:
2. mkdir note-taking-app
3. cd note-taking-app
4. mkdir client server
5. **Initialize Node.js in Each Subdirectory**:

- Navigate into each folder (client and server) and run:
- npm init -y

- This command creates a package.json file for managing project metadata and dependencies.

2.4 Installing Dependencies

1. **Server Dependencies**:
   - o Inside the server directory:
   - o npm install express mongoose cors dotenv
   - o
   - o express: Framework for building the server.
   - o mongoose: ODM for interacting with MongoDB.
   - o cors: Middleware to enable CORS (Cross-Origin Resource Sharing).
   - o dotenv: Module to load environment variables from a .env file.
2. **Client Dependencies**:
   - o Inside the client directory:
   - o npm install react react-dom react-router-dom axios
   - o react and react-dom: Libraries for building the UI.
   - o react-router-dom: For routing in the React application.
   - o axios: For making HTTP requests to the server.
3. **Development Tools**:
   - o Install Webpack, Babel, and other development tools in the client directory:
   - o npm install --save-dev webpack webpack-cli webpack-dev-server babel-loader @babel/core @babel/preset-env @babel/preset-react html-webpack-plugin css-loader style-loader

2.5 Configuring Webpack and Babel

Create a webpack.config.js file in the client folder with the following configuration:

```
const path = require('path');
const HtmlWebpackPlugin = require('html-webpack-plugin');

module.exports = {
  entry: './src/index.js',
  output: {
    path: path.resolve(__dirname, 'dist'),
    filename: 'bundle.js'
  },
  module: {
    rules: [
      {
        test: /\\\\.jsx?$/,
        exclude: /node_modules/,
```

```
      use: {
        loader: 'babel-loader',
        options: {
          presets: ['@babel/preset-env', '@babel/preset-react']
        }
      }
    },
    {
      test: /\\\\.css$/,
      use: ['style-loader', 'css-loader']
    }
  ]
},
plugins: [
  new HtmlWebpackPlugin({
    template: './public/index.html'
  })
],
devServer: {
  historyApiFallback: true,
}
};
```

This setup ensures that your front-end and back-end are well-prepared for development, with all necessary tools and dependencies installed.

3. Building the Backend

The backend of our note-taking application will handle CRUD operations for notes, manage user authentication (optional), and interact with the MongoDB database to store and retrieve data. This section will guide you through setting up the Express server, defining the database schema with Mongoose, and implementing API routes.

3.1 Server Initialization

1. **Create the Main Server File**:
    o   In your server directory, create a file named server.js.
    o   This file will be the entry point for your server.
2. **Basic Server Setup**:

- Set up an Express server with initial configurations:
- const express = require('express');
- const mongoose = require('mongoose');

- const cors = require('cors');
- const dotenv = require('dotenv');
-
- dotenv.config(); // Load environment variables from .env file
-
- const app = express();
- const PORT = process.env.PORT || 5000;
-
- app.use(cors());
- app.use(express.json()); // Middleware to parse JSON
-
- app.listen(PORT, () => {
-   console.log(`Server running on port ${PORT}`);
- });

3.2 Database Connection

1. **Configure MongoDB with Mongoose**:
   - Ensure you have the MongoDB connection URI in your .env file (e.g., from MongoDB Atlas or your local MongoDB setup).
   - Connect to MongoDB using Mongoose:
   - const dbURI = process.env.MONGODB_URI;
   - mongoose.connect(dbURI, { useNewUrlParser: true, useUnifiedTopology: true })
   - .then(() => console.log('Database connected successfully'))
   - .catch(err => console.error('MongoDB connection error:', err));

3.3 Models

1. **Define a Mongoose Schema for Notes**:
   - In the server/models directory, create a file named Note.js.
   - Define the schema and model for a note:
   - const mongoose = require('mongoose');
   -
   - const noteSchema = new mongoose.Schema({
   -   title: {
   -     type: String,
   -     required: true,
   -     trim: true
   -   },
   -   content: {

- o   type: String,
- o   required: true
- o   },
- o   date: {
- o   type: Date,
- o   default: Date.now
- o   }
- o   });
- o
- o   const Note = mongoose.model('Note', noteSchema);
- o   module.exports = Note;

3.4 API Routes

1. **Set Up Express Routes for CRUD Operations**:

- Create a routes directory and a file for notes routes (notes.js):
- const express = require('express');
- const router = express.Router();
- const Note = require('../models/Note');
-
- // GET all notes
- router.get('/', async (req, res) => {
-   try {
-    const notes = await Note.find();
-    res.json(notes);
-   } catch (err) {
-    res.status(500).json({ message: err.message });
-   }
- });
-
- // POST a new note
- router.post('/', async (req, res) => {
-   const note = new Note({
-    title: req.body.title,
-    content: req.body.content
-   });
-   try {
-    const newNote = await note.save();
-    res.status(201).json(newNote);

- ```} catch (err) {```
- ```   res.status(400).json({ message: err.message });```
- ```   }```
- ```});```
-
- ```// Additional routes for PUT and DELETE```
-
- ```module.exports = router;```

1. **Integrate Routes into the Server**:

- In server.js, import and use the routes:
- ```const notesRouter = require('./routes/notes');```
- ```app.use('/api/notes', notesRouter);```

With the backend setup complete, your server is now capable of handling requests to manage notes, including creating, reading, updating, and deleting them. This robust backend architecture ensures that your application can efficiently process and store data, serving as the backbone for the note-taking functionality.

4. Designing the Frontend

The frontend of our note-taking application will provide a user-friendly interface for interacting with the notes. We will use React to build a dynamic and responsive SPA (Single Page Application). This section will guide you through setting up the React environment, creating the necessary components, and integrating them with the backend API.

4.1 Setting Up React

1. **Create React App**:
   o Navigate to the client directory and initialize a new React application:
   o npx create-react-app .
   o This command sets up a new React project with all necessary configurations.
2. **Clean Up**:
   o Remove unnecessary files and code to start with a clean slate, simplifying the initial setup and ensuring that you begin development with only what you need.

4.2 Component Structure

1. **Designing Components**:
   - o Plan and create the necessary components for the application:
     - ▪ App: The main component that houses the overall layout.
     - ▪ NoteList: Displays a list of all notes.
     - ▪ NoteItem: Represents a single note in the list.
     - ▪ NoteEditor: Used for creating a new note or editing an existing one.
     - ▪ SearchBar: Allows users to filter notes based on search criteria.
2. **Routing Setup**:

- Use react-router-dom to manage navigation within the application:
- npm install react-router-dom
- Set up basic routes in App.js:
- import React from 'react';
- import { BrowserRouter as Router, Route, Switch } from 'react-router-dom';
- import NoteList from './components/NoteList';
- import NoteEditor from './components/NoteEditor';
-
- function App() {
-   return (
-    <Router>
-
-      <Switch>
-       <Route path="/" exact component={NoteList} />
-       <Route path="/edit/:id" component={NoteEditor} />
-       <Route path="/create" component={NoteEditor} />
-      </Switch>
-
-    </Router>
-   );
- }
-
- export default App;

4.3 Styling

1. **CSS and Frameworks**:

- Decide whether to use plain CSS, a CSS preprocessor like SASS, or a CSS framework such as Bootstrap or Material-UI:
- npm install @material-ui/core
- Utilize the chosen style method to create responsive and aesthetically pleasing components.

4.4 Connecting to the Backend

1. **API Integration**:

- Use axios for making HTTP requests to your backend:
- npm install axios
- Implement API calls in NoteList and NoteEditor for CRUD operations:
- import axios from 'axios';
-
- // Example in NoteList for fetching notes
- useEffect(() => {
- const fetchNotes = async () => {
- try {
- const response = await axios.get('/api/notes');
- setNotes(response.data);
- } catch (error) {
- console.error('Error fetching notes:', error);
- }
- };
-
- fetchNotes();
- }, []);

4.5 Testing and Validation

1. **Component Testing**:

- Write tests using Jest and React Testing Library to ensure components render correctly and functionality works as expected:
- npm install --save-dev @testing-library/react

With the frontend designed and integrated with the backend, your application now has a functional, responsive user interface that allows users to manage their notes effectively. The next steps include finalizing features, refining the user interface, and preparing for deployment.

5. Integrating Frontend with Backend

Integrating the frontend with the backend is a critical step in full-stack application development. This process ensures that the user interface interacts effectively with server-side functionalities, enabling a dynamic and responsive user experience. In this section, we will cover how to connect the React frontend of our note-taking application with the Express backend, focusing on data fetching, state management, and handling updates.

5.1 API Integration

1. **Using Axios for HTTP Requests**:
   - Install Axios in the client project to handle HTTP requests to the backend server:
   - npm install axios
   - Create an Axios instance configured with the base URL of your backend:
   - import axios from 'axios';
   -
   - const api = axios.create({
   -   baseURL: '<http://localhost:5000/api>',
   -   headers: {
   -     'Content-Type': 'application/json'
   -   }
   - });
2. **Fetching Data from the Backend**:
   - Implement data fetching in the NoteList component to retrieve notes from the backend:
   - import React, { useEffect, useState } from 'react';
   - import NoteItem from './NoteItem';
   - import api from './api';
   -
   - function NoteList() {
   -   const [notes, setNotes] = useState([]);
   -
   -   useEffect(() => {
   -     const fetchNotes = async () => {
   -       try {
   -         const response = await api.get('/notes');
   -         setNotes(response.data);
   -       } catch (error) {
   -         console.error('Error fetching notes:', error);
   -       }

```
      };

      fetchNotes();
    }, []);

    return (
      <div>
        {notes.map(note => (
          <NoteItem key={note._id} note={note} />
        ))}
      </div>
    );
  }

  export default NoteList;
```

3. **Handling Create, Update, and Delete Operations**:
   - In the NoteEditor component, implement the functionality to add or update notes:

```
  function NoteEditor({ history, match }) {
    const [note, setNote] = useState({ title: '', content: '' });

    const handleChange = (e) => {
      const { name, value } = e.target;
      setNote(prevNote => ({
        ...prevNote,
        [name]: value
      }));
    };

    const handleSubmit = async (e) => {
      e.preventDefault();
      try {
        if (match.params.id) {
          await api.put(`/notes/${match.params.id}`, note);
        } else {
          await api.post('/notes', note);
        }
        history.push('/');
      } catch (error) {
        console.error('Error saving the note:', error);
      }
```

```
o    };
o
o    return (
o      <form onSubmit={handleSubmit}>
o        <input name="title" value={note.title} onChange={handleChange} />
o        <textarea          name="content"          value={note.content}
     onChange={handleChange} />
o        <button type="submit">Save</button>
o      </form>
o    );
o  }
```

5.2 State Management

1. **Using Context API for Global State**:
   o Optionally, implement React's Context API to manage the state globally across components, which is particularly useful for handling authentication states or shared data across components.
   o Define a context for notes and wrap your component hierarchy in this context provider to make notes accessible throughout the component tree.

5.3 Error Handling and User Feedback

1. **Implementing Error Handling**:
   o Provide feedback to the user when API calls fail, using error messages displayed in the UI.
   o Use try-catch blocks in your asynchronous operations to catch and handle errors.
2. **Loading States**:
   o Manage loading states in your components to inform users when data is being fetched or saved. Display loaders or progress indicators to enhance user experience.

Integrating the frontend with the backend is a pivotal phase in full-stack development, requiring careful attention to API interactions, state management, and user feedback mechanisms. By following the guidelines and examples provided, your application will be capable of handling real-time data operations efficiently, providing a seamless and interactive user experience.

6. Features Implementation

Now that the basic integration between the frontend and backend of our note-taking application is complete, it's time to focus on implementing specific features that will enhance the functionality and user experience. This section will cover the addition of search and filter capabilities, the implementation of authentication, and other essential features.

6.1 CRUD Operations

Ensure that the basic CRUD operations are functioning correctly across your application. This includes:

1. **Creating Notes**: Users should be able to create new notes through a form.
2. **Reading Notes**: Display all notes in a list or grid view.
3. **Updating Notes**: Enable editing of existing notes.
4. **Deleting Notes**: Allow users to delete notes they no longer need.

6.2 Search Functionality

Implementing a search feature allows users to quickly find specific notes based on keywords or content.

1. **Search Bar Component**:
   o   Add a search bar to the NoteList component that lets users input search terms.
2. function SearchBar({ setSearchTerm }) {
3.   return (
4.   <input
5.   type="text"
6.   onChange={(e) => setSearchTerm(e.target.value)}
7.   placeholder="Search notes..."
8.   />
9.   );
10. }
11. **Filter Notes Based on Search**:
    o   Use the search term to filter the displayed notes.
12. const [searchTerm, setSearchTerm] = useState('');
13. const filteredNotes = notes.filter(note =>
14.   note.title.toLowerCase().includes(searchTerm.toLowerCase()) ||
15.   note.content.toLowerCase().includes(searchTerm.toLowerCase())
16.);

```
17.
18.  return (
19.    <div>
20.      <SearchBar setSearchTerm={setSearchTerm} />
21.      {filteredNotes.map(note => (
22.        <NoteItem key={note._id} note={note} />
23.      ))}
24.    </div>
25.  );
```

6.3 User Authentication

If your application requires users to log in:

1. **Setup Authentication Routes**:
   - Implement routes in the backend for user registration and login using Express.
   - Use libraries such as bcrypt for password hashing and jsonwebtoken for issuing JWTs.
2. **Authentication in the Frontend**:
   - Create Login and Register components.
   - Manage authentication state using React Context or Redux to store user information and tokens.
   - Protect routes that require authentication using higher-order components or hooks that redirect unauthenticated users.

6.4 Additional Features

Consider implementing additional features that can improve the usability and functionality of the application:

1. **Note Organization**:
   - Allow users to tag notes or organize them into categories or folders.
   - Implement drag-and-drop functionality to rearrange notes.
2. **Rich Text Editing**:
   - Integrate a rich text editor like react-quill or draft-js to allow users to format their notes, add links, lists, and other rich text features.
3. **Sharing and Collaboration**:
   - Enable sharing of notes with other users or the ability to collaborate on a single note in real-time.
4. **Notifications and Reminders**:

- o   Add the ability to set reminders for notes and send notifications to the user via email or web notifications.

The implementation of these features will transform the basic note-taking app into a robust, full-featured application that caters to a variety of user needs. Each feature not only enhances the user experience but also adds complexity and learning opportunities to your project. By carefully planning and executing these features, your application will stand out in terms of functionality and usability.

7. Testing

Thorough testing is crucial in ensuring that your full-stack note-taking application functions correctly and provides a reliable user experience. This section will guide you through setting up and conducting various types of tests, covering both the frontend and backend components of your application.

7.1 Unit Testing

1. **Backend Testing**:
   - o   Use testing frameworks such as Mocha and Chai for the backend. These tools will help you test your Express routes and database operations.
   - o   Example of a basic test for a GET route in an Express app:
   - o   const chai = require('chai');
   - o   const chaiHttp = require('chai-http');
   - o   const server = require('../server');
   - o   const should = chai.should();
   - o
   - o   chai.use(chaiHttp);
   - o
   - o   describe('Notes', () => {
   - o   describe('/GET notes', () => {
   - o   it('it should GET all the notes', (done) => {
   - o   chai.request(server)
   - o   .get('/api/notes')
   - o   .end((err, res) => {
   - o   res.should.have.status(200);
   - o   res.body.should.be.a('array');
   - o   done();
   - o   });
   - o   });
   - o   });

o   });
2. **Frontend Testing**:
    o   Use Jest and React Testing Library to test your React components. These tools are ideal for ensuring your components render correctly and handle state management as expected.
    o   Example of a test for a React component that displays a note:
    o   import { render, screen } from '@testing-library/react';
    o   import NoteItem from './NoteItem';
    o
    o   test('displays the correct note content', () => {
    o   const note = { title: 'Test Note', content: 'This is a test note' };
    o   render(<NoteItem note={note} />);
    o
    o   expect(screen.getByText('Test Note')).toBeInTheDocument();
    o   expect(screen.getByText('This is a test note')).toBeInTheDocument();
    o   });

7.2 Integration Testing

Integration tests help ensure that the various parts of your application work well together, from the frontend interacting with the backend APIs to database integration.

1. **API and Database Integration**:
    o   Test the integration between your API routes and the database to verify that operations such as creating, retrieving, updating, and deleting notes are performed correctly.
    o   These tests typically involve making requests to your API endpoints and checking the responses and database state.

7.3 End-to-End (E2E) Testing

End-to-end testing simulates real user scenarios from start to finish. Tools like Cypress or Selenium can be used for E2E testing to automate interactions with the actual UI and backend.

1. **Setting up Cypress**:
    o   Install Cypress in your frontend project:
    o   npm install cypress --save-dev
    o   Add a script to your package.json to open Cypress:
    o   "scripts": {
    o   "cypress:open": "cypress open"
    o   }

- o   Write tests that interact with your application as a user would:
- o   describe('Note management', () => {
- o   it('creates a new note', () => {
- o   cy.visit('/');
- o   cy.contains('New Note').click();
- o   cy.get('[data-testid="note-title-input"]').type('New Note');
- o   cy.get('[data-testid="note-content-input"]').type('Note content here');
- o   cy.contains('Save').click();
- o   cy.contains('New Note').should('exist');
- o   cy.contains('Note content here').should('exist');
- o   });
- o   });

7.4 Performance Testing

Consider performance testing for your application to ensure it handles load efficiently, especially if you expect high traffic or data-intensive operations.

Load Testing:

Tools like JMeter or Artillery can simulate multiple users or requests to your application to test how it handles increased load.

Comprehensive testing is integral to developing a reliable and robust application. By implementing unit, integration, E2E, and performance tests, you ensure that each component of your application performs as expected and that they work seamlessly together. This approach not only minimizes bugs and issues in production but also boosts confidence in the quality of your application.

8. Deployment

Deploying your full-stack note-taking application is the final step in making your app accessible to users on the web. This section will guide you through the processes of preparing your application for production, choosing a hosting solution, and ensuring a smooth deployment.

8.1 Preparing for Deployment

1.   **Environment Variables**:

- o Ensure that all sensitive information and environment-specific settings (like database URLs) are stored in environment variables and not hard-coded into your codebase.
- o Create .env files for different environments (e.g., .env.production, .env.development).

2. **Optimization**:
   - o Minimize and optimize your frontend assets. This can be done using Webpack for bundling your JavaScript, CSS, and other assets.
   - o Ensure that images and other media are compressed without losing quality.

3. **Security Enhancements**:
   - o Implement security best practices such as HTTPS, data validation, and CORS settings.
   - o Use security-related HTTP headers like Strict-Transport-Security or Content-Security-Policy.

8.2 Choosing a Hosting Solution

1. **Backend (Node.js + Express)**:
   - o **Heroku**: A popular choice for Node.js applications. Heroku simplifies deployment processes and offers a free tier for small projects.
   - o **DigitalOcean** or **AWS Elastic Beanstalk**: These services offer more control over the server and are suitable for scaling.

2. **Frontend (React)**:
   - o **Netlify**: Ideal for hosting static sites and SPA built with React. It offers continuous deployment from Git repositories, automated HTTPS, and many more features out of the box.
   - o **Vercel**: Similar to Netlify, it provides excellent support for React applications with benefits like SSR (Server-Side Rendering) and SSG (Static Site Generation).

3. **Database**:
   - o **MongoDB Atlas**: A cloud database service that seamlessly integrates with any application. It's easy to set up and connect with Node.js.

8.3 Deployment Steps

1. **Backend Deployment**:
   - o **Heroku**:
     - ▪ Create a Heroku account and install the Heroku CLI.
     - ▪ Log in to your Heroku CLI and create a new app.
     - ▪ Set environment variables in Heroku dashboard.
     - ▪ Deploy your app using Git:
     - ▪ git add .

- git commit -m "Prepare for deployment"
- git push heroku master
- Heroku automatically detects a Node.js app and builds your project accordingly.

2. **Frontend Deployment**:
   - **Netlify**:
     - Push your code to a Git repository (GitHub, GitLab, or Bitbucket).
     - Connect your repository to Netlify from the "New site from Git" option.
     - Configure your build settings and publish directory (build/ for create-react-app).
     - Netlify will deploy your site and provide a URL upon successful deployment.

8.4 Post-Deployment

1. **Monitoring**:
   - Monitor your application's performance and stability using tools like New Relic or Logentries.
   - Set up alerts for downtime or critical errors.
2. **Analytics**:
   - Integrate Google Analytics or a similar service to understand user behavior and traffic patterns.
3. **Continuous Integration/Continuous Deployment (CI/CD)**:
   - If not already set up, configure CI/CD pipelines to automate your build and deployment processes. This ensures that updates to your codebase trigger automatic deployments.

Deployment is a critical phase that makes your application available to users worldwide. By choosing the right hosting solutions and following the detailed steps for deploying both frontend and backend components, you ensure that your application is robust, secure, and scalable. This setup not only serves current users efficiently but also provides a strong foundation for future growth and enhancements.

9. Documentation and Maintenance

Thorough documentation and diligent maintenance are crucial for the long-term success and scalability of your full-stack note-taking application. This section will guide you through best practices for creating effective documentation and strategies for maintaining your application to ensure its continuous improvement and reliability.

9.1 Creating Documentation

1. **User Documentation**:
   - **Purpose and Audience**: Target end-users who will interact with the application. Explain how to use the application, detailing features like how to create, edit, delete, and search for notes.
   - **Format**: Consider user-friendly formats such as online help pages, PDF guides, or interactive tutorials. Tools like Adobe FrameMaker, MadCap Flare, or simpler options like a Git repository's wiki can be effective.
2. **Developer Documentation**:
   - **Code Documentation**: Use inline comments and tools like JSDoc to annotate your source code. This helps developers understand complex parts of the code and the purpose of specific functions and classes.
   - **API Documentation**: Document your backend API endpoints if your application exposes an API. Tools like Swagger (OpenAPI) can automatically generate interactive API documentation that helps other developers understand and use your API correctly.
   - **Architecture Overview**: Provide a high-level overview of the application architecture, including the frontend and backend setups, database schema designs, and interactions between different parts of the application.
3. **Maintenance Guidelines**:
   - Include guidelines for updating libraries and dependencies, procedures for testing after updates, and best practices for ensuring compatibility and security with each new release.

9.2 Maintenance Strategies

1. **Regular Updates and Dependency Management**:
   - Regularly update your application's dependencies to leverage improvements and security patches in the libraries you use.
   - Use tools like Dependabot or Snyk to automate dependency updates and security vulnerability checks.
2. **Bug Tracking and Issue Resolution**:
   - Implement a system for tracking and managing bugs and issues, using platforms like Jira, Trello, or GitHub Issues.
   - Encourage users to report issues and provide feedback through built-in support features or external platforms like email or a dedicated support portal.
3. **Performance Monitoring**:
   - Utilize tools like Google Analytics for user interaction, and New Relic or Datadog for backend performance monitoring.

- o Regularly review performance reports to identify and resolve bottlenecks or scalability issues.
4. **Backup and Disaster Recovery**:
    - o Implement regular backup procedures for your database and server environments.
    - o Develop a disaster recovery plan that includes steps for restoring data and services in case of hardware failure, data corruption, or security breaches.
5. **Security Practices**:
    - o Continuously monitor security advisories related to the technologies you use.
    - o Conduct regular security audits and penetration testing to identify and address vulnerabilities.
6. **Community and Open Source Engagement** (if applicable):
    - o If your project is open-source, encourage community contributions by clearly documenting how to set up development environments, submit changes, and communicate with your project team.
    - o Manage pull requests and community contributions effectively to ensure that they align with the project goals and quality standards.

Effective documentation and proactive maintenance are pivotal for the smooth operation and future growth of your application. By providing clear and comprehensive documentation, you enable users and developers to understand and effectively use or contribute to your application. Additionally, by adhering to robust maintenance practices, you ensure that your application remains secure, performant, and relevant to its users, thereby extending its lifecycle and enhancing user satisfaction.

10. Extensions and Improvements

After deploying your full-stack note-taking application, it's important to consider potential enhancements and extensions to keep the application relevant, improve user experience, and meet emerging needs. This section will outline possible improvements and new features that can be integrated into your application to extend its functionality and maintain its competitiveness in the market.

10.1 Feature Enhancements

1. **Tagging System**:
    - o Implement a tagging feature to allow users to tag their notes for better organization and retrieval. Users can filter notes by tags, making it easier to find related information quickly.
2. **Collaborative Editing**:

- o Introduce real-time collaborative editing features, similar to Google Docs, allowing multiple users to edit the same note simultaneously. This can be achieved using WebSockets or technologies like Firebase.

3. **Rich Text Editing**:
   - o Upgrade the note editor to support rich text features, including bold, italic, underline, bullet points, and custom fonts. Consider integrating a rich text editor library like Quill or CKEditor.

4. **Mobile App**:
   - o Develop a mobile version of the application using React Native or another mobile framework to provide users with on-the-go access to their notes.

5. **Integration with Third-party Services**:
   - o Allow users to integrate their notes with other services such as Google Calendar for reminders, Dropbox for backups, or Slack for sharing.

6. **Export and Import Notes**:
   - o Provide functionality for users to export their notes to formats like PDF or Markdown and import notes from other platforms.

7. **Voice Notes and Transcription**:
   - o Implement voice note capabilities where users can record voice memos that are automatically transcribed into text.

10.2 Performance Improvements

1. **Optimize Load Times**:
   - o Analyze and optimize load times using tools like Google Lighthouse. Minimize the size of assets, use lazy loading for images and components, and ensure efficient server response times.

2. **Database Optimization**:
   - o Optimize database queries to improve response times and scalability. Implement indexing for quicker searches, especially if the application handles a large volume of notes.

3. **Caching Strategies**:
   - o Implement caching mechanisms on the client-side and server-side to store frequently accessed data temporarily, reducing load times and server requests.

10.3 Scalability

1. **Microservices Architecture**:
   - o If the application grows significantly, consider breaking down the server architecture into microservices. This approach can help manage complexities in the application, improving scalability and maintenance.

2. **Cloud Auto-scaling**:
   - o   Utilize cloud auto-scaling features to handle varying loads efficiently, ensuring that the application remains responsive under heavy usage.

10.4 Security Enhancements

1. **Regular Security Audits**:
   - o   Conduct regular security audits and update security practices to protect against new vulnerabilities and threats.
2. **Enhanced Data Encryption**:
   - o   Implement enhanced encryption measures for data at rest and in transit, particularly for sensitive user data.
3. **Two-Factor Authentication (2FA)**:
   - o   Offer two-factor authentication for user accounts to provide an additional layer of security.

10.5 User Experience (UX) Improvements

1. **User Feedback Loop**:
   - o   Establish a continuous user feedback loop to gather and analyze user suggestions and complaints. Use this data to prioritize new features and improvements.
2. **Personalization**:
   - o   Implement personalization options such as customizable themes and layouts to enhance user engagement.

The process of extending and improving your application is ongoing. By continuously introducing new features, optimizing performance, and enhancing security and usability, you ensure that your application adapts to user needs and technological advancements. These improvements not only retain existing users but also attract new users, fostering a growing and engaged user base for your application.

Conclusion

As we conclude "JavaScript from Zero to Superhero: Unlock Your Web Development Superpowers," it is essential to reflect on the journey we have undertaken together. From the fundamental concepts of JavaScript to the complexities of deploying a full-stack application, this book has covered a wide range of topics designed to equip you with the skills necessary to excel in the world of web development.

The Journey Through JavaScript

We began our exploration with the basics of JavaScript, understanding its syntax, operators, data types, and structures. These foundational skills are crucial for any developer and serve as the building blocks for more advanced topics. As we progressed, we delved into the functionalities that make JavaScript a powerful tool in both client-side and server-side programming.

We explored how JavaScript handles asynchronous operations—a critical concept in modern web applications. Understanding callbacks, promises, and async/await patterns not only demystifies how JavaScript deals with operations that take time to complete but also illustrates the language's robust handling of such scenarios, making our applications more efficient and responsive.

Diving into the DOM and Beyond

The document object model (DOM) was another cornerstone topic. By learning how to manipulate the DOM, we gained the ability to create dynamic content and interactive user experiences. This knowledge is vital for any web developer looking to build engaging and interactive websites.

As we moved into more advanced topics, we tackled modern JavaScript frameworks and libraries such as React, Vue, and Angular. These tools are indispensable in the current web development landscape, offering powerful solutions for building scalable and maintainable applications. Understanding these frameworks allows developers to keep pace with the industry and meet the demands of complex project requirements.

The Server-Side of JavaScript with Node.js

Our journey also took us through the server-side aspects of JavaScript with Node.js, enriching your toolkit by enabling you to develop full-stack capabilities. This knowledge allows you to handle server-side scripting, APIs, and databases, making you a versatile asset in any development team.

The chapter on deploying JavaScript applications encapsulated the crucial final steps needed to launch a web application. By covering deployment platforms, optimizations, and best practices, we ensured that you are well-prepared to take your projects from development to production, showcasing your applications to the world.

Practical Application and Real-World Skills

Throughout this book, practical exercises and projects were integrated to provide hands-on experience. The projects, ranging from simple scripts to a comprehensive full-stack note-taking application, were designed to challenge you and enhance your learning through real-world application of the concepts discussed. These exercises are not merely academic; they are stepping stones to building your portfolio and improving your problem-solving skills in web development.

The Future of JavaScript and Web Development

Looking forward, the landscape of web development and JavaScript is ever-evolving. New frameworks, tools, and best practices are continuously emerging. As a developer, staying updated with these changes is crucial. Engage with the community, contribute to open-source projects, and never stop learning. Technologies like WebAssembly and progressive web apps (PWAs) are on the horizon, promising to further blur the lines between desktop and web applications.

The Role of Continuous Learning

The field of technology is one of perpetual learning. What you have learned from this book is a solid foundation, but the architecture of your career in web development will be built on continuous education and adaptation. Participate in coding bootcamps, online courses, and developer meetups. Read blogs, watch tutorials, and keep coding. Every line of code you write, every bug you fix, and every project you complete propels you further in your journey.

In Conclusion

This book was crafted not only to teach you JavaScript but also to inspire you to explore the vast possibilities it presents. Whether you aspire to be a front-end developer, a Node.js expert, or a full-stack engineer, the skills you have acquired here are invaluable. JavaScript is more than just a programming language; it is a gateway to fulfilling your creative and professional aspirations in the digital world.

As you close this book, remember that the end of this reading is just the beginning of your adventure in web development. With your new skills, a proactive attitude, and a passion for building and creating, you are now equipped to take on the world of web development. Embrace challenges, celebrate your successes, and continue to grow. Your journey as a JavaScript developer is just getting started.

Where to continue?

If you've completed this book, and are hungry for more programming knowledge, we'd like to recommend some other books from our software company that you might find useful. These books cover a wide range of topics and are designed to help you continue to expand your programming skills.

1. **"ChatGPT API Bible: Mastering Python Programming for Conversational AI"**: Provide a hands-on, step-by-step guide to utilizing ChatGPT, covering everything from API integration to fine-tuning the model for specific tasks or industries.
2. **"Natural Language Processing with Python: Building your Own Customer Service ChatBot"**: This expansive book offers an in-depth exploration of NLP. It successfully simplifies complex concepts using engaging explanations and intuitive examples.
3. **"Data Analysis with Python"** - Python is a powerful language for data analysis, and this book will help you unlock its full potential. It covers topics such as data cleaning, data manipulation, and data visualization, and provides you with practical exercises to help you apply what you've learned.
4. **"Machine Learning with Python"** - Machine learning is one of the most exciting fields in computer science, and this book will help you get started with building your own machine learning models using Python. It covers topics such as linear regression, logistic regression, and decision trees.
5. **"Mastering ChatGPT and Prompt Engineering"** - In this book, we will take you on a comprehensive journey through the world of prompt engineering, covering everything from the fundamentals of AI language models to advanced strategies and real-world applications.

All of these books are designed to help you continue to expand your programming skills and deepen your understanding of the Python language. We believe that programming is a skill that can be learned and developed over time, and we are committed to providing resources to help you achieve your goals.

We'd also like to take this opportunity to thank you for choosing our software company as your guide in your programming journey. We hope that you have found this book of Python for beginners to be a valuable resource, and we look forward to continuing to provide you with high-quality programming resources in the future. If you have any feedback or suggestions for future books or resources, please don't hesitate to get in touch with us. We'd love to hear from you!

Know more about us

At Cuantum Technologies, we specialize in building web applications that deliver creative experiences and solve real-world problems. Our developers have expertise in a wide range of programming languages and frameworks, including Python, Django, React, Three,js, and Vue.js, among others. We are constantly exploring new technologies and techniques to stay at the forefront of the industry, and we pride ourselves on our ability to create solutions that meet our clients' needs.

If you are interested in learning more about our Cuantum Technologies and the services that we offer, please visit our website at books.cuantum.tech. We would be happy to answer any questions that you may have and to discuss how we can help you with your software development needs.

CUANTUM
TECHNOLOGIES

www.cuantum.tech

www.ingramcontent.com/pod-product-compliance
Lightning Source LLC
Chambersburg PA
CBHW082117210326
41599CB00031B/5790